High Energy Physics

Volume I

High Energy Physics
Volume I

Edited by **Lucy Flynn**

NY RESEARCH PRESS

New York

Published by NY Research Press,
23 West, 55th Street, Suite 816,
New York, NY 10019, USA
www.nyresearchpress.com

High Energy Physics: Volume I
Edited by Lucy Flynn

International Standard Book Number: 978-1-63238-287-0 (Hardback)

Printed in China.

Contents

Preface

High Energy Physics is a field of physics that is focused on the search for the fundamental and foundational forces and particles that constitute and build the world the around us. It is also more commonly known by the name particle physics. Anyone with a basic foundation in advanced physics would know that as atoms are made up of even smaller particles. High energy physics thus becomes a scientific field that explores the questions about the basic and elementary constituents that make up matter and energy along with a focus on the interactions between them, and the nature of space and time. High energy physics researches also often focus on the creation of subatomic particles that are deviated from the normal part of matter that one encounters daily. Such particles have high energy, that is, compared to various natural processes like lightning or nuclear decay, the energy per articles of such particles are of magnitudes many times higher. That is why this field can be termed as that of high energy physics. Research is essential and constantly being done in this field to discover newer facets of the universe that one did not know before and therefore the demand for skilled graduates in this field has never been higher.

This book is an attempt to compile and collate all available research on high energy physics under one umbrella. I am grateful to those who put their hard work, effort and expertise into these research projects as well as those who were supportive in this endeavour.

Editor

1

Electromagnetic Properties of Neutrinos

C. Broggini,[1] C. Giunti,[2] and A. Studenikin[3,4]

[1] *INFN, Sezione di Padova, Via F. Marzolo 8, 35131 Padova, Italy*
[2] *INFN, Sezione di Torino, Via P. Giuria 1, 10125 Torino, Italy*
[3] *Department of Theoretical Physics, Moscow State University, Moscow 119991, Russia*
[4] *Joint Institute for Nuclear Research, Dubna 141980, Moscow, Russia*

Correspondence should be addressed to C. Giunti, giunti@to.infn.it

Academic Editor: Jose Bernabeu

In this paper, we discuss the main theoretical aspects and experimental effects of neutrino electromagnetic properties. We start with a general description of the electromagnetic form factors of Dirac and Majorana neutrinos. Then, we discuss the theory and phenomenology of the magnetic and electric dipole moments, summarizing the experimental results and the theoretical predictions. We discuss also the phenomenology of a neutrino charge radius and radiative decay. Finally, we describe the theory of neutrino spin and spin-flavor precession in a transverse magnetic field and we summarize its phenomenological applications.

1. Introduction

The investigation of neutrino properties is one of the most active fields of research in current high-energy physics. Neutrinos are special particles, because they interact very weakly and their masses are much smaller than those of the other fundamental fermions (charged leptons and quarks). In the Standard Model neutrinos are massless and have only weak interactions. However, the observation of neutrino oscillations by many experiments (see [1–4]) imply that neutrinos are massive and mixed. Therefore, the Standard Model must be extended to account for neutrino masses. In many extensions of the Standard Model neutrinos acquire also electromagnetic properties through quantum loops effects. Hence, the theoretical and experimental study of neutrino electromagnetic interactions is a powerful tool in the search for the fundamental theory beyond the Standard Model. Moreover, the electromagnetic interactions of neutrinos can generate important effects, especially in astrophysical environments, where neutrinos propagate for long distances in magnetic fields both in vacuum and in matter.

In this paper, we review the theory and phenomenology of neutrino electromagnetic interactions. After a derivation of all the possible types of electromagnetic interactions of Dirac and Majorana neutrinos, we discuss their effects in terrestrial and astrophysical environments and the corresponding experimental results. In spite of many efforts in the search of neutrino electromagnetic interactions, up to now there is no positive experimental indication in favor of their existence. However, the existence of neutrino masses and mixing implies that nontrivial neutrino electromagnetic properties are plausible and experimentalists and theorists are eagerly looking for them.

In this review, we use the notation and conventions in [1]. When we consider neutrino mixing, we have the relation:

$$\nu_{\alpha L} = \sum_{k=1}^{3} U_{\alpha k} \nu_{kL} \quad (\alpha = e, \mu, \tau) \tag{1.1}$$

between the left-handed components of the three flavor neutrino fields ν_e, ν_μ, ν_τ and the left-handed components of three massive neutrino fields ν_k with masses m_k $(k = 1, 2, 3)$. The 3×3 mixing matrix U is unitary $(U^\dagger = U^{-1})$.

Neutrino electromagnetic properties are discussed in [4–7], and in the previous reviews in [8–14].

The structure of this paper is as follows. In Section 2, we discuss the general form of the electromagnetic interaction of Dirac and Majorana neutrinos, which is expressed in terms of form factors, and the derivation of the form factors in gauge models. In Section 3, we discuss the phenomenology of the neutrino magnetic and electric dipole moments in laboratory experiments. These are the most studied electromagnetic properties of neutrinos, both experimentally and theoretically. In Section 4, we review the theory and experimental constraints on the neutrino charge radius. In Section 5, we discuss neutrino radiative decay and the astrophysical bounds on a neutrino magnetic moment obtained from the study of plasmon decay in stars. In Section 6 we discuss neutrino spin and spin-flavor precession. In conclusion, in Section 7, we summarize the status of our knowledge of neutrino electromagnetic properties and we discuss the prospects for future research.

2. Electromagnetic Form Factors

The importance of neutrino electromagnetic properties was first mentioned by Pauli in 1930, when he postulated the existence of this particle and discussed the possibility that the neutrino might have a magnetic moment. Systematic theoretical studies of neutrino electromagnetic properties started after it was shown that in the extended Standard Model with right-handed neutrinos the magnetic moment of a massive neutrino is, in general, nonvanishing and that its value is determined by the neutrino mass [15–23].

Neutrino electromagnetic properties are important because they are directly connected to fundamentals of particle physics. For example, neutrino electromagnetic properties can be used to distinguish Dirac and Majorana neutrinos (see [21, 22, 24–27]) and also as probes of new physics that might exist beyond the Standard Model (see [28–30]).

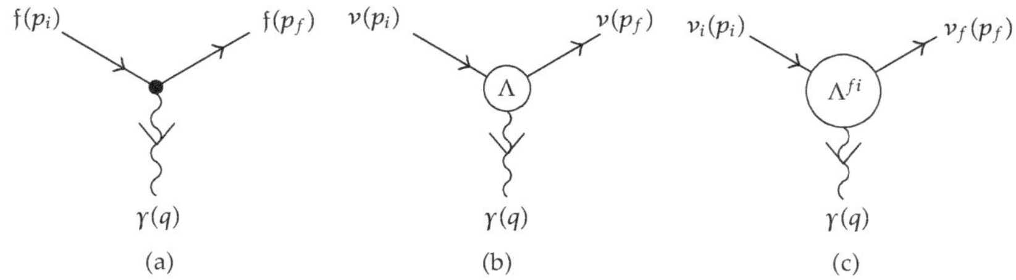

Figure 1: Tree-level coupling of a charged fermion $\mathfrak{f}$ with a photon γ (a), effective coupling of a neutrino ν with a photon (b), and effective coupling of neutrinos with a photon taking into account possible transitions between two different initial and final massive neutrinos ν_i and ν_f (c).

2.1. Dirac Neutrinos

In the Standard Model, the interaction of a fermionic field $\mathfrak{f}(x)$ with the electromagnetic field $A^{\mu}(x)$ is given by the interaction Hamiltonian:

$$\mathscr{H}_{\mathrm{em}}(x) = j_{\mu}(x)A^{\mu}(x) = q_{\mathfrak{f}}\overline{\mathfrak{f}}(x)\gamma_{\mu}\mathfrak{f}(x)A^{\mu}(x), \tag{2.1}$$

where $q_{\mathfrak{f}}$ is the charge of the fermion $\mathfrak{f}$. Figure 1(a) shows the corresponding tree-level Feynman diagram (the photon γ is the quantum of the electromagnetic field $A^{\mu}(x)$).

For neutrinos, the electric charge is zero and there are no electromagnetic interactions at tree-level (however, in some theories beyond the Standard Model neutrinos can be millicharged particles (see [13])). However, such interactions can arise at the quantum level from loop diagrams at higher order of the perturbative expansion of the interaction. In the one-photon approximation, the electromagnetic interactions of a neutrino field $\nu(x)$ can be described by the effective interaction Hamiltonian:

$$\mathscr{H}_{\mathrm{eff}}(x) = j_{\mu}^{\mathrm{eff}}(x)A^{\mu}(x) = \overline{\nu}(x)\Lambda_{\mu}\nu(x)A^{\mu}(x), \tag{2.2}$$

where $j_{\mu}^{\mathrm{eff}}(x)$ is the effective neutrino electromagnetic current four-vector and Λ_{μ} is a 4×4 matrix in spinor space which can contain space-time derivatives, such that $j_{\mu}^{\mathrm{eff}}(x)$ transforms as a four-vector. Since radiative corrections are generated by weak interactions which are not invariant under a parity transformation, $j_{\mu}^{\mathrm{eff}}(x)$ can be a sum of polar and axial parts. The corresponding diagram for the interaction of a neutrino with a photon is shown in Figure 1(b), where the blob represents the quantum loop contributions.

We are interested in the neutrino part of the amplitude corresponding to the diagram in Figure 1(b), which is given by the matrix element:

$$\left\langle \nu(p_f, h_f) \middle| j_{\mu}^{\mathrm{eff}}(x) \middle| \nu(p_i, h_i) \right\rangle, \tag{2.3}$$

where $p_i(p_f)$ and $h_i(h_f)$ are the four-momentum and helicity of the initial (final) neutrino. Taking into account that

$$\partial^{\mu} j_{\mu}^{\mathrm{eff}}(x) = i\left[\mathcal{P}^{\mu}, j_{\mu}^{\mathrm{eff}}(x)\right], \tag{2.4}$$

where $\mathcal{P}^\mu$ is the four-momentum operator which generate translations, the effective current can be written as

$$j_\mu^{\text{eff}}(x) = e^{i\mathcal{P}\cdot x} j_\mu^{\text{eff}}(0) e^{-i\mathcal{P}\cdot x}. \tag{2.5}$$

Since $\mathcal{P}^\mu|\nu(p)\rangle = p^\mu|\nu(p)\rangle$, we have

$$\left\langle \nu(p_f) \middle| j_\mu^{\text{eff}}(x) \middle| \nu(p_i) \right\rangle = e^{i(p_f - p_i)\cdot x} \left\langle \nu(p_f) \middle| j_\mu^{\text{eff}}(0) \middle| \nu(p_i) \right\rangle, \tag{2.6}$$

where we suppressed for simplicity the helicity labels which are not of immediate relevance. Here we see that the unknown quantity which determines the neutrino-photon interaction is $\langle \nu(p_f)|j_\mu^{\text{eff}}(0)|\nu(p_i)\rangle$. Considering that the incoming and outgoing neutrinos are free particles which are described by free Dirac fields with the standard Fourier expansion in (2.139) of [1], we have

$$\left\langle \nu(p_f) \middle| j_\mu^{\text{eff}}(0) \middle| \nu(p_i) \right\rangle = \overline{u}(p_f) \Lambda_\mu(p_f, p_i) u(p_i). \tag{2.7}$$

The electromagnetic properties of neutrinos are embodied by $\Lambda_\mu(p_f, p_i)$, which is a matrix in spinor space and can be decomposed in terms of linearly independent products of Dirac γ matrices and the available kinematical four-vectors p_i and p_f. The most general decomposition can be written as (see [11])

$$\begin{aligned} \Lambda_\mu(p_f, p_i) = {} & f_1\left(q^2\right) q_\mu + f_2\left(q^2\right) q_\mu \gamma_5 + f_3\left(q^2\right) \gamma_\mu + f_4\left(q^2\right) \gamma_\mu \gamma_5 \\ & + f_5\left(q^2\right) \sigma_{\mu\nu} q^\nu + f_6\left(q^2\right) \epsilon_{\mu\nu\rho\gamma} q^\nu \sigma^{\rho\gamma}, \end{aligned} \tag{2.8}$$

where $f_k(q^2)$ are six Lorentz-invariant form factors ($k = 1, \ldots, 6$) and q is the four-momentum of the photon, which is given by

$$q = p_i - p_f, \tag{2.9}$$

from energy-momentum conservation. Notice that the form factors depend only on q^2, which is the only available Lorentz-invariant kinematical quantity, since $(p_i + p_f)^2 = 4m^2 - q^2$. Therefore, $\Lambda_\mu(p_f, p_i)$ depends only on q and from now on we will denote it as $\Lambda_\mu(q)$.

Since the Hamiltonian and the electromagnetic field are Hermitian ($\mathcal{H}_{\text{eff}}^\dagger = \mathcal{H}_{\text{eff}}$ and $A^{\mu\dagger} = A^\mu$), the effective current must be Hermitian, $j_\mu^{\text{eff}\dagger} = j_\mu^{\text{eff}}$. Hence, we have

$$\left\langle \nu(p_f) \middle| j_\mu^{\text{eff}}(0) \middle| \nu(p_i) \right\rangle = \left\langle \nu(p_i) \middle| j_\mu^{\text{eff}}(0) \middle| \nu(p_f) \right\rangle^*, \tag{2.10}$$

which leads to

$$\Lambda_\mu(q) = \gamma^0 \Lambda_\mu^\dagger(-q) \gamma^0. \tag{2.11}$$

This constraint implies that

$$f_2, f_3, f_4 \quad \text{are real,} \tag{2.12}$$

$$f_1, f_5, f_6 \quad \text{are imaginary.} \tag{2.13}$$

The number of independent form factors can be reduced by imposing current conservation, $\partial^\mu j_\mu^{\text{eff}}(x) = 0$, which is required by gauge invariance (i.e., invariance of $\mathcal{H}_{\text{eff}}(x)$ under the transformation $A^\mu(x) \longrightarrow A^\mu(x) + \partial^\mu \varphi(x)$ for any $\varphi(x)$, which leaves invariant the electromagnetic tensor $F^{\mu\nu} = \partial^\mu A^\nu - \partial^\nu A^\mu$). Using (2.4), current conservation implies that

$$\left\langle \nu(p_f) \right| \left[\mathcal{P}^\mu, j_\mu^{\text{eff}}(0) \right] \left| \nu(p_i) \right\rangle = 0. \tag{2.14}$$

Hence, in momentum space, we have the constraint:

$$q^\mu \, \overline{u}(p_f) \Lambda_\mu(q) u(p_i) = 0, \tag{2.15}$$

which implies that

$$f_1\left(q^2\right)q^2 + f_2\left(q^2\right)q^2\gamma_5 + 2mf_4\left(q^2\right)\gamma_5 = 0. \tag{2.16}$$

Since γ_5 and the unity matrix are linearly independent, we obtain the constraints:

$$f_1\left(q^2\right) = 0, \quad f_4\left(q^2\right) = -\frac{f_2(q^2)q^2}{2m}. \tag{2.17}$$

Therefore, in the most general case, consistent with Lorentz and electromagnetic gauge invariance, the vertex function is defined in terms of four form factors [25–27]:

$$\Lambda_\mu(q) = f_Q\left(q^2\right)\gamma_\mu - f_M\left(q^2\right)i\sigma_{\mu\nu}q^\nu + f_E\left(q^2\right)\sigma_{\mu\nu}q^\nu\gamma_5 + f_A\left(q^2\right)\left(q^2\gamma_\mu - q_\mu \not{q}\right)\gamma_5. \tag{2.18}$$

where $f_Q = f_3$, $f_M = if_5$, $f_E = -2if_6$, and $f_A = -f_2/2m$ are the real charge, dipole magnetic and electric, and anapole neutrino form factors. For the coupling with a real photon ($q^2 = 0$),

$$f_Q(0) = q, \qquad f_M(0) = \mu, \qquad f_E(0) = \epsilon, \qquad f_A(0) = a, \tag{2.19}$$

where q, μ, ϵ, and a are, respectively, the neutrino charge, magnetic moment, electric moment and anapole moment. Although above we stated that $q = 0$, here we did not enforce this equality because in some theories beyond the Standard Model neutrinos can be millicharged particles (see [13]).

Now, it is interesting to study the properties of $\mathcal{H}_{\text{eff}}(x)$ under a CP transformation, in order to find which of the terms in (2.18) violate CP. Let us consider the active CP transformation:

$$\mathrm{U}_{\mathrm{CP}}\nu(x)\mathrm{U}_{\mathrm{CP}}^\dagger = \xi^{\mathrm{CP}}\gamma^0 \mathcal{C}\overline{\nu}^T(x_{\mathrm{P}}), \tag{2.20}$$

where ξ^{CP} is a phase, $\mathcal{C}$ is the charge-conjugation matrix (such that $\mathcal{C}\gamma_\mu^T\mathcal{C}^{-1} = -\gamma_\mu$, $\mathcal{C}^\dagger = \mathcal{C}^{-1}$ and $\mathcal{C}^T = -\mathcal{C}$), and $x_{\mathrm{P}}^\mu = x_\mu$. For the Standard Model electric current $j_\mu(x)$ in (2.1), we have

$$j_\mu(x) \xrightarrow{\mathrm{CP}} \mathrm{U}_{\mathrm{CP}} j_\mu(x) \mathrm{U}_{\mathrm{CP}}^\dagger = -j^\mu(x_{\mathrm{P}}). \tag{2.21}$$

Hence, the Standard Model electromagnetic interaction Hamiltonian $\mathcal{H}_{\mathrm{em}}(x)$ is left invariant by (the transformation $x \longrightarrow x_{\mathrm{P}}$ is irrelevant since all amplitudes are obtained by integrating over d^4x)

$$A_\mu(x) \xrightarrow{\mathrm{CP}} -A^\mu(x_{\mathrm{P}}). \tag{2.22}$$

CP is conserved in neutrino electromagnetic interactions (in the one-photon approximation) if $j_\mu^{\mathrm{eff}}(x)$ transforms as $j_\mu(x)$:

$$\mathrm{CP} \Longleftrightarrow \mathrm{U}_{\mathrm{CP}} j_\mu^{\mathrm{eff}}(x) \mathrm{U}_{\mathrm{CP}}^\dagger = -j_{\mathrm{eff}}^\mu(x_{\mathrm{P}}). \tag{2.23}$$

For the matrix element (2.7), we obtain

$$\mathrm{CP} \Longleftrightarrow \Lambda_\mu(q) \xrightarrow{\mathrm{CP}} -\Lambda^\mu(q). \tag{2.24}$$

One can find that under a CP transformation we have

$$\Lambda_\mu(q) \xrightarrow{\mathrm{CP}} \gamma^0 \mathcal{C} \Lambda_\mu^T(q_{\mathrm{P}}) \mathcal{C}^\dagger \gamma^0, \tag{2.25}$$

with $q_{\mathrm{P}}^\mu = q_\mu$. Using the form factor expansion in (2.18), we obtain

$$\Lambda_\mu(q) \xrightarrow{\mathrm{CP}} -\left[f_Q\left(q^2\right)\gamma^\mu - f_M\left(q^2\right)i\sigma^{\mu\nu}q_\nu - f_E\left(q^2\right)\sigma^{\mu\nu}q_\nu\gamma_5 + f_A\left(q^2\right)\left(q^2\gamma^\mu - q^\mu \not{q}\right)\gamma_5\right]. \tag{2.26}$$

Therefore, only the electric dipole form factor violates CP:

$$\mathrm{CP} \Longleftrightarrow f_E\left(q^2\right) = 0. \tag{2.27}$$

So far, we have considered only one massive neutrino field $\nu(x)$, but according to the mixing relation (1.1), the three flavor neutrino fields ν_e, ν_μ, ν_τ are unitary linear combinations of three massive neutrinos ν_k ($k = 1, 2, 3$). Therefore, we must generalize the discussion to the case of more than one massive neutrino field. The effective electromagnetic interaction Hamiltonian in (2.2) is generalized to

$$\mathcal{H}_{\mathrm{eff}}(x) = j_\mu^{\mathrm{eff}}(x) A^\mu(x) = \sum_{k,j=1}^{3} \overline{\nu_k}(x) \Lambda_\mu^{kj} \nu_j(x) A^\mu(x), \tag{2.28}$$

where we take into account possible transitions between different massive neutrinos. The physical effect of $\mathcal{H}_{\text{eff}}$ is described by the effective electromagnetic vertex in Figure 1(c), with the neutrino matrix element:

$$\left\langle \nu_f(p_f) \middle| j_\mu^{\text{eff}}(0) \middle| \nu_i(p_i) \right\rangle = \overline{u_f}(p_f) \Lambda_\mu^{fi}(p_f, p_i) u_i(p_i). \tag{2.29}$$

As in the case of one massive neutrino field, $\Lambda_\mu^{fi}(p_f, p_i)$ depends only on the four-momentum q transferred to the photon and can be expressed in terms of six Lorentz-invariant form factors:

$$\begin{aligned} \Lambda_\mu^{fi}(q) = {} & f_1^{fi}\left(q^2\right) q_\mu + f_2^{fi}\left(q^2\right) q_\mu \gamma_5 + f_3^{fi}\left(q^2\right) \gamma_\mu + f_4^{fi}\left(q^2\right) \gamma_\mu \gamma_5 \\ & + f_5^{fi}\left(q^2\right) \sigma_{\mu\nu} q^\nu + f_6^{fi}\left(q^2\right) \epsilon_{\mu\nu\rho\gamma} q^\nu \sigma^{\rho\gamma}. \end{aligned} \tag{2.30}$$

The Hermitian nature of j_μ^{eff} implies that $\langle \nu_f(p_f) | j_\mu^{\text{eff}}(0) | \nu_i(p_i) \rangle = \langle \nu_i(p_i) | j_\mu^{\text{eff}}(0) | \nu_f(p_f) \rangle^*$, leading to the constraint:

$$\Lambda_\mu^{fi}(q) = \gamma^0 \left[\Lambda_\mu^{if}(-q)\right]^\dagger \gamma^0. \tag{2.31}$$

Considering the 3×3 form factor matrices f_k in the space of massive neutrinos with components f_k^{fi} for $k = 1, \ldots, 6$, we find that

$$f_2, f_3, f_4 \quad \text{are Hermitian,} \tag{2.32}$$

$$f_1, f_5, f_6 \quad \text{are anti-Hermitian.} \tag{2.33}$$

Following the same method used in (2.4)–(2.16), one can find that current conservation implies the constraints:

$$f_1^{fi}\left(q^2\right) q^2 + f_3^{fi}\left(q^2\right)(m_f - m_i) = 0, \qquad f_2^{fi}\left(q^2\right) q^2 + f_4^{fi}\left(q^2\right)(m_f + m_i) = 0. \tag{2.34}$$

Therefore, we obtain

$$\Lambda_\mu^{fi}(q) = \left(\gamma_\mu - q_\mu \frac{\not{q}}{q^2}\right)\left[f_Q^{fi}\left(q^2\right) + f_A^{fi}\left(q^2\right) q^2 \gamma_5\right] - i\sigma_{\mu\nu} q^\nu \left[f_M^{fi}\left(q^2\right) + i f_E^{fi}\left(q^2\right) \gamma_5\right], \tag{2.35}$$

where $f_Q^{fi} = f_3^{fi}$, $f_M^{fi} = i f_5^{fi}$, $f_E^{fi} = -2i f_6^{fi}$ and $f_A^{fi} = -f_2^{fi}/(m_f + m_i)$, with

$$f_\Omega^{fi} = \left(f_\Omega^{if}\right)^* \quad (\Omega = Q, M, E, A). \tag{2.36}$$

Note that since $\overline{u_f}(p_f) \not{q} u_i(p_i) = (m_f - m_i) \overline{u_f}(p_f) u_i(p_i)$, if $f = i$, (2.35) correctly reduces to (2.18).

The form factors with $f = i$ are called "diagonal," whereas those with $f \neq i$ are called "off-diagonal" or "transition form factors." This terminology follows from the expression:

$$\Lambda_\mu(q) = \left(\gamma_\mu - q_\mu \frac{q}{q^2}\right)\left[f_Q\left(q^2\right) + f_A\left(q^2\right)q^2\gamma_5\right] - i\sigma_{\mu\nu}q^\nu\left[f_M\left(q^2\right) + i f_E\left(q^2\right)\gamma_5\right], \tag{2.37}$$

in which $\Lambda_\mu(q)$ is a 3×3 matrix in the space of massive neutrinos expressed in terms of the four Hermitian 3×3 matrices of form factors:

$$f_\Omega = f_\Omega^\dagger \quad (\Omega = Q, M, E, A). \tag{2.38}$$

For the coupling with a real photon ($q^2 = 0$), we have

$$f_Q^{fi}(0) = q_{fi}, \qquad f_M^{fi}(0) = \mu_{fi}, \qquad f_E^{fi}(0) = \epsilon_{fi}, \qquad f_A^{fi}(0) = a_{fi}, \tag{2.39}$$

where q_{fi}, μ_{fi}, ϵ_{fi}, and a_{fi} are, respectively, the neutrino charge, magnetic moment, electric moment, and anapole moment of diagonal ($f = i$) and transition ($f \neq i$) types.

Considering now CP invariance, the transformation (2.23) of $j_\mu^{\text{eff}}(x)$ implies the constraint in (2.24) for the $N \times N$ matrix $\Lambda_\mu(q)$ in the space of massive neutrinos. Using (2.20), we obtain

$$\Lambda_\mu^{fi}(q) \xrightarrow{\text{CP}} \xi_f^{\text{CP}} \xi_i^{\text{CP}*} \gamma^0 \mathcal{C}\left[\Lambda_\mu^{if}(q_{\text{P}})\right]^T \mathcal{C}^\dagger \gamma^0, \tag{2.40}$$

where ξ_k^{CP} is the CP phase of ν_k. Since the three massive neutrinos take part to standard charged-current weak interactions, their CP phases are equal if CP is conserved (see [1]). Hence, we have

$$\Lambda_\mu^{fi}(q) \xrightarrow{\text{CP}} \gamma^0 \mathcal{C}\left[\Lambda_\mu^{if}(q_{\text{P}})\right]^T \mathcal{C}^\dagger \gamma^0. \tag{2.41}$$

Using the form factor expansion in (2.35), we obtain

$$\Lambda_\mu^{fi}(q) \xrightarrow{\text{CP}} -\left\{\left(\gamma^\mu - q^\mu \frac{q}{q^2}\right)\left[f_Q^{if}\left(q^2\right) + f_A^{if}\left(q^2\right)q^2\gamma_5\right] - i\sigma^{\mu\nu}q_\nu\left[f_M^{if}\left(q^2\right) - i f_E^{if}\left(q^2\right)\gamma_5\right]\right\}. \tag{2.42}$$

Imposing the constraint in (2.24), for the form factors, we obtain

$$\text{CP} \Longleftrightarrow \begin{cases} f_\Omega^{fi} = f_\Omega^{if} = \left(f_\Omega^{fi}\right)^* & (\Omega = Q, M, A), \\ f_E^{fi} = -f_E^{if} = -\left(f_E^{fi}\right)^*, & \end{cases} \tag{2.43}$$

where, in the last equalities, we took into account the constraints (2.36). For the Hermitian 3×3 form factor matrices, we obtain that if CP is conserved f_Q, f_M and f_A, are real and symmetric and f_E is imaginary and antisymmetric:

$$\mathrm{CP} \Longleftrightarrow \begin{cases} f_\Omega = f_\Omega^{\mathrm{T}} = f_\Omega^* & (\Omega = Q, M, A), \\ f_E = -f_E^{\mathrm{T}} = -f_E^*. \end{cases} \tag{2.44}$$

Let us now consider antineutrinos. Using for the massive neutrino fields, the Fourier expansion in (2.139) of [1], the effective antineutrino matrix element for $\overline{\nu}_i(p_i) \to \overline{\nu}_f(p_f)$ transitions is given by

$$\left\langle \overline{\nu}_f(p_f) \middle| j_\mu^{\mathrm{eff}}(0) \middle| \overline{\nu}_i(p_i) \right\rangle = -\overline{v_i}(p_i)\Lambda_\mu^{if}(q)v_f(p_f). \tag{2.45}$$

Using the relation:

$$u(p) = C\overline{v}^T(p), \tag{2.46}$$

we can write it as

$$\left\langle \overline{\nu}_f(p_f) \middle| j_\mu^{\mathrm{eff}}(0) \middle| \overline{\nu}_i(p_i) \right\rangle = \overline{u_f}(p_f)C\left[\Lambda_\mu^{if}(q)\right]^T C^\dagger u_i(p_i), \tag{2.47}$$

where transposition operates in spinor space. Therefore, the effective form factor matrix in spinor space for antineutrinos is given by

$$\overline{\Lambda}_\mu^{fi}(q) = C\left[\Lambda_\mu^{if}(q)\right]^T C^\dagger. \tag{2.48}$$

Using the properties of the charge-conjugation matrix and the expression (2.35) for $\Lambda_\mu^{if}(q)$, we obtain the antineutrino form factors:

$$\overline{f}_\Omega^{fi} = -f_\Omega^{if} \quad (\Omega = Q, M, E), \tag{2.49}$$

$$\overline{f}_A^{fi} = f_A^{if}. \tag{2.50}$$

Therefore, in particular the diagonal magnetic and electric moments of neutrinos and antineutrinos have the same size with opposite signs, as the charge, if it exists. On the other hand, the diagonal neutrino and antineutrino anapole moments are equal.

2.2. Majorana Neutrinos

A massive Majorana neutrino is a neutral spin 1/2 particle which coincides with its antiparticle. The four degrees of freedom of a massive Dirac field (two helicities and two particle-antiparticle) are reduced to two (two helicities) by the Majorana constraint:

$$\nu_k = \nu_k^c = \mathcal{C}\overline{\nu_k}^T. \tag{2.51}$$

Since a Majorana field has half the degrees of freedom of a Dirac field, it is possible that its electromagnetic properties are reduced. From the relations (2.49) and (2.50) between neutrino and antineutrino form factors in the Dirac case, we can infer that in the Majorana case the charge, magnetic, and electric form factor matrices are antisymmetric and the anapole form factor matrix is symmetric. In order to confirm this deduction, let us calculate the neutrino matrix element corresponding to the effective electromagnetic vertex in Figure 1(c), with the effective interaction Hamiltonian in (2.28), which takes into account possible transitions between two different initial and final massive Majorana neutrinos ν_i and ν_f. Using the neutrino Majorana fields, the Fourier expansion in (6.99) of [1], we obtain

$$\left\langle \nu_f(p_f) \left| j_\mu^{\text{eff}}(0) \right| \nu_i(p_i) \right\rangle = \overline{u_f}(p_f)\Lambda_\mu^{fi}(p_f, p_i)u_i(p_i) - \overline{v_i}(p_i)\Lambda_\mu^{if}(p_f, p_i)v_f(p_f). \tag{2.52}$$

Using (2.46), we can write it as

$$\overline{u_f}(p_f)\left\{\Lambda_\mu^{fi}(p_f, p_i) + \mathcal{C}\left[\Lambda_\mu^{if}(p_f, p_i)\right]^T \mathcal{C}^\dagger\right\}u_i(p_i), \tag{2.53}$$

where transposition operates in spinor space. Therefore, the effective form factor matrix in spinor space for Majorana neutrinos is given by

$$\widetilde{\Lambda}_\mu^{fi}(p_f, p_i) = \Lambda_\mu^{fi}(p_f, p_i) + \mathcal{C}\left[\Lambda_\mu^{if}(p_f, p_i)\right]^T \mathcal{C}^\dagger. \tag{2.54}$$

As in the case of Dirac neutrinos, $\Lambda_\mu^{fi}(p_f, p_i)$ depends only on $q = p_f - p_i$ and can be expressed in terms of six Lorentz-invariant form factors according to (2.30). Hence, we can write the 3×3 matrix $\widetilde{\Lambda}_\mu(p_f, p_i)$ in the space of massive Majorana neutrinos as

$$\begin{aligned}\widetilde{\Lambda}_\mu(q) = {} & \widetilde{f}_1\left(q^2\right)q_\mu + \widetilde{f}_2\left(q^2\right)q_\mu\gamma_5 + \widetilde{f}_3\left(q^2\right)\gamma_\mu + \widetilde{f}_4\left(q^2\right)\gamma_\mu\gamma_5 \\ & + \widetilde{f}_5\left(q^2\right)\sigma_{\mu\nu}q^\nu + \widetilde{f}_6\left(q^2\right)\epsilon_{\mu\nu\rho\gamma}q^\nu\sigma^{\rho\gamma},\end{aligned} \tag{2.55}$$

with

$$\begin{aligned}\widetilde{f}_k = f_k + f_k^T \Longrightarrow \widetilde{f}_k = \widetilde{f}_k^T \quad \text{for } k = 1, 2, 4, \\ \widetilde{f}_k = f_k - f_k^T \Longrightarrow \widetilde{f}_k = -\widetilde{f}_k^T \quad \text{for } k = 3, 5, 6.\end{aligned} \tag{2.56}$$

Now, we can follow the discussion in Section 2.1 for Dirac neutrinos taking into account the additional constraints (2.56) for Majorana neutrinos. The hermiticity of j_μ^{eff} and current conservation lead to an expression similar to that in (2.37):

$$\widetilde{\Lambda}_\mu(q) = \left(\gamma_\mu - q_\mu \frac{\not{q}}{q^2}\right)\left[\widetilde{f}_Q(q^2) + \widetilde{f}_A(q^2)q^2\gamma_5\right] - i\sigma_{\mu\nu}q^\nu\left[\widetilde{f}_M(q^2) + i\widetilde{f}_E(q^2)\gamma_5\right], \quad (2.57)$$

with $\widetilde{f}_Q^{fi} = \widetilde{f}_3^{fi}$, $\widetilde{f}_M^{fi} = i\widetilde{f}_5^{fi}$, $\widetilde{f}_E^{fi} = -2i\widetilde{f}_6^{fi}$, and $\widetilde{f}_A^{fi} = -\widetilde{f}_2^{fi}/(m_f + m_i)$. For the Hermitian 3×3 form factor matrices in the space of massive neutrinos,

$$\widetilde{f}_\Omega = \widetilde{f}_\Omega^\dagger \quad (\Omega = Q, M, E, A), \quad (2.58)$$

the Majorana constraints (2.56) imply that

$$\widetilde{f}_\Omega = -\widetilde{f}_\Omega^T \quad (\Omega = Q, M, E), \quad (2.59)$$

$$\widetilde{f}_A = \widetilde{f}_A^T. \quad (2.60)$$

These relations confirm the expectation discussed above that for Majorana neutrinos the charge, magnetic and electric form factor matrices are antisymmetric and the anapole form factor matrix is symmetric.

Since $\widetilde{f}_Q$, $\widetilde{f}_M$, and $\widetilde{f}_E$ are antisymmetric, a Majorana neutrino does not have diagonal charge and dipole magnetic, and electric form factors. It can only have a diagonal anapole form factor. On the other hand, Majorana neutrinos can have as many off-diagonal (transition) form factors as Dirac neutrinos.

Since the form factor matrices are Hermitian as in the Dirac case, $\widetilde{f}_Q$, $\widetilde{f}_M$, and $\widetilde{f}_E$ are imaginary, whereas $\widetilde{f}_A$ is real:

$$\widetilde{f}_\Omega = -\widetilde{f}_\Omega^* \quad (\Omega = Q, M, E), \quad (2.61)$$

$$\widetilde{f}_A = \widetilde{f}_A^*. \quad (2.62)$$

Considering now CP invariance, the case of Majorana neutrinos is rather different from that of Dirac neutrinos, because the CP phases of the massive Majorana fields ν_k are constrained by the CP invariance of the Lagrangian Majorana mass term:

$$\mathcal{L}_M = \frac{1}{2}\sum_k m_k \nu_k^T\, \mathcal{C}^\dagger \nu_k. \quad (2.63)$$

In order to prove this statement, let us first notice that since a massive Majorana neutrino field ν_k is constrained by the Majorana relation in (2.51), only the parity transformation part is effective in a CP transformation:

$$\mathrm{U}_{\mathrm{CP}}\nu_k(x)\mathrm{U}_{\mathrm{CP}}^\dagger = \xi_k^{\mathrm{CP}}\gamma^0\nu_k(x_{\mathrm{P}}). \quad (2.64)$$

Considering the mass term in (2.63), we have

$$U_{\rm CP}\nu_k^T\, \mathcal{C}^\dagger \nu_k U_{\rm CP}^\dagger = -\xi_k^{{\rm CP}2} \nu_k^T \mathcal{C}^\dagger \nu_k. \tag{2.65}$$

Therefore,

$$\text{CP} \Longleftrightarrow \xi_k^{\rm CP} = \eta_k i, \tag{2.66}$$

with $\eta_k = \pm 1$. These CP signs can be different for the different massive neutrinos, even if they all take part to the standard charged-current weak interactions through neutrino mixing, because they can be compensated by the Majorana CP phases in the mixing matrix (see [1]). Therefore, from (2.40), we have

$$\widetilde{\Lambda}_\mu^{fi}(q) \xrightarrow{\rm CP} \eta_f \eta_i \gamma^0 \mathcal{C} \left[\widetilde{\Lambda}_\mu^{if}(q_{\rm P})\right]^T \mathcal{C}^\dagger \gamma^0. \tag{2.67}$$

Imposing a CP constraint analogous to that in (2.24), we obtain

$$\text{CP} \Longleftrightarrow \begin{cases} f_\Omega^{fi} = \eta_f \eta_i f_\Omega^{if} = \eta_f \eta_i \left(f_\Omega^{fi}\right)^*, \\ f_E^{fi} = -\eta_f \eta_i f_E^{if} = -\eta_f \eta_i \left(f_E^{fi}\right)^*, \end{cases} \tag{2.68}$$

with $\Omega = Q, M, A$. Taking into account the constraints (2.61) and (2.62), we have two cases:

$$\text{CP}, \eta_f = \eta_i \Longleftrightarrow f_Q^{fi} = f_M^{fi} = 0, \tag{2.69}$$

$$\text{CP}, \eta_f = -\eta_i \Longleftrightarrow f_E^{fi} = f_A^{fi} = 0. \tag{2.70}$$

Therefore, if CP is conserved, two massive Majorana neutrinos can have either a transition electric form factor or a transition magnetic form factor, but not both, and the transition electric form factor can exist only together with a transition anapole form factor, whereas the transition magnetic form factor can exist only together with a transition charge form factor. In the diagonal case $f = i$, (2.69) does not give any constraint, because only diagonal anapole form factors are allowed for Majorana neutrinos.

2.3. *Form Factors in Gauge Models*

From the demand that the form factors at zero momentum transfer, $q^2 = 0$, are elements of the scattering matrix, it follows that in any consistent theoretical model the form factors in the matrix element (2.7) should be gauge independent and finite. Then, the form factors values at $q^2 = 0$ determine the static electromagnetic properties of the neutrino that can be probed or measured in the direct interaction with external electromagnetic fields. This is the case for charge, dipole magnetic, and electric neutrino form factors in the minimally extended Standard Model.

In non-Abelian gauge theories, the form factors in the matrix element (2.7) at nonzero momentum transfer, $q^2 \neq 0$, can be noninvariant under gauge transformations. This happens because in general the off-shell photon propagator is gauge dependent. Therefore, the one-photon approximation is not enough to get physical quantities. In this case, the form factors in the matrix element (2.7) cannot be directly measured in an experiment with an external electromagnetic field. However, they can contribute to higher-order diagrams describing some processes that are accessible for experimental observation (see [31]).

Note that there is an important difference between the electromagnetic vertex function of massive and massless neutrinos [32, 33]. For the case of a massless neutrino, the matrix element (2.7) of the electromagnetic current can be expressed in terms of only one Dirac form factor $f_D(q^2)$ (see also [30]):

$$\overline{u}(p')\Lambda_\mu(q)u(p) = f_D\left(q^2\right)\overline{u}(p')\gamma_\mu(1+\gamma_5)u(p). \tag{2.71}$$

It follows that the electric charge and anapole form factors for a massless neutrino are related to the Dirac form factor $f_D(q^2)$, and hence to each other:

$$f_Q\left(q^2\right) = f_D\left(q^2\right), \qquad f_A\left(q^2\right) = \frac{f_D(q^2)}{q^2}. \tag{2.72}$$

In the case of a massive neutrino, there is no such simple relation between electric charge and anapole form factors since the $q_\mu \not{q}\gamma_5$ term in the anapole part of the vertex function (2.18) cannot be neglected.

Moreover, a direct calculation of the massive neutrino electromagnetic vertex function, taking into account all the diagrams in Figures 15–17 of [13], reveals that each of the Feynman diagrams gives nonzero contribution to the term proportional to $\gamma_\mu\gamma_5$ [32, 33]. These contributions are not vanishing even at $q^2 = 0$. Therefore, in addition to the usual four terms in (2.18) an extra term proportional to $\gamma_\mu\gamma_5$ appears and a corresponding additional form factor $f_5(q^2)$ must be introduced. This problem is related to the decomposition of the massive neutrino electromagnetic vertex function. The calculation of the contributions of the proper vertex diagrams (Figure 15 of [13]) and $\gamma - Z$ self-energy diagrams (Figures 16 and 17 of [13]) for arbitrary gauge fixing parameter $\alpha = 1/\xi$ in the general R_ξ gauge and arbitrary mass parameter $a = m_l^2/m_W^2$ shows that at least in the zeroth and first orders of the expansion over the small neutrino mass parameter $b = (m_\nu/m_W)^2$, the corresponding "charge" $f_5(q^2 = 0)$ is zero. The cancellation of contributions from the proper vertex and self-energy diagrams to the form factor $f_5(q^2)$ at $q^2 \neq 0$,

$$f_5\left(q^2\right) = f_5^{(\gamma-Z)}\left(q^2\right) + f_5^{(\text{prop.vert.})}\left(q^2\right) = 0, \tag{2.73}$$

was also shown [32, 33] for arbitrary mass parameters a and b in the 't Hooft-Feynman gauge $\alpha = 1$.

Hence, in the minimally extended Standard Model one can perform a direct calculation of the neutrino vertex function leading to the four terms in (2.18) with gauge-invariant electric charge, magnetic, electric, and anapole moments.

3. Magnetic and Electric Dipole Moments

The neutrino dipole magnetic and electric form factors (and the corresponding magnetic and electric dipole moments) are theoretically the most well-studied and understood among the form factors. They also attract a reasonable attention from experimentalists, although the neutrino magnetic moment predicted in the extended Standard Model with right-handed neutrinos is proportional to the neutrino mass and, therefore, it is many orders of magnitude smaller than the present experimental limits obtained in terrestrial experiments.

3.1. *Theoretical Predictions*

The first calculations of the neutrino dipole moments within the minimal extension of the Standard Model with right-handed neutrinos were performed in [15–22]. The explicit evaluation of the one-loop contributions to the neutrino dipole moments in the leading approximation over the small parameters $b_i = m_i^2/m_W^2$ (where m_i are the neutrino masses, $i = 1,2,3$), that in addition exactly accounts for the dependence on the small parameters $a_l = m_l^2/m_W^2$ (with $l = e, \mu, \tau$), yields, for Dirac neutrinos [7, 18–23],

$$\left.\begin{matrix} \mu_{ij}^{\mathrm{D}} \\ \epsilon_{ij}^{\mathrm{D}} \end{matrix}\right\} = \frac{eG_F}{8\sqrt{2}\pi^2}(m_i \pm m_j) \sum_{l=e,\mu,\tau} f(a_l) U_{li}^* U_{lj}, \tag{3.1}$$

where

$$f(a_l) = \frac{3}{4}\left[1 + \frac{1}{1-a_l} - \frac{2a_l}{(1-a_l)^2} - \frac{2a_l^2 \ln a_l}{(1-a_l)^3}\right]. \tag{3.2}$$

All the charged lepton parameters a_l are small. In the limit $a_l \ll 1$, one has

$$f(a_l) \simeq \frac{3}{2}\left(1 - \frac{a_l}{2}\right). \tag{3.3}$$

From (3.1) and (3.3), the diagonal magnetic moments of Dirac neutrinos are given by

$$\mu_{ii}^{\mathrm{D}} \simeq \frac{3eG_F m_i}{8\sqrt{2}\pi^2}\left(1 - \frac{1}{2}\sum_{l=e,\mu,\tau} a_l |U_{li}|^2\right). \tag{3.4}$$

This result exhibits the following important features. The magnetic moment of a Dirac neutrino is proportional to the neutrino mass and for a massless Dirac neutrino in the Standard Model (in the absence of right-handed charged currents), the magnetic moment is zero. The magnetic moment of a massive Dirac neutrino, at the leading order in a_l, is independent of the neutrino mixing matrix and of the values of the charged lepton masses. The numerical value of the Dirac neutrino magnetic moment is

$$\mu_{ii}^{\mathrm{D}} \simeq 3.2 \times 10^{-19}\left(\frac{m_i}{\mathrm{eV}}\right)\mu_B. \tag{3.5}$$

Taking into account the existing constraint of the order of 1 eV on the neutrino masses (see [1–4]), this value is several orders of magnitude smaller than the present experimental limits, which are discussed in Section 3.4.

From (3.1), it can be clearly seen that in the extended Standard Model with right-handed neutrinos the static (diagonal) electric dipole moment of a Dirac neutrino vanishes, $\epsilon_{ii}^{\mathrm{D}} = 0$, in spite of possible CP violations generated by the Dirac phase in the mixing matrix (as shown in (2.27), Dirac neutrinos may have nonzero diagonal electric moments only in theories where CP invariance is violated). For a Majorana neutrino, both the diagonal magnetic and electric moments are zero, $\mu_{ii}^{M} = \epsilon_{ii}^{M} = 0$, as shown in Section 2.2.

Let us consider now the neutrino transition moments, which are given by (3.1) for $i \neq j$. Considering only the leading term $f(a_l) \simeq 3/2$ in the expansion (3.3), one gets vanishing transition moments, because of the unitarity relation:

$$\sum_{l} U_{li}^{*} U_{lj} = \delta_{ij}. \tag{3.6}$$

Therefore, the first nonvanishing contribution comes from the second term in the expansion (3.3) of $f(a_l)$, which contains the additional small factor $a_l = m_l^2/m_W^2$:

$$\left.\begin{matrix} \mu_{ij}^{\mathrm{D}} \\ \epsilon_{ij}^{\mathrm{D}} \end{matrix}\right\} \simeq -\frac{3eG_F}{32\sqrt{2}\pi^2}(m_i \pm m_j) \sum_{l=e,\mu,\tau} \left(\frac{m_l}{m_W}\right)^2 U_{li}^{*} U_{lj}, \tag{3.7}$$

for $i \neq j$. Thus, the transition moments are suppressed with respect to the diagonal magnetic moments in (3.4). This suppression is called "GIM mechanism," in analogy with the suppression of flavor-changing neutral currents in hadronic processes discovered in [34]. Numerically, the Dirac transition moments are given by

$$\left.\begin{matrix} \mu_{ij}^{\mathrm{D}} \\ \epsilon_{ij}^{\mathrm{D}} \end{matrix}\right\} \simeq -4 \times 10^{-23} \left(\frac{m_i \pm m_j}{\mathrm{eV}}\right) f_{ij} \mu_B, \tag{3.8}$$

with

$$f_{ij} = \sum_{l=e,\mu,\tau} \left(\frac{m_l}{m_\tau}\right)^2 U_{li}^{*} U_{lj}. \tag{3.9}$$

Also, Majorana neutrinos can have nonvanishing transition magnetic and electric moments, as discussed in Section 2.2. Assuming CP conservation and neglecting model-dependent Feynman diagrams depending on the details of the scalar sector [7, 21, 22, 24], if ν_i and ν_j have the same CP phase,

$$\mu_{ij}^{M} = 0, \qquad \epsilon_{ij}^{M} = 2\epsilon_{ij}^{\mathrm{D}}, \tag{3.10}$$

whereas if ν_i and ν_j have opposite CP phases,

$$\mu_{ij}^{M} = 2\mu_{ij}^{D}, \qquad \epsilon_{ij}^{D} = 0, \tag{3.11}$$

with ϵ_{ij}^{D} and μ_{ij}^{D} given by (3.1). Hence, although the nonvanishing Majorana transition moments are twice the Dirac ones, they are equally suppressed by the GIM mechanism. However, the model-dependent contributions of the scalar sector can enhance the Majorana transition moments (see [21, 35, 36]).

In recent studies, the value of the diagonal magnetic moment of a massive Dirac neutrino was calculated in the one-loop approximation in the extended Standard Model with right-handed neutrinos, accounting for the dependence on the neutrino mass parameter $b_i = m_i^2/m_W^2$ [37] and accounting for the exact dependence on both mass parameters b_i and $a_l = m_l^2/m_W^2$ [32, 33]. The calculations of the neutrino magnetic moment which take into account exactly the dependence on the masses of all particles can be useful in the case of a heavy neutrino with a mass comparable or even exceeding the values of the masses of other known particles. Note that the LEP data require that the number of light neutrinos coupled to the Z boson is three [38]. Therefore, any additional active neutrino must be heavier than $m_Z/2$. This possibility is not excluded by current data (see [39]).

For a heavy neutrino with mass m_i, much larger than the charged lepton masses but smaller than the W-boson mass (2 GeV $\ll m_i \ll$ 80 GeV), the authors of [32, 33] obtained the diagonal magnetic moment:

$$\mu_{ii} \simeq \frac{3eG_F}{8\pi^2\sqrt{2}} m_i \left(1 + \frac{5}{18} b_i\right), \tag{3.12}$$

whereas for a heavy neutrino with mass m_i much larger than the W-boson mass, they got

$$\mu_{ii} \simeq \frac{eG_F}{8\pi^2\sqrt{2}} m_i. \tag{3.13}$$

Note that in both cases the Dirac neutrino magnetic moment is proportional to the neutrino mass. This is an expected result, because the calculations have been performed within the extended Standard Model with right-handed neutrinos.

At this point, a question arises: "Is a neutrino magnetic moment always proportional to the neutrino mass?." The answer is "No." For example, much larger values of the Dirac neutrino magnetic moment can be obtained in $SU(2)_L \times SU(2)_R \times U(1)$ left-right symmetric models with direct right-handed neutrino interactions (see [15, 40–42]). The massive gauge bosons states W_1 and W_2 have, respectively, predominant left-handed and right-handed coupling, since

$$W_1 = W_L \cos\xi - W_R \sin\xi, \qquad W_2 = W_L \sin\xi + W_R \cos\xi, \tag{3.14}$$

where ξ is a small mixing angle and the fields W_L and W_R have pure $V \pm A$ interactions. The magnetic moment of a neutrino ν_l calculated in this model, neglecting neutrino mixing, is

$$\mu_{\nu_l} = \frac{eG_F}{2\sqrt{2}\pi^2}\left[m_l\left(1 - \frac{m_{W_1}^2}{m_{W_2}^2}\right)\sin 2\xi + \frac{3}{4}m_{\nu_l}\left(1 + \frac{m_{W_1}^2}{m_{W_2}^2}\right)\right], \tag{3.15}$$

where the term proportional to the charged lepton mass m_l is due to the left-right mixing. This term can exceed the second term in (3.15), which is proportional to the neutrino mass m_{ν_l}.

3.2. Neutrino-Electron Elastic Scattering

The most sensitive and widely used method for the experimental investigation of the neutrino magnetic moment is provided by direct laboratory measurements of low-energy elastic scattering of neutrinos and antineutrinos with electrons in reactor, accelerator and solar experiments. Detailed descriptions of several experiments can be found in [12, 43].

Extensive experimental studies of the neutrino magnetic moment, performed during many years, are stimulated by the hope to observe a value much larger than the prediction in (3.5) of the minimally extended Standard Model with right-handed neutrinos. It would be a clear indication of new physics beyond the extended Standard Model. For example, the effective magnetic moment in $\overline{\nu}_e$-e elastic scattering in a class of extradimension models can be as large as $\sim 10^{-10}\,\mu_B$ [44]. Future higher precision reactor experiments can, therefore, be used to provide new constraints on large extradimensions.

The possibility for neutrino-electron elastic scattering due to neutrino magnetic moment was first considered in [45] and the cross-section of this process was calculated in [46, 47]. Discussions on the derivation of the cross-section and on the optimal conditions for bounding the neutrino magnetic moment, as well as a collection of cross-section formulas for elastic scattering of neutrinos (antineutrinos) on electrons, nucleons, and nuclei can be found in [48, 49].

Let us consider the elastic scattering:

$$\nu + e^- \longrightarrow \nu + e^- \tag{3.16}$$

of a neutrino with energy E_ν with an electron at rest in the laboratory frame. There are two observables: the kinetic energy T of the recoil electron and the recoil angle χ with respect to the neutrino beam, which are related by

$$\cos\chi = \frac{E_\nu + m_e}{E_\nu}\left[\frac{T}{T + 2m_e}\right]^{1/2}. \tag{3.17}$$

The electron kinetic energy is constrained from the energy-momentum conservation by

$$T \le \frac{2E_\nu^2}{2E_\nu + m_e}. \tag{3.18}$$

Since, in the ultrarelativistic limit, the neutrino magnetic moment interaction changes the neutrino helicity and the Standard Model weak interaction conserves the neutrino helicity, the two contributions add incoherently in the cross-section which can be written as [49]

$$\frac{d\sigma}{dT} = \left(\frac{d\sigma}{dT}\right)_{\text{SM}} + \left(\frac{d\sigma}{dT}\right)_{\mu}. \tag{3.19}$$

The small interference term due to neutrino masses has been derived in [50].

The weak-interaction cross-section is given by

$$\left(\frac{d\sigma}{dT}\right)_{\text{SM}} = \frac{G_F^2 m_e}{2\pi}\left[(g_V + g_A)^2 + (g_V - g_A)^2\left(1 - \frac{T}{E_\nu}\right)^2 + \left(g_A^2 - g_V^2\right)\frac{m_e T}{E_\nu^2}\right], \tag{3.20}$$

with the standard coupling constants g_V and g_A given by

$$g_V = \begin{cases} 2\sin^2\theta_W + \dfrac{1}{2} & \text{for } \nu_e, \\ 2\sin^2\theta_W - \dfrac{1}{2} & \text{for } \nu_\mu, \nu_\tau, \end{cases} \tag{3.21}$$

$$g_A = \begin{cases} \dfrac{1}{2} & \text{for } \nu_e, \\ -\dfrac{1}{2} & \text{for } \nu_\mu, \nu_\tau. \end{cases} \tag{3.22}$$

In antineutrino-electron elastic scattering, one must substitute $g_A \to -g_A$.

The neutrino magnetic-moment contribution to the cross section is given by [49]

$$\left(\frac{d\sigma}{dT}\right)_{\mu} = \frac{\pi\alpha^2}{m_e^2}\left(\frac{1}{T} - \frac{1}{E_\nu}\right)\left(\frac{\mu_\nu}{\mu_B}\right)^2, \tag{3.23}$$

where μ_ν is the effective magnetic moment discussed in Section 3.3.

The two terms $(d\sigma/dT)_{\text{SM}}$ and $(d\sigma/dT)_\mu$ exhibit quite different dependences on the experimentally observable electron kinetic energy T, as illustrated in Figure 2, where the values of the two terms averaged over the typical antineutrino reactor spectrum are plotted for six values of the neutrino magnetic moment, $\mu_\nu^{(N)} = N \times 10^{-11}\,\mu_B$, with $N = 1, 2, 3, 4, 5, 6$ [43] (see also [49]). One can see that small values of the neutrino magnetic moment can be probed by lowering the electron recoil energy threshold. In fact, from (3.20) and (3.23), one can find that $(d\sigma/dT)_\mu$ exceeds $(d\sigma/dT)_{\text{SM}}$ for

$$T \lesssim \frac{\pi^2\alpha^2}{G_F^2 m_e^3}\left(\frac{\mu_\nu}{\mu_B}\right)^2. \tag{3.24}$$

It was proposed in [51] that electron binding in atoms (the "atomic ionization" effect in neutrino interactions on Ge target) can significantly increase the electromagnetic contribution

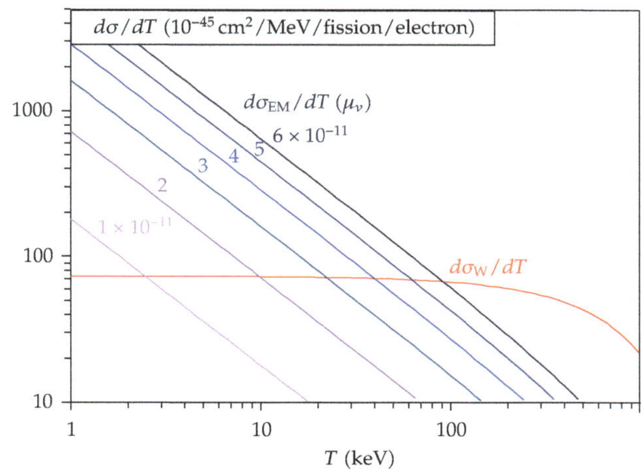

Figure 2: Standard Model weak (W) and magnetic moment electromagnetic (EM) contributions to the cross section for several values of the neutrino magnetic moment [43].

to the differential cross-section with respect to the free electron approximation. However, detailed considerations of the atomic ionization effect in (anti)neutrino atomic electron scattering experiments presented in [52–57] show that the effect is by far too small to have measurable consequences even in the case of the low energy threshold of 2.8 keV reached in the GEMMA experiment [58].

3.3. *Effective Dipole Moments*

In scattering experiments, the neutrino is created at some distance from the detector as a flavor neutrino, which is a superposition of massive neutrinos. Therefore, the magnetic moment that is measured in these experiment is not that of a massive neutrino, but it is an effective magnetic moment which takes into account neutrino mixing and the oscillations during the propagation between source and detector [50, 59]:

$$\mu_\nu^2(\nu_\alpha, L, E) = \sum_j \left| \sum_k U_{\alpha k}^* e^{-im_k^2 L/2E} (\mu_{jk} - i\epsilon_{jk}) \right|^2, \tag{3.25}$$

where we have written explicitly the dependence from the initial neutrino flavor ν_α, the distance L, and the energy E. In this expression of the effective μ_ν one can see that in general both the magnetic and electric dipole moments contribute to the elastic scattering. Note that $\mu_\nu(\nu_\alpha, L, E)$ depends only on the neutrino squared-mass differences: considering for simplicity only the magnetic moment contribution, we have

$$\mu_\nu^2(\nu_\alpha, L, E) = \sum_j \sum_{kk'} U_{\alpha k}^* U_{\alpha k'} e^{-i\Delta m_{kk'}^2 L/2E} \mu_{jk} \mu_{jk'}. \tag{3.26}$$

In the case of Majorana neutrinos, there are no diagonal magnetic and electric dipole moments and μ_ν^2 receives contributions only from the transition dipole moments.

Furthermore, if CP is conserved, there are either only magnetic or electric transition dipole moments (see Section 2.2).

The general expression for $\mu_\nu^2(\nu_\alpha, L, E)$ can be simplified in some cases [59]. For instance, for Dirac neutrinos with only diagonal magnetic moments $\mu_{ij} = \mu_i\delta_{ij}$, we have the effective flavor magnetic moment:

$$\mu_\nu^2(\nu_\alpha, L, E) \longrightarrow \left(\mu_\alpha^{\mathrm{D}}\right)^2 = \sum_i |U_{\alpha i}|^2 \mu_i^2. \tag{3.27}$$

Since in this case there is no dependence on the distance L and the neutrino energy, the magnetic cross-section is characterized by the initial neutrino flavor rather than by the composition of mass states in the detector. In this case, measurements of all flavor magnetic moments and mixing parameters can allow the extraction of all the fundamental moments μ_i.

3.4. Experimental Limits

The constraints on the neutrino magnetic moment in direct laboratory experiments have been obtained so far from the lack of any observable distortion of the recoil electron energy spectrum. Experiments of this type have started about 40 years ago. The strategy applied during all these years in reactor experiments is rather simple: minimize the threshold on the recoil energy for the detection of the scattered electron, keeping at the same time a reasonable background level. Since the region of interest coincides with the energy region dominated by radioactivity, low intrinsic radioactivity detectors have been employed together with active and passive shields. In addition, all the experiments were running inside laboratories with significant overburden of concrete, corresponding to several meters of water. This was enough to suppress the soft component of cosmic rays.

In all the experiments, with one exception only, the signal due to the antineutrinos from the reactor (about 2×10^{20} s^{-1} GW thermal^{-1}) is obtained from the difference between the reactor-on and reactor-off rate. Clearly, this requires the same background with the reactor on as with the reactor off.

$\overline{\nu}_e$-e elastic scattering was first observed in the pioneering experiment [60] at the Savannah River Laboratory. The setup was made of a 15.9 kg plastic scintillator target divided into 16 optically isolated elements, totally enclosed inside a 300 kg NaI crystal shielded with lead and cadmium. Finally, the entire setup was immersed into 2200 liters of liquid scintillator. Both the NaI and the liquid scintillator detectors were working as veto against cosmics and gamma-rays from the laboratory. In the electron kinetic energy range from 1.5 MeV to 4.5 MeV, a reactor-on rate of 47.5 ± 1 events/day was measured, to be compared with a reactor-off rate of 40.4 ± 0.9 events/day. A revised analysis of the Savannah River Laboratory data [49] with an improved reactor neutrino spectrum and a more precise value of $\sin^2\theta_W$ gave hints for a neutrino magnetic moment on the order of (2–4)$\times 10^{-10}$ μ_B.

However, lower limits were then obtained by two experiments performed at nuclear reactors in Russia. The Krasnoyarsk experiment [61] had a 103 kg target of liquid organofluoric scintillator contained into seven scintillation chambers. The absence of hydrogen in the target was a significant improvement, since the $\overline{\nu}_e$-p charged-current reaction, which has a much larger cross-section than $\overline{\nu}_e$-e, had been an important background source in the Savannah River experiment. The active target was then surrounded by a passive shield of steel, copper, lead, and borated polyethylene. Finally, two layers of plastic scintillators were vetoing the cosmic muons. With a count rate of 8.3 ± 0.3 (reactor-on) and 7.1 ± 0.4 (reactor-off) in the electron energy range 3.15–5.18 MeV, the Krasnoyarsk collaboration obtained the limit:

$$\mu_{\bar{\nu}_e} \leq 2.4 \times 10^{-10}\, \mu_B \quad (90\%\ \text{C.L.}). \tag{3.28}$$

A completely different detector was built for the Rovno [62] experiment: 600 silicon detectors, for a total mass of 75 kg, with a passive shield of mercury, copper, cadmium absorber, and graphite. Finally, the setup was enclosed inside a veto made of plastic scintillators. With a rate of 4963 ± 12 events/day (reactor-on) and 4921 ± 16 events/day (reactor-off) in the electron recoil energy range 0.6–2 MeV, it has been possible to obtain a limit on the neutrino magnetic moment of

$$\mu_{\bar{\nu}_e} \leq 1.9 \times 10^{-10}\, \mu_B \quad (90\%\ \text{C.L.}). \tag{3.29}$$

Finally, more stringent limits have been obtained in the two most recent experiments at rectors. TEXONO [63] has been performed at the Kuo-Sheng nuclear power station. The detector, a 1.06 kg high-purity germanium, was completely surrounded by NaI (Tl) and CsI (Tl) crystals working as anti-Compton. The whole setup was contained inside a shield made of copper, boron-loaded polyethylene, stainless steel, lead, and plastic scintillators. A background of about 1 event/keV·kg·day could be achieved above the threshold of 12 keV, giving the limit:

$$\mu_{\bar{\nu}_e} \leq 7.4 \times 10^{-11}\, \mu_B \quad (90\%\ \text{C.L.}). \tag{3.30}$$

At the moment, the world best limit is coming from the GEMMA experiment at the Kalinin nuclear power plant. A 1.5 kg high-purity germanium detector is placed inside a cup-shaped NaI crystal and surrounded by copper, lead, and plastic scintillators. With an energy threshold as low as 2.8 keV, the GEMMA collaboration obtained [58]

$$\mu_{\bar{\nu}_e} \leq 2.9 \times 10^{-11}\, \mu_B \quad (90\%\ \text{C.L.}). \tag{3.31}$$

The experiment which followed a strategy different from the study of the reactor-on and reactor-off rate was MUNU. As a matter of fact, the detector [64] was able to provide not only the energy but also the topology of events. As a consequence, the initial direction of an electron track could be measured and the electron scattering angle reconstructed. This allowed to look for the reactor signal by comparing forward electrons, having as reference the reactor to detector axis, with the backward ones. In this way, the background is measured online, which eliminates problems from detector instabilities, as well as from a possible time dependence of the background itself. The central component of the detector consisted of an acrylic vessel time-projected chamber (a cylinder 90 cm in diameter and 162 cm long) filled with CF_4 at 3 bar pressure and immersed in a steel tank (2 m diameter and 3.8 m long) filled with 10 m^3 liquid scintillator viewed by 48 photomultipliers. The total target mass of CF_4 was 11.4 kg. Finally, the setup was surrounded by boron-loaded polyethylene and lead. With a total rate of 6.8 ± 0.3 events/day in the forward direction and a background of 5.8 ± 0.17 events/day, the following upper bounds have been obtained [65]:

$$\mu_{\overline{\nu}_e} \le 9 \times 10^{-11}\, \mu_B \quad (90\%\ \text{C.L.}). \tag{3.32}$$

Several experiments at accelerators have searched for an effect due to the magnetic moment of ν_μ in ν_μ–e and $\overline{\nu}_\mu$–e elastic scattering (see [66]). The current best limit has been obtained in the LSND experiment [67]:

$$\mu_{\nu_\mu} \le 6.8 \times 10^{-10}\, \mu_B \quad (90\%\ \text{C.L.}). \tag{3.33}$$

The DONUT collaboration has investigated ν_τ–e and $\overline{\nu}_\tau$–e elastic scattering, finding the limit [68]:

$$\mu_{\nu_\tau} \le 3.9 \times 10^{-7}\, \mu_B \quad (90\%\ \text{C.L.}). \tag{3.34}$$

Solar neutrino experiments as Super-Kamiokande and Borexino can also search for a neutrino magnetic moment signal by studying the shape of the electron spectrum. Since the neutrino magnetic moment depends both on the mixing and on the propagation properties of the neutrino, then oscillations are here relevant.

The analysis of the recoil electron spectrum generated by solar neutrinos in the Super-Kamiokande experiment experiment gave [69]

$$\mu_\nu \le 1.1 \times 10^{-10}\, \mu_B \quad (90\%\ \text{C.L.}), \tag{3.35}$$

where μ_ν is not the same as $\mu_{\overline{\nu}_e}$ since it is given by a different combination of the magnetic moment of the neutrino mass eigenstates (see Section 3.3).

The limit

$$\mu_\nu \le 5.4 \times 10^{-11}\, \mu_B \quad (90\%\ \text{C.L.}) \tag{3.36}$$

has been recently obtained in the Borexino solar neutrino scattering experiment [70]. An upper limit on the neutrino magnetic moment $\mu_\nu \le 8.4 \times 10^{-11}\, \mu_B$ has been found in an independent analysis of the first release of the Borexino experiment data performed in [71]. It was also shown that with reasonable assumptions on the oscillation probability this limit translates into the conservative upper limits on the magnetic moments of ν_μ and ν_τ:

$$\mu_{\nu_\mu} \le 1.5 \times 10^{-10}\, \mu_B, \qquad \mu_{\nu_\tau} \le 1.9 \times 10^{-10}\, \mu_B \quad (90\%\ \text{C.L.}). \tag{3.37}$$

The limit on μ_{ν_τ} is three order of magnitude stronger than the direct limit (3.34).

The global fit [72, 73] of the magnetic moment data from the reactor and solar neutrino experiments for the Majorana neutrinos produces limits on the neutrino transition moments:

$$\mu_{23}, \mu_{31}, \mu_{12} < 1.8 \times 10^{-10}\, \mu_B \quad (90\%\ \text{C.L.}). \tag{3.38}$$

Finally, an interesting new possibility for providing more stringent constraints on the neutrino magnetic moment from $\overline{\nu}_e$-e scattering experiments was discussed in [74] on the basis of an observation [75] that "dynamical zeros" appear in the Standard Model contribution to the scattering cross-section.

3.5. Theoretical Considerations

As it was already mentioned before, there is a gap of many orders of magnitude between the present experimental limits $\sim 10^{-11}\,\mu_B$ on neutrino magnetic moments (discussed in Section 3.4) and the prediction (3.5) of the minimal extension of the Standard Model with right-handed neutrinos. At the same time, the experimental sensitivity of reactor $\overline{\nu}_e$-e elastic scattering experiments has improved by only one order of magnitude during a period of about twenty years (see [49], where a sensitivity of $\sim 10^{-10}\,\mu_B$ is discussed). However, the experimental studies of neutrino magnetic moments are stimulated by the hope that new physics beyond the minimally extended Standard Model with right-handed neutrinos might give much stronger contributions. One of the examples in which it is possible to avoid the neutrino magnetic moment being proportional to a (small) neutrino mass, that would in principle make a neutrino magnetic moment accessible for experimental observations, is realized in the left-right symmetric models considered at the end of Section 3.1.

Other interesting possibilities of obtaining neutrino magnetic moments lager than the prediction (3.5) of the minimal extension the Standard Model with right-handed neutrinos have been considered recently. In this concern, we note that it was proposed in [44] to probe a class of large extradimensions models with future reactors searches for neutrino magnetic moments. The results obtained within the Minimal Supersymmetric Standard Model with R-parity violating interactions [76, 77] show that the Majorana transition magnetic moment might be significantly above the scale of (3.5).

Considering the problem of large neutrino magnetic moments, one can write down a generic relation between the size of a neutrino magnetic moment μ_ν and the corresponding neutrino mass m_ν [28, 29, 35, 78–80]. Suppose that a large neutrino magnetic moment is generated by physics beyond a minimal extension of the Standard Model at an energy scale characterized by Λ. For a generic diagram corresponding to this contribution to μ_ν, one can again use the Feynman graph in Figure 1(b); the shaded circle in this case denotes effects of new physics beyond the Standard Model. The contribution of this diagram to the magnetic moment is

$$\mu_\nu \sim \frac{eG}{\Lambda}, \tag{3.39}$$

where e is the electric charge and G is a combination of coupling constants and loop factors. The same diagram of Figure 1(b) but without the photon line gives a new physics contribution to the neutrino mass:

$$\delta m_\nu \sim G\Lambda. \tag{3.40}$$

Combining the estimates (3.39) and (3.40), one can get the relation:

$$\delta m_\nu \sim \frac{\Lambda^2}{2m_e}\frac{\mu_\nu}{\mu_B} = \frac{\mu_\nu}{10^{-18}\mu_B}\left(\frac{\Lambda}{1\ \text{TeV}}\right)^2 \text{eV} \tag{3.41}$$

between the one-loop contribution to the neutrino mass and the neutrino magnetic moment.

It follows that, generally, in theoretical models that predict large values for the neutrino magnetic moment, simultaneously large contributions to the neutrino mass arise. Therefore, a particular fine tuning is needed to get a large value for the neutrino magnetic moment while keeping the neutrino mass within experimental bounds. One of the possibilities [78] is based on the idea of suppressing the ratio m_ν/μ_ν with a symmetry: if a $SU(2)_\nu$ symmetry is an exact symmetry of the Lagrangian of a model, because of different symmetry properties of the mass and magnetic moment even a massless neutrino can have a nonzero magnetic moment. If, as it happens in a realistic model, the $SU(2)_\nu$ symmetry is broken and if this breaking is small, the ratio m_ν/μ_ν is also small, giving a natural way to obtain a magnetic moment of the order of $\sim 10^{-11}\mu_B$ without contradictions with the neutrino mass experimental constraints. Several possibilities based on the general idea of [78] were considered in [81–86].

Another idea of neutrino mass suppression without suppression of the neutrino magnetic moment was discussed in [35] within the Zee model [87], which is based on the Standard Model gauge group $SU(2)_L \times U(1)_Y$ and contains at least three Higgs doublets and a charged field which is a singlet of $SU(2)_L$. For this kind of models, there is a suppression of the neutrino mass diagram, while the magnetic moment diagram is not suppressed.

It is possible to show with more general and rigorous considerations [28, 29, 80] that the Λ^2 dependence in (3.41) arises from the quadratic divergence in the renormalization of the dimension-four neutrino mass operator. A general and model-independent upper bound on the Dirac neutrino magnetic moment, which can be generated by an effective theory beyond the Standard Model, has been derived [28, 29, 80] from the demand of absence of fine-tuning of effective operator coefficients and from the current experimental information on neutrino masses. A model with Dirac fermions, scalars, and gauge bosons that is valid below the scale Λ and respects the Standard Model $SU(2)_L \times U(1)_Y$ symmetry was considered. Integrating out the physics above the scale Λ, the following effective Lagrangian that involves right-handed neutrinos ν_R, lepton isodoublets and the Higgs doublet can be obtained:

$$\mathcal{L}_{\text{eff}} = \sum_{n,j} \frac{C_j^n(\mu)}{\Lambda^{n-4}} \mathcal{O}_j^{(n)}(\mu) + \text{H.c.}, \tag{3.42}$$

where μ is the renormalization scale, $n \geq 4$ denotes the operator dimension and j runs over independent operators of a given dimension. For $n = 4$, a neutrino mass arises from the operator $\mathcal{O}_1^{(4)} = \overline{L}\tilde{\Phi}\nu_R$, where $\tilde{\Phi} = i\sigma_2\Phi^*$. In addition, if the scale Λ is not extremely large with respect to the electroweak scale, an important contribution to the neutrino mass can arise also from higher dimension operators. At this point, it is important to note that the combination of the $n = 6$ operators appearing in the Lagrangian (3.42) contains the magnetic moment operator $\overline{\nu}\sigma_{\mu\nu}\nu F^{\mu\nu}$ and also generates a contribution δm_ν to the neutrino mass [28, 29, 80]. Solving the renormalization group equation from the scale Λ to the electroweak scale, one finds that the contributions to the neutrino magnetic moment and to the neutrino mass are connected to each other by

$$\left|\mu_\nu^{\mathrm{D}}\right| = \frac{16\sqrt{2}G_F m_e \delta m_\nu \sin^4\theta_W}{9\alpha^2 |f| \ln(\Lambda/v)} \mu_B, \tag{3.43}$$

where α is the fine structure constant, v is the vacuum expectation value of the Higgs doublet,

$$f = 1 - r - \frac{2}{3}\tan^2\theta_W - \frac{1}{3}(1+r)\tan^4\theta_W, \tag{3.44}$$

and r is a ratio of effective operator coefficients defined at the scale Λ which is of order unity without fine tuning. If the neutrino magnetic moment is generated by new physics at a scale $\Lambda \sim 1\,\mathrm{TeV}$ and the corresponding contribution to the neutrino mass is $\delta m_\nu \lesssim 1\,\mathrm{eV}$, then the bound $\mu_\nu \lesssim 10^{-14}\,\mu_B$ can be obtained. This bound is some orders of magnitude stronger than the constraints from reactor and solar neutrino scattering experiments discussed before.

The model-independent limit on a Majorana neutrino transition magnetic moment μ_ν^{M} was also discussed in [28, 29, 80]. However, the limit in the Majorana case is much weaker than that in the Dirac case, because for a Majorana neutrino, the magnetic moment contribution to the mass is Yukawa suppressed. The limit on μ_ν^{M} is also weaker than the present experimental limits if μ_ν^{M} is generated by new physics at the scale $\Lambda \sim 1$ TeV. An important conclusion of [28, 29, 80], based on model-independent considerations of the contributions to μ_ν, is that if a neutrino magnetic moment of order $\mu_\nu \geq 10^{-15}\,\mu_B$ were observed in an experiment, it would give a proof that neutrinos are Majorana rather than Dirac particles.

4. Neutrino Charge Radius

Even if the electric charge of a neutrino is vanishing, the electric form factor $f_Q(q^2)$ can still contain nontrivial information about neutrino electromagnetic properties. Considering $f_Q(0) = 0$, in the static limit ($q^2 \to 0$), the electric form factor is given by

$$f_Q\left(q^2\right) = q^2 \left.\frac{df_Q(q^2)}{dq^2}\right|_{q^2=0} + \cdots. \tag{4.1}$$

The leading contribution can be expressed in terms of a neutrino charge radius considering a static spherically symmetric charge distribution of density $\rho(r)$ (with $r = |\vec{x}|$) in the so-called "Breit frame," where $q_0 = 0$. In this approximation, we have

$$f_Q\left(q^2\right) = \int \rho(r) e^{i\vec{q}\cdot\vec{x}} d^3x = 4\pi \int \rho(r) \frac{\sin(qr)}{qr} r^2 dr, \tag{4.2}$$

where $q = |\vec{q}|$. Since $df_Q/dq^2|_{q^2=0} = -\langle r^2\rangle/6$, with $\langle r^2\rangle = \int r^2\rho(r)dr$, the neutrino charge radius is defined by

$$\left\langle r_\nu^2 \right\rangle = -6\frac{df_Q(q^2)}{dq^2}\bigg|_{q^2=0}. \tag{4.3}$$

Note that $\langle r_\nu^2\rangle$ can be negative, because the charge density $\rho(r)$ is not a positively defined function of r.

In one of the first studies [31], it was claimed that in the Standard Model and in the unitary gauge, the neutrino charge radius is ultraviolet-divergent and so it is not a physical quantity. A direct one-loop calculation [32, 33] of proper vertices and $\gamma - Z$ self-energy (Figures 15 and 16 of [13]) contributions to the neutrino charge radius performed in a general R_ξ gauge for a massive Dirac neutrino gave also a divergent result. However, it was shown [88], using the unitary gauge, that, by including in addition to the usual terms also contributions from diagrams of the neutrino-lepton neutral current scattering (Z boson diagrams), it is possible to obtain for the neutrino charge radius a gauge-dependent but finite quantity. Later on, it was also shown [16] that in order to define the neutrino charge radius as a physical quantity, one has also to consider box diagrams (see Figure 18 of [13]), which contribute to the scattering process $\nu + \ell \to \nu + \ell$, and that in combination with contributions from the proper diagrams it is possible to obtain a finite and gauge-independent value for the neutrino charge radius. In this way, the neutrino electroweak radius was defined [89, 90] and an additional set of diagrams that give contribution to its value was discussed in [91]. Finally, in a series of papers [92–94] the neutrino electroweak radius as a physical observable has been introduced. In the corresponding calculations, performed in the one-loop approximation including additional terms from the $\gamma - Z$ boson mixing and the box diagrams involving W and Z bosons, the following gauge-invariant result for the neutrino charge radius has been obtained:

$$\left\langle r_{\nu_\alpha}^2 \right\rangle = \frac{G_F}{4\sqrt{2}\pi^2}\left[3 - 2\log\left(\frac{m_\alpha^2}{m_W^2}\right)\right], \tag{4.4}$$

where m_W and m_α are the W boson and lepton masses ($\alpha = e, \mu, \tau$). This result, however, revived the discussion [95–98] on the definition of the neutrino charge radius. Numerically, for the electron neutrino electroweak radius, it yields [92–94]

$$\left\langle r_{\nu_e}^2 \right\rangle = 4 \times 10^{-33}\ \text{cm}^2, \tag{4.5}$$

which is very close to the numerical estimations obtained much earlier in [89, 90].

Note that the neutrino charge radius can be considered as an effective scale of the particle's "size," which should influence physical processes such as, for instance, neutrino scattering off electron. To incorporate the neutrino charge radius contribution in the cross-section (3.20), the following substitution [49, 99, 100] can be used:

$$g_V \longrightarrow \frac{1}{2} + 2\sin^2\theta_W + \frac{2}{3}m_W^2\left\langle r_{\nu_e}^2 \right\rangle \sin^2\theta_W. \tag{4.6}$$

Using this method, the TEXONO collaboration obtained [101]

$$-2.1 \times 10^{-32}\,\text{cm}^2 < \left\langle r_{\bar{\nu}_e}^2 \right\rangle < 3.3 \times 10^{-32}\,\text{cm}^2 \quad (90\%\ \text{C.L.}). \tag{4.7}$$

Other available bounds on the electron neutrino charge radius are from primordial nucleosynthesis [102]

$$\left\langle r_{\nu_e}^2 \right\rangle \lesssim 7 \times 10^{-33}\,\text{cm}^2, \tag{4.8}$$

from SN 1987A [103]

$$\left\langle r_{\bar{\nu}_e}^2 \right\rangle \lesssim 2 \times 10^{-33}\,\text{cm}^2, \tag{4.9}$$

from neutrino neutral-current reactions [104]

$$-2.74 \times 10^{-32}\,\text{cm}^2 < \left\langle r_{\nu_e}^2 \right\rangle < 4.88 \times 10^{-32}\,\text{cm}^2 \quad (90\%\ \text{C.L.}), \tag{4.10}$$

from solar experiments (Kamiokande II and Homestake) [105, 106]

$$\left\langle r_{\nu_e}^2 \right\rangle < 2.3 \times 10^{-32}\,\text{cm}^2 \quad (95\%\ \text{C.L.}), \tag{4.11}$$

from an evaluation of the weak mixing angle $\sin^2\theta_W$ by a combined fit of all electron neutrino elastic scattering data [107]

$$-0.13 \times 10^{-32}\,\text{cm}^2 < \left\langle r_{\nu_e}^2 \right\rangle < 3.32 \times 10^{-32}\,\text{cm}^2 \quad (90\%\ \text{C.L.}). \tag{4.12}$$

Comparing the theoretical value in (4.5) with the experimental limits in (4.7)–(4.12), one can see that they differ at most by one order of magnitude. Therefore, one may expect that the experimental accuracy will soon reach the value needed to probe the theoretical predictions for the neutrino effective charge radius.

The effects of new physics beyond the Standard Model can also contribute to the neutrino charge radius. Let us only mention that the anomalous $WW\gamma$ vertex contribution to the neutrino effective charge radius has been studied in [30], and shown to correspond to a contribution $\lesssim 10^{-34}\,\text{cm}^2$ to $|\langle r_{\nu_e}^2 \rangle|$. Note that this is only one order of magnitude lower than the expected value of the charge radius in the Standard Model.

A detailed discussion on the possibilities to constrain the ν_τ and ν_μ charge radii from astrophysical and cosmological observations and from terrestrial experiments can be found in [108].

5. Radiative Decay and Plasmon Decay

If the masses of neutrinos are nondegenerate, the radiative decay of a heavier neutrino ν_i into a lighter neutrino ν_f (with $m_i > m_f$) with emission of a photon,

$$\nu_i \longrightarrow \nu_f + \gamma, \tag{5.1}$$

may proceed in vacuum [15, 16, 18–21, 23, 109, 110]. Early discussions of the possible role of neutrino radiative decay in different astrophysical and cosmological settings can be found in [111–116].

For the case of a Dirac neutrino, the decay rate in the minimal extension of the Standard Model with right-handed neutrinos is [15, 16, 18–21, 23, 109, 110]

$$\Gamma_{\nu_i^D \to \nu_j^D + \gamma} = \frac{\alpha G_F^2}{128\pi^4}\left(\frac{m_i^2 - m_j^2}{m_j}\right)^3 \left(m_i^2 + m_j^2\right)\left|\sum_{l=e,\mu,\tau} f(a_l) U_{lj} U_{li}^*\right|^2, \tag{5.2}$$

where $f(a_l)$ is given by (3.2). Recalling the results for the Dirac neutrino magnetic and electric transition moments μ_{ij} and ϵ_{ij}, given in (3.1), one can rewrite (5.2) as (see [117, 118])

$$\Gamma_{\nu_i \to \nu_j + \gamma} = \frac{\left|\mu_{ij}\right|^2 + \left|\epsilon_{ij}^2\right|}{8\pi}\left(\frac{m_i^2 - m_j^2}{m_j}\right)^3. \tag{5.3}$$

For degenerate neutrino masses ($m_i = m_j$), the process is kinematically forbidden in vacuum.

Note that there are models (see for instance [119]) in which the neutrino radiative decay rates (as well as the magnetic moment discussed above) of a nonstandard Dirac neutrinos are much larger than those predicted in the minimally extended Standard Model.

For Majorana neutrinos, if CP is violated, the decay rate is given by (5.3). If CP is conserved, we have two cases: if the Majorana neutrinos ν_i and ν_j have the same CP eigenvalues, (2.69) and (3.10) imply that the decay process is induced purely by the neutrino electric transition dipole moment, because $\mu_{ij} = 0$; on the other hand, if the two Majorana neutrinos have opposite CP eigenvalues, from (2.70) and (3.11), one can see that the transition is purely of magnetic dipole type ($\epsilon_{ij} = 0$).

For numerical estimations, it is convenient to express (5.3) in the following form:

$$\Gamma_{\nu_i \to \nu_j + \gamma} = 5.3\left(\frac{\mu_{\text{eff}}}{\mu_B}\right)^2 \left(\frac{m_i^2 - m_j^2}{m_j^2}\right)^3 \left(\frac{m_i}{1\,\text{eV}}\right)^3 s^{-1}, \tag{5.4}$$

with the effective neutrino magnetic moment $\mu_{\text{eff}} = \sqrt{|\mu_{ij}|^2 + |\epsilon_{ij}^2|}$.

The neutrino radiative decay can be constrained by the absence of decay photons in reactor $\overline{\nu}_e$ and solar ν_e fluxes. The limits on μ_{eff} that have been obtained from these considerations are much weaker than those obtained from neutrino scattering terrestrial experiments. Stronger constraints on μ_{eff} (though still weaker than the terrestrial ones) have been obtained from the neutrino decay limit set by SN 1987A and from the limits on the

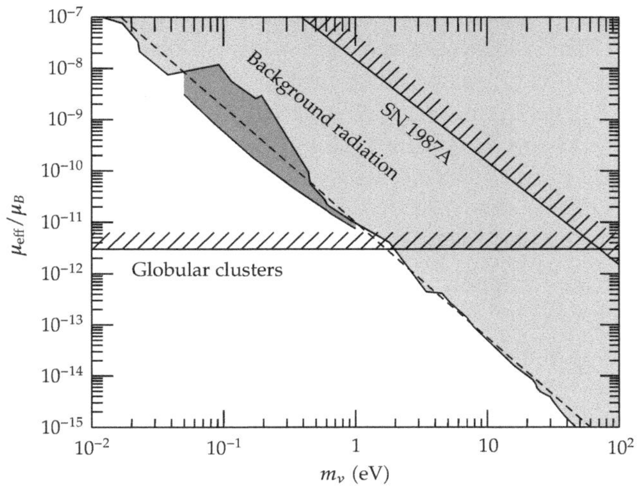

Figure 3: Astrophysical limits on neutrino transition moments [117, 118].

distortion of the Cosmic Microwave Background Radiation (CMBR). These limits can be expressed as (see [117, 118] and references therein)

$$\frac{\mu_{\text{eff}}}{\mu_{\text{B}}} < \begin{cases} 0.9 \times 10^{-1}\left(\dfrac{\text{eV}}{m_\nu}\right)^2 & \text{Reactor } (\overline{\nu}_e), \\ 0.5 \times 10^{-5}\left(\dfrac{\text{eV}}{m_\nu}\right)^2 & \text{Sun } (\nu_e), \\ 1.5 \times 10^{-8}\left(\dfrac{\text{eV}}{m_\nu}\right)^2 & \text{SN1987A (all flavors)}, \\ 1.0 \times 10^{-11}\left(\dfrac{\text{eV}}{m_\nu}\right)^{9/4} & \text{CMBR (all flavors)}. \end{cases} \tag{5.5}$$

Detailed discussions (and corresponding references) on the astrophysical constraints on the neutrino magnetic and electric transition moments, summarized in Figure 3, can be found in [117, 118].

For completeness, we would like to mention that other processes characterized by the same signature of (5.1) have been considered (for a review of the literature see [120–124]).

(i) The photon radiation by a massless neutrino ($\nu_i \rightarrow \nu_j + \gamma, i = j$) due to the vacuum polarization loop diagram in the presence of an external magnetic field [125].

(ii) The photon radiation by a massive neutrino with nonvanishing magnetic moment in constant magnetic and electromagnetic wave fields [126–129].

(iii) The Cherenkov radiation due to the nonvanishing neutrino magnetic moment in an homogeneous and infinitely extended medium, which is only possible if the speed of the neutrino is larger than the speed of light in the medium [130, 131].

(iv) The transition radiation due to a nonvanishing neutrino magnetic moment which would be produced when the neutrino crosses the interface of two media with different refractive indices [132–134].

(v) The Cherenkov radiation of a massless neutrino due to its induced charge in a medium [135, 136]. (Note that the neutrino electromagnetic properties are in general affected by the external environment. In particular, a neutrino can acquire an electric charge in magnetized matter [135, 136] and the neutrino magnetic moment depends on the strength of external electromagnetic fields [127, 137, 138]. A recent study of the neutrino electromagnetic vertex in magnetized matter can be found in [139]. See also [122, 123] for a review of neutrino interactions in external electromagnetic fields.)

(vi) The Cherenkov radiation of massive and massless neutrinos in a magnetized medium [120, 140].

(vii) The neutrino radiative decay ($\nu_i \longrightarrow \nu_j + \gamma, i \neq j$) in external fields and media (see [141–146] and references therein).

(viii) The spin light of neutrino in matter ($SL\nu$) that is a mechanism of electromagnetic radiation due to the precession or transition of magnetic or electric (transition) moments of massive neutrinos when they propagate in background matter [121, 147–151].

A very interesting process, for the purpose of constraining neutrino electromagnetic properties, is the photon (plasmon) decay into a neutrino-antineutrino pair:

$$\gamma^* \longrightarrow \nu + \overline{\nu}. \tag{5.6}$$

This process becomes kinematically allowed in media, because a photon with the dispersion relation $\omega_\gamma^2 + \vec{k}_\gamma^2 > 0$ roughly behaves as a particle with an effective mass.

Plasmon decay generated by the neutrino coupling to photons due to a magnetic moment μ_ν (and/or to a neutrino electric millicharge q_ν) was first considered in [152] as a possible source of energy loss of the Sun. The requirement that the energy loss does not exceed the solar luminosity gave [117, 118]

$$\mu_\nu \lesssim 4 \times 10^{-10}\ \mu_B, \tag{5.7}$$

and $q_\nu \lesssim 6 \times 10^{-14}\, e$.

The tightest astrophysical bound on a neutrino magnetic moment is provided by the observed properties of globular cluster stars. The plasmon decay (5.6) inside the star liberates the energy ω_γ in the form of neutrinos that freely escape the stellar environment. This nonstandard energy loss cools a red giant star so fast that it can delay helium ignition. From the lack of observational evidence of this effect, the following limit has been found [153]:

$$\mu_\nu \leq 3 \times 10^{-12}\ \mu_B, \tag{5.8}$$

and $q_\nu \lesssim 2 \times 10^{-14}\, e$. This is the most stringent astrophysical constraint on a neutrino magnetic moment, applicable to both Dirac and Majorana neutrinos. The same limit applies for the neutrino magnetic transition moments as well as for the electric (transition) moments.

Recently, it has been shown that the additional cooling due to neutrino magnetic moments generates qualitative changes to the structure and evolution of stars with masses between 7 and 18 solar masses, rather than simply changing the time scales of their burning [154]. The resulting sensitivity to the neutrino magnetic moment has been estimated to be at the level of $(2-4) \times 10^{-11}\ \mu_B$.

6. Spin-Flavor Precession

If neutrinos have magnetic moments, the spin can precess in a transverse magnetic field [155–157].

Let us first consider the spin precession of a Dirac neutrino generated by its diagonal magnetic moment μ. The spatial evolution of the left-handed and right-handed helicity amplitudes $\varphi_L(x)$ and $\varphi_R(x)$ in a transverse magnetic field $B_\perp(x)$ is given by

$$i\frac{\mathrm{d}}{\mathrm{d}x}\begin{pmatrix}\varphi_L(x)\\ \varphi_R(x)\end{pmatrix} = \begin{pmatrix}0 & \mu B_\perp(x)\\ \mu B_\perp(x) & 0\end{pmatrix}\begin{pmatrix}\varphi_L(x)\\ \varphi_R(x)\end{pmatrix}. \tag{6.1}$$

The differential equation (6.1) can be solved through the transformation:

$$\begin{pmatrix}\varphi_L(x)\\ \varphi_R(x)\end{pmatrix} = \frac{1}{\sqrt{2}}\begin{pmatrix}1 & 1\\ -1 & 1\end{pmatrix}\begin{pmatrix}\varphi_-(x)\\ \varphi_+(x)\end{pmatrix}. \tag{6.2}$$

The new amplitudes $\varphi_-(x)$ and $\varphi_+(x)$ satisfy decoupled differential equations, whose solutions are

$$\varphi_\mp(x) = \exp\left[\pm i\int_0^x \mathrm{d}x'\mu B_\perp(x')\right]\varphi_\mp(0). \tag{6.3}$$

If we consider an initial left-handed neutrino, we have

$$\begin{pmatrix}\varphi_L(0)\\ \varphi_R(0)\end{pmatrix} = \begin{pmatrix}1\\ 0\end{pmatrix} \Longrightarrow \begin{pmatrix}\varphi_-(0)\\ \varphi_+(0)\end{pmatrix} = \frac{1}{\sqrt{2}}\begin{pmatrix}1\\ 1\end{pmatrix}. \tag{6.4}$$

Then, the probability of $\nu_L \to \nu_R$ transitions is given by

$$P_{\nu_L\to\nu_R}(x) = |\varphi_R(x)|^2 = \sin^2\left(\int_0^x \mathrm{d}x'\mu B_\perp(x')\right). \tag{6.5}$$

Note that the transition probability is independent from the neutrino energy (contrary to the case of flavor oscillations) and the amplitude of the oscillating probability is unity. Hence, when the argument of the sine is equal to $\pi/2$, there is complete $\nu_L \to \nu_R$ conversion.

The precession $\nu_{eL} \to \nu_{eR}$ in the magnetic field of the Sun was considered in 1971 [155] as a possible solution of the solar neutrino problem. If neutrinos are Dirac particles, right-handed neutrinos are sterile and a $\nu_{eL} \to \nu_{eR}$ conversion could explain the disappearance of active solar ν_{eL}'s.

In 1986, it was realized [156, 157] that the matter effect during neutrino propagation inside of the Sun suppresses $\nu_{eL} \to \nu_{eR}$ transition by lifting the degeneracy of ν_{eL} and ν_{eR}.

Indeed, taking into account matter effects, the evolution equation (6.1) for a Dirac neutrino becomes

$$i\frac{\mathrm{d}}{\mathrm{d}x}\begin{pmatrix}\varphi_L(x)\\ \varphi_R(x)\end{pmatrix} = \begin{pmatrix}V(x) & \mu B_\perp(x)\\ \mu B_\perp(x) & 0\end{pmatrix}\begin{pmatrix}\varphi_L(x)\\ \varphi_R(x)\end{pmatrix}, \tag{6.6}$$

with the appropriate potential $V(x)$ which depends on the neutrino flavor:

$$V_\alpha(x) = V_{\mathrm{CC}}(x)\ \delta_{\alpha e} + V_{\mathrm{NC}}(x). \tag{6.7}$$

Here, V_{CC} and V_{NC} are the charged-current and neutral-current potentials given by

$$V_{\mathrm{CC}}(x) = \sqrt{2}G_F N_e(x), \qquad V_{\mathrm{NC}}(x) = -\frac{1}{2}\sqrt{2}G_F N_n(x), \tag{6.8}$$

where $N_e(x)$ and $N_n(x)$ are the electron and neutron number densities in the medium. For antineutrinos, $\overline{V}_\alpha(x) = -V_\alpha(x)$.

In the case of a constant matter density, the differential equation (6.6) can be solved analytically with the orthogonal transformation:

$$\begin{pmatrix}\varphi_L(x)\\ \varphi_R(x)\end{pmatrix} = \begin{pmatrix}\cos\xi & \sin\xi\\ -\sin\xi & \cos\xi\end{pmatrix}\begin{pmatrix}\varphi_-(x)\\ \varphi_+(x)\end{pmatrix}. \tag{6.9}$$

The angle ξ is chosen in order to diagonalize the matrix operator in (6.6):

$$\sin 2\xi = \frac{2\mu B_\perp}{\Delta E_M}, \tag{6.10}$$

with the effective energy splitting in matter

$$\Delta E_{\mathrm{M}} = \sqrt{V^2 + (2\mu B_\perp)^2}. \tag{6.11}$$

The decoupled evolution of $\varphi_\mp(x)$ is given by:

$$\varphi_\mp(x) = \exp\left[-\frac{i}{2}(V \mp \Delta E_M)\right]\varphi_\mp(0). \tag{6.12}$$

For an initial left-handed neutrino,

$$\begin{pmatrix}\varphi_-(0)\\ \varphi_+(0)\end{pmatrix} = \begin{pmatrix}\cos\xi\\ \sin\xi\end{pmatrix}, \tag{6.13}$$

leading to the oscillatory transition probability:

$$P_{\nu_L\to\nu_R}(x) = |\varphi_R(x)|^2 = \sin^2 2\xi \sin^2\left(\frac{1}{2}\ \Delta E_M x\right). \tag{6.14}$$

Since in matter $\Delta E_M > 2\mu B_\perp$, the matter effect suppresses the amplitude of $\nu_L \to \nu_R$ transitions. However, these transitions are still independent from the neutrino energy, which does not enter in the evolution equation (6.6).

When it was known, in 1986 [156, 157], that the matter potential has the effect of suppressing $\nu_L \to \nu_R$ transitions because it breaks the degeneracy of left-handed and right-handed states, it did not take long to realize, in 1988 [158, 159], that the matter potentials can cause resonant spin-flavor precession if different flavor neutrinos have transition magnetic moments (spin-flavor precession in vacuum was previously discussed in [24]).

Let us denote with $\varphi_{\alpha h}(x)$ the flavor and helicity amplitudes (with $\alpha = e, \mu, \tau$ and $h = \pm 1$), that is, $\varphi_{\alpha -1}(x) \equiv \varphi_{\alpha L}(x)$ and $\varphi_{\alpha +1}(x) \equiv \varphi_{\alpha R}(x)$. Considering neutrino mixing, the evolution of the flavor and helicity amplitudes is given by

$$i\,\frac{d\varphi_{\alpha h}(x)}{dx} = \sum_{\beta} \sum_{h'=\pm 1} \left[\left(\sum_k U_{\alpha k} \frac{m_k^2}{2E} U_{\beta k}^* + V_\alpha(x) \delta_{\alpha\beta} \right) \delta_{hh'} + \mu_{\alpha\beta} B_\perp(x) \delta_{-hh'} \right] \varphi_{\beta h'}(x), \tag{6.15}$$

with the effective magnetic moments in the flavor basis:

$$\mu_{\alpha\beta} = \sum_{k,j} U_{\alpha k} \mu_{kj} U_{\beta j}^*. \tag{6.16}$$

For a Dirac neutrino, from (2.36), we have

$$\mu_{jk} = {\mu_{kj}}^* \Longrightarrow \mu_{\beta\alpha} = \mu_{\alpha\beta}^*. \tag{6.17}$$

If CP is conserved, from (2.43), and the reality of the mixing matrix, for a Dirac neutrino, we obtain

$$\text{CP} \Longrightarrow \mu_{jk} = \mu_{kj} \Longrightarrow \mu_{\beta\alpha} = \mu_{\alpha\beta}. \tag{6.18}$$

For a Majorana neutrino, from (2.59) and (2.61), we have

$$\mu_{jk} = -\mu_{kj}, \qquad \mu_{kj} = -{\mu_{kj}}^*. \tag{6.19}$$

Hence, in the mass basis of Majorana neutrinos, there are no diagonal magnetic moments and the transition magnetic moments are imaginary. If CP is not conserved, the mixing matrix is not real and the constraints (6.19) do not imply similar relations between the effective magnetic moments in the flavor basis, for which we have only the relation in (6.17) as for Dirac neutrinos. In particular, Majorana neutrinos can have diagonal effective magnetic moments in the flavor basis if CP is not conserved. Let us emphasize that both Dirac and Majorana phases contribute to this effect. Therefore, it occurs also in the case of two-neutrino mixing, in which there is one Majorana phase.

On the other hand, if CP is conserved, there is no additional constraint on the magnetic moments of Majorana neutrinos in the mass basis, as we have seen in Section 2.2. However, in this case, the mixing matrix is real and we have

$$\text{CP} \Longrightarrow \mu_{\beta\alpha} = -\mu_{\alpha\beta}, \qquad \mu_{\alpha\beta} = -{\mu_{\alpha\beta}}^*. \tag{6.20}$$

Hence, only if CP is conserved, there are no diagonal magnetic moments of Majorana neutrinos in the flavor basis as in the mass basis.

In the following, we discuss the spin-flavor evolution equation in the two-neutrino mixing approximation, which is interesting for understanding the relevant features of neutrino spin-flavor precession.

Considering Dirac neutrinos, from (6.15), it follows that the generalization of (6.1) to two-neutrino $\nu_e - \nu_\mu$ mixing is, using an analogous notation,

$$i\frac{\mathrm{d}}{\mathrm{d}x}\begin{pmatrix} \varphi_{eL}(x) \\ \varphi_{\mu L}(x) \\ \varphi_{eR}(x) \\ \varphi_{\mu R}(x) \end{pmatrix} = \mathrm{H}\begin{pmatrix} \varphi_{eL}(x) \\ \varphi_{\mu L}(x) \\ \varphi_{eR}(x) \\ \varphi_{\mu R}(x) \end{pmatrix}, \tag{6.21}$$

with the effective Hamiltonian matrix:

$$\mathrm{H} = \begin{pmatrix} -\frac{\Delta m^2}{4E}\cos 2\vartheta + V_e & \frac{\Delta m^2}{4E}\sin 2\vartheta & \mu_{ee}B_\perp(x) & \mu_{e\mu}B_\perp(x) \\ \frac{\Delta m^2}{4E}\sin 2\vartheta & \frac{\Delta m^2}{4E}\cos 2\vartheta + V_\mu & \mu_{e\mu}^* B_\perp(x) & \mu_{\mu\mu}B_\perp(x) \\ \mu_{ee}B_\perp(x) & \mu_{e\mu}B_\perp(x) & -\frac{\Delta m^2}{4E}\cos 2\vartheta & \frac{\Delta m^2}{4E}\sin 2\vartheta \\ \mu_{e\mu}^* B_\perp(x) & \mu_{\mu\mu}B_\perp(x) & \frac{\Delta m^2}{4E}\sin 2\vartheta & \frac{\Delta m^2}{4E}\cos 2\vartheta \end{pmatrix}, \tag{6.22}$$

where we have used the constraint (6.17) for the transition magnetic moments. The matter potential can generate resonances, which occur when two diagonal elements of H become equal. Besides the standard MSW resonance in the $\nu_{eL} \leftrightarrows \nu_{\mu L}$ channel for $V_{\mathrm{CC}} = \Delta m^2 \cos 2\vartheta / 2E$ (see [1–4]), there are two possibilities

(1) There is a resonance in the $\nu_{eL} \leftrightarrows \nu_{\mu R}$ channel for

$$V_e = \frac{\Delta m^2}{2E}\cos 2\vartheta. \tag{6.23}$$

The density at which this resonance occurs is not the same as that of the MSW resonance, because of the neutral-current contribution to $V_e = V_{\mathrm{CC}} + V_{\mathrm{NC}}$. The location of this resonance depends on both N_e and N_n.

(2) There is a resonance in the $\nu_{\mu L} \leftrightarrows \nu_{eR}$ channel for

$$V_\mu = -\frac{\Delta m^2}{2E}\cos 2\vartheta. \tag{6.24}$$

If $\cos 2\vartheta > 0$, this resonance is possible in normal matter, since the sign of $V_\mu = V_{\mathrm{NC}}$ is negative, as one can see from (6.8).

In practice, the effect of these resonances could be the disappearance of active ν_{eL} or $\nu_{\mu L}$ into sterile right-handed states.

Let us consider now the more interesting case of Majorana neutrinos, which presents two fundamental differences with respect to the Dirac case.

(A) If CP is conserved, Majorana neutrinos can have only a transition magnetic moment $\mu_{e\mu} = -\mu_{\mu e} = -\mu^*_{e\mu}$ in the flavor basis.

(B) The right-handed states are not sterile, but they interact as right-handed Dirac antineutrinos.

Assuming CP conservation, the evolution equation of the amplitudes is given by (6.21) with the effective Hamiltonian matrix:

$$
H = \begin{pmatrix}
-\frac{\Delta m^2}{4E}\cos 2\vartheta + V_e & \frac{\Delta m^2}{4E}\sin 2\vartheta & 0 & \mu_{e\mu}B_\perp(x) \\
\frac{\Delta m^2}{4E}\sin 2\vartheta & \frac{\Delta m^2}{4E}\cos 2\vartheta + V_\mu & -\mu_{e\mu}B_\perp(x) & 0 \\
0 & \mu_{e\mu}B_\perp(x) & -\frac{\Delta m^2}{4E}\cos 2\vartheta - V_e & \frac{\Delta m^2}{4E}\sin 2\vartheta \\
-\mu_{e\mu}B_\perp(x) & 0 & \frac{\Delta m^2}{4E}\sin 2\vartheta & \frac{\Delta m^2}{4E}\cos 2\vartheta - V_\mu
\end{pmatrix}. \tag{6.25}
$$

Again, besides the standard MSW resonance in the $\nu_{eL} \leftrightarrows \nu_{\mu L}$ channel, there are two possible resonances

(1) There is a resonance in the $\nu_{eL} \leftrightarrows \nu_{\mu R}$ channel for

$$
V_{\mathrm{CC}} + 2V_{\mathrm{NC}} = \frac{\Delta m^2}{2E}\cos 2\vartheta. \tag{6.26}
$$

(2) There is a resonance in the $\nu_{\mu L} \leftrightarrows \nu_{eR}$ channel for

$$
V_{\mathrm{CC}} + 2V_{\mathrm{NC}} = -\frac{\Delta m^2}{2E}\cos 2\vartheta. \tag{6.27}
$$

The location of both resonances depend on both N_e and N_n. If $\cos 2\vartheta > 0$, only the first resonance can occur in normal matter, where $N_n \simeq N_e/6$. A realization of the second resonance requires a large neutron number density, as that in a neutron star.

The neutrino spin oscillations in a transverse magnetic field with a possible rotation of the field-strength vector in a plane orthogonal to the neutrino-propagation direction (such rotating fields may exist in the convective zone of the Sun) have been considered in [160–163]. The effect of the magnetic-field rotation may substantially shift the resonance point of neutrino oscillations. Neutrino spin oscillations in electromagnetic fields with other different configurations, including a longitudinal magnetic field and the field of an electromagnetic wave, were examined in [164–169].

It is possible to formulate a criterion [163] for finding out if the neutrino spin and spin-flavor precession is significant for given neutrino and background medium properties. The probability of oscillatory transitions between two neutrino states $\nu_{\alpha L} \leftrightarrows \nu_{\beta R}$ can be expressed in terms of the elements of the effective Hamiltonian matrices (6.22) and (6.25) as

$$P_{\nu_{\alpha L} \leftrightarrows \nu_{\beta R}} = \sin^2\vartheta_{\text{eff}} \sin^2 \frac{x\pi}{L_{\text{eff}}}, \tag{6.28}$$

where

$$\sin^2\vartheta_{\text{eff}} = \frac{4\text{H}_{\alpha\beta}^2}{4\text{H}_{\alpha\beta}^2 + (\text{H}_{\beta\beta} - \text{H}_{\alpha\alpha})^2}, \qquad L_{\text{eff}} = \frac{2\pi}{\sqrt{4\text{H}_{\alpha\beta}^2 + (\text{H}_{\beta\beta} - \text{H}_{\alpha\alpha})^2}}. \tag{6.29}$$

The transition probability can be of order unity if the following two conditions hold simultaneously: (1) the amplitude of the transition probability must be sizable (at least $\sin^2\vartheta_{\text{eff}} \gtrsim 1/2$); (2) the neutrino path length in a medium with a magnetic field should be longer than half the effective length of oscillations L_{eff}. In accordance with this criterion, it is possible to introduce the critical strength of a magnetic field B_{cr} which determines the region of field values $B_{\perp} > B_{\text{cr}}$ at which the probability amplitude is not small ($\sin^2\vartheta_{\text{eff}} > 1/2$):

$$B_{\text{cr}} = \frac{1}{2\tilde{\mu}} \sqrt{(\text{H}_{\beta\beta} - \text{H}_{\alpha\alpha})^2}, \tag{6.30}$$

where $\tilde{\mu}$ is μ_{ee}, $\mu_{\mu\mu}$, $\mu_{e\mu}$, or $\mu_{\mu e}$ depending on the type of neutrino transition process in question.

Consider, for instance, the case of $\nu_{eL} \leftrightarrows \nu_{\mu R}$ transitions of Majorana neutrinos. From (6.25) and (6.30), it follows [163] that

$$B_{\text{cr}} = \left| \frac{1}{2\tilde{\mu}} \left(\frac{\Delta m^2 \cos 2\vartheta}{2E} - \sqrt{2} G_{\text{F}} N_{\text{eff}} \right) \right|, \tag{6.31}$$

where $N_{\text{eff}} = N_e - N_n$. For getting numerical estimates of B_{cr}, it is convenient to rewrite (6.31) in the following form:

$$B_{\text{cr}} \approx 43 \frac{\mu_B}{\tilde{\mu}} \left| \cos 2\vartheta \left(\frac{\Delta m^2}{\text{eV}^2} \right) \left(\frac{\text{MeV}}{E} \right) - 2.5 \times 10^{-31} \left(\frac{N_{\text{eff}}}{\text{cm}^{-3}} \right) \right| \text{Gauss.} \tag{6.32}$$

An interesting feature of the evolution equation (6.21) in the case of Majorana neutrinos is that the interplay of spin precession and flavor oscillations can generate $\nu_{eL} \rightarrow \nu_{eR}$ transitions [170]. Since ν_{eR} interacts as right-handed Dirac antineutrinos, it is often denoted by $\overline{\nu}_{eR}$, or only $\overline{\nu}_e$, and called "electron antineutrino." This state can be detected through the inverse β-decay reaction:

$$\overline{\nu}_e + p \longrightarrow n + e^+, \tag{6.33}$$

having a threshold $E_{\text{th}} = 1.8\,\text{MeV}$.

The possibility of $\nu_e \to \overline{\nu}_e$ transitions generated by spin-flavor precession of Majorana neutrinos is particularly interesting for solar neutrinos, which experience matter effects in the interior of the Sun in the presence of the solar magnetic field (see [10, 171]). Taking into account the dominant $\nu_e \to \nu_a$ transitions due to neutrino oscillations (see [1–4]), with $\nu_a = \cos\vartheta_{23}\nu_\mu - \sin\vartheta_{23}\nu_\tau$ and $\sin^2 2\vartheta_{23} > 0.95$ (90% C.L.) [66], the probability of solar $\nu_e \to \overline{\nu}_e$ transitions is given by [172]

$$P_{\nu_e \to \overline{\nu}_e} \simeq 1.8 \times 10^{-10} \sin^2 2\vartheta_{12} \left(\frac{\mu_{ea}}{10^{-12}\ \mu_B} \frac{B_\perp(0.05 R_\odot)}{10\ \text{kG}} \right)^2, \tag{6.34}$$

where μ_{ea} is the transition magnetic moment between ν_e and ν_a, $\sin^2 2\vartheta_{12} = 0.857^{+0.023}_{-0.025}$ [66] and $R_\odot$ is the radius of the Sun.

It is also possible that spin-flavor precession occurs in the convective zone of the Sun, where there can be random turbulent magnetic fields [173–175]. In this case [176],

$$\begin{aligned} P_{\nu_e \to \overline{\nu}_e} \approx 10^{-7} S^2 &\left(\frac{\mu_{ea}}{10^{-12}\ \mu_B} \right)^2 \left(\frac{B}{20\ \text{kG}} \right)^2 \left(\frac{3 \times 10^4\ \text{km}}{L_{\max}} \right)^{p-1} \\ \times &\left(\frac{8 \times 10^{-5}\ \text{eV}^2}{\Delta m^2_{21}} \right)^p \left(\frac{E}{10\,\text{MeV}} \right)^p \left(\frac{\cos^2\vartheta_{12}}{0.7} \right)^p, \end{aligned} \tag{6.35}$$

where S is a factor of order unity depending on the spatial configuration of the magnetic field, B is the average strength of the magnetic field at the spatial scale $L_{\max}$, which is the largest scale of the turbulence, p is the power of the turbulence scaling, $\Delta m^2_{21} = 7.50^{+0.19}_{-0.20} \times 10^{-5}\,\text{eV}^2$ [66], and E is the neutrino energy. A possible value of p is 5/3 [173–175], corresponding to Kolmogorov turbulence. Conservative values for the other parameters are $B = 20$ kG and $L_{\max} = 3 \times 10^4$ km.

In 2002, the Super-Kamiokande Collaboration established for the flux of solar $\overline{\nu}_e$'s a 90% C.L. an upper limit of 0.8% of the Standard Solar Model (SSM) neutrino flux in the range of energy from 8 to 20 MeV [177] by taking as a reference the BP00 SSM prediction $\phi^{\text{BP00}}_{^8\text{B}} = 5.05 \times 10^6\,\text{cm}^{-2}\,\text{s}^{-1}$ for the solar ^{8}B flux [178] and assuming an undistorted ^{8}B spectrum for the $\overline{\nu}_e$'s. This limit was improved in 2003 by the KamLAND Collaboration [179] to 2.8×10^{-4} of the BP00 SSM prediction at 90% C.L. by measuring $\phi_{\overline{\nu}_e} < 370\,\text{cm}^{-2}\,\text{s}^{-1}$ (90% C.L.) in the energy range 8.3–14.8 MeV, which corresponds to $\phi_{\overline{\nu}_e} < 1250\,\text{cm}^{-2}\,\text{s}^{-1}$ (90% C.L.) in the entire ^{8}B energy range assuming an undistorted spectrum.

Recently, the Borexino collaboration established the best limit on the probability of solar $\nu_e \to \overline{\nu}_e$ transitions [180]:

$$P_{\nu_e \to \overline{\nu}_e} < 1.3 \times 10^{-4} \quad \text{(90\% C.L.)}, \tag{6.36}$$

by taking as a reference $\phi^{\text{SSM}}_{^8\text{B}} = 5.88 \times 10^6\,\text{cm}^{-2}\,\text{s}^{-1}$ [181] and assuming an undistorted ^{8}B spectrum for the $\overline{\nu}_e$'s. They measured $\phi_{\overline{\nu}_e} < 320\,\text{cm}^{-2}\,\text{s}^{-1}$ (90% C.L.) for $E_{\overline{\nu}_e} > 7.3\,\text{MeV}$, which corresponds to $\phi_{\overline{\nu}_e} < 760\,\text{cm}^{-2}\,\text{s}^{-1}$ (90% C.L.) in the entire ^{8}B energy range assuming an undistorted spectrum.

The implications of the limits on the flux of solar $\overline{\nu}_e$'s on Earth for the spin-flavor precession of solar neutrinos have been studied in several papers [172–175, 182–185], taking into account the dominant $\nu_e \to \nu_\mu, \nu_\tau$ transitions due to neutrino oscillations (see [1–4]). Using (6.34) and (6.36), we obtain

$$\mu_{ea} \lesssim 1.3 \times 10^{-12} \frac{7\ \mathrm{MG}}{B_\perp(0.05R_\odot)} \mu_\mathrm{B}, \tag{6.37}$$

with 600 G $\lesssim B_\perp(0.05R_\odot) \lesssim$ 7 MG [180]. In the case of spin-flavor precession in the convective zone of the Sun with random turbulent magnetic fields, (6.35) and (6.36) give, assuming $p = 5/3$,

$$\mu_{ea} \lesssim 4 \times 10^{-11}\,\mathrm{S}^{-1} \frac{20\ \mathrm{kG}}{B} \left(\frac{L_{\max}}{3 \times 10^4\ \mathrm{km}}\right)^{1/3} \mu_\mathrm{B}. \tag{6.38}$$

The spin-flavor mechanism was also considered [186] in order to describe time variations of solar-neutrino fluxes in Gallium experiments. The effect of a nonzero neutrino magnetic moment is also of interest in connection with the analysis of helioseismological observations [187].

The idea that the neutrino magnetic moment may solve the supernova problem, that is, that the neutrino spin-flip transitions in a magnetic field provide an efficient mechanism of energy transfer from a protoneutron star, was first discussed in [188] and then investigated in some detail in [189–191]. The possibility of a loss of up to half of the active left-handed neutrinos because of their transition to sterile right-handed neutrinos in strong magnetic fields at the boundary of the neutron star (the so-called boundary effect) was considered in [163].

7. Summary and Perspectives

In this review, we have considered the electromagnetic properties of neutrinos with focus on the most important issues related to the problem. The main results discussed in the paper can be summed up as follows.

In the most general case, the neutrino electromagnetic vertex function is defined in terms of four form factors: the charge, dipole magnetic, and electric and anapole form factors. This decomposition is consistent with Lorentz and electromagnetic gauge invariance. The four form factors at zero momentum transfer $q^2 = 0$ are, respectively, the neutrino charge, magnetic moment, electric moment, and anapole moment. These quantities contribute to elements of the scattering matrix and describe neutrino interactions with real photons.

An important characteristic of neutrino electromagnetic properties is that they are different for Dirac and Majorana neutrinos. In particular, Majorana neutrinos cannot have diagonal magnetic or electric moments. Thus, studies of neutrino electromagnetic interactions can be used as a procedure to distinguish whether a neutrino is a Dirac or Majorana particle.

Moreover, CP invariance in the lepton sector puts additional constraints on the neutrino form factors and can be tested with experimental probes of neutrino electromagnetic interactions.

Up to now, no effect of neutrino electromagnetic properties has been found in terrestrial laboratory experiments and in the analyses of astrophysical and cosmological data. However, massive neutrinos have nontrivial electromagnetic properties in a wide set of theoretical frameworks, including the simplest extension of the Standard Model with inclusion of singlet right-handed neutrinos. Therefore, the search for nonvanishing neutrino electromagnetic properties is of great interest for experimentalist and theorists.

The neutrino dipole magnetic (and also electric) moment is theoretically the most well studied and understood among the neutrino electromagnetic moments. In models which extend the Standard Model with the addition of singlet right-handed neutrinos, a Dirac neutrino has a nonzero magnetic moment proportional to the neutrino mass, which yields a very small value for the magnetic moment, less than about $3 \times 10^{-19}\,\mu_B$ for a neutrino mass smaller than 1 eV (3.5). Extra terms contributing to the magnetic moment of the neutrino that are not proportional to the neutrino mass may exist, for example, in the framework of left-right symmetric models. In this type of models, as well as in other generalizations of the Standard Model, as, for instance, in supersymmetric and extradimension models, values of the neutrino magnetic moment much larger than $10^{-19}\,\mu_B$ can be obtained.

The most severe terrestrial experimental upper bound for the effective electron antineutrino magnetic moment, $\mu_{\bar{\nu}_e} \leq 2.9 \times 10^{-11}\,\mu_B$ at 90% C.L. (3.31), has been recently obtained in the direct antineutrino-electron scattering experiment performed by the GEMMA collaboration [58]. This value is still about an order of magnitude weaker than the upper bound $\mu_\nu \lesssim 3 \times 10^{-12}\,\mu_B$ (5.8) obtained from astrophysics (considering the cooling of red giant stars) [153].

There is a gap of some orders of magnitude between the present experimental limits $\sim 10^{-11} \div 10^{-12}\,\mu_B$ on neutrino magnetic moments and the predictions of different extensions of the Standard Model which hint at a range $\sim 10^{-14} \div 10^{-15}\,\mu_B$ [28, 29, 80] (see Section (3.5)). The terrestrial experimental constraints have been improved by only one order of magnitude during a period of about twenty years. Further improvements are very important, but unfortunately at the moment there is no new idea which could lead to fast improvements in the near future. On the other hand, astrophysical studies could allow significant improvements of the sensitivity to nontrivial neutrino electromagnetic properties and maybe find a positive indication in their favor. In particular, neutrino flows in extreme astrophysical environments with very strong magnetic fields are sensitive to small values of the neutrino electromagnetic moments. An example is the modelling of neutrinos propagation during core-collapse supernovae (see [192] and references therein) where very strong magnetic fields are believed to exist and in which the influence of neutrino electromagnetic properties has not yet been taken into account.

Acknowledgments

The work of A. Studenikin on this paper has been partially supported by the Russian Foundation for Basic Research (Grants no. 11-02-01509 and no. 12-02-06833) and the Ministry of Education and Science of Russia (State Contract no. 12.741.11.0180 and Projects no. 2012-1.2.1-12-000-1012-1958 and no. 2012-1.1-12-000-1011-6097). C. Giunti would like to thank the Department of Physics of the University of Torino for hospitality and support.

References

[1] C. Giunti and C. W. Kim, *Fundamentals of Neutrino Physics and Astrophysics*, Oxford University Press, Oxford, UK, 2007.

[2] M. C. Gonzalez-Garcia and M. Maltoni, "Phenomenology with massive neutrinos," *Physics Reports*, vol. 460, no. 1–3, pp. 1–129, 2008.

[3] S. Bilenky, *Introduction to the Physics of Massive and Mixed Neutrinos*, vol. 817 of *Lecture Notes in Physics*, Springer, New York, NY, USA, 2010.
[4] Z. Z. Xing and S. Zhou, *Neutrinos in Particle Physics, Astronomy and Cosmology*, Zhejiang University Press, 2011.
[5] C. W. Kim and A. Pevsner, *Neutrinos in Physics and Astrophysics*, vol. 8 of *Contemporary Concepts in Physics*, Harwood Academic Press, 1993.
[6] M. Fukugita and T. Yanagida, *Physics of Neutrinos and Applications to Astrophysics*, Springer, New York, NY, USA, 2003.
[7] R. N. Mohapatra and P. B. Pal, *Massive Neutrinos in Physics and Astrophysics*, vol. 72 of *Lecture Notes in Physics*, World Scientific, River Edge, NJ, USA, 3rd edition, 2004.
[8] G. G. Raffelt, "Astrophysical methods to constrain axions and other novel particle phenomena," *Physics Reports*, vol. 198, no. 1-2, pp. 1–113, 1990.
[9] G. G. Raffelt, "Astrophysics probes of particle physics," *Physics Reports*, vol. 333-334, pp. 593–618, 2000.
[10] J. Pulido, "The solar neutrino problem and the neutrino magnetic moment," *Physics Reports*, vol. 211, no. 4, pp. 167–199, 1992.
[11] M. Nowakowski, E. A. Paschos, and J. M. Rodríguez, "All electromagnetic form factors," *European Journal of Physics*, vol. 26, no. 4, pp. 545–560, 2005.
[12] H. T. Wong and H. B. Li, "Neutrino magnetic moment," *Modern Physics Letters A*, vol. 20, no. 15, article 1103, 2005.
[13] C. Giunti and A. Studenikin, "Neutrino electromagnetic properties," *Physics of Atomic Nuclei*, vol. 72, no. 12, pp. 2089–2125, 2009.
[14] A. Studenikin, "Neutrino magnetic moment: a window to new physics," *Nuclear Physics B*, vol. 188, pp. 220–222, 2009, Proceedings of the Neutrino Oscillation Workshop, Otranto, Italy, September 2008.
[15] W. J. Marciano and A. I. Sanda, "Exotic decays of the muon and heavy leptons in gauge theories," *Physics Letters B*, vol. 67, no. 3, pp. 303–305, 1977.
[16] B. W. Lee and R. E. Shrock, "Natural suppression of symmetry violation in gauge theories: muon- and electron-lepton-number nonconservation," *Physical Review D*, vol. 16, no. 5, pp. 1444–1473, 1977.
[17] K. Fujikawa and R. Shrock, "Magnetic moment of a massive neutrino and neutrino-spin rotation," *Physical Review Letters*, vol. 45, no. 12, pp. 963–966, 1980.
[18] S. Petcov, "Neutrino gamma in the Weinberg-Salam Model with Neutrino Mixing," *Soviet Journal of Nuclear Physics*, vol. 25, p. 340, 1977.
[19] S. Petcov, "Neutrino gamma in the Weinberg-Salam Model with Neutrino Mixing," *Soviet Journal of Nuclear Physics*, vol. 25, p. 698, 1977.
[20] S. Petcov, "Neutrino gamma in the Weinberg-Salam Model with Neutrino Mixing," *Soviet Journal of Nuclear Physics*, vol. 25, p. 1336, 1977.
[21] P. B. Pal and L. Wolfenstein, "Radiative decays of massive neutrinos," *Physical Review D*, vol. 25, no. 3, pp. 766–773, 1982.
[22] R. E. Shrock, "Electromagnetic properties and decays of Dirac and Majorana neutrinos in a general class of gauge theories," *Nuclear Physics B*, vol. 206, no. 3, pp. 359–379, 1982.
[23] S. M. Bilenky and S. T. Petcov, "Massive neutrinos and neutrino oscillations," *Reviews of Modern Physics*, vol. 59, no. 3, pp. 671–754, 1987.
[24] J. Schechter and J. W. F. Valle, "Majorana neutrinos and magnetic fields," *Physical Review D*, vol. 24, no. 7, pp. 1883–1889, 1981.
[25] J. F. Nieves, "Electromagnetic properties of Majorana neutrinos," *Physical Review D*, vol. 26, no. 11, pp. 3152–3158, 1982.
[26] B. Kayser, "Majorana neutrinos and their electromagnetic properties," *Physical Review D*, vol. 26, no. 7, pp. 1662–1670, 1982.
[27] B. Kayser, "CPT, CP, and C phases, and their effects, in Majorana-particle processes," *Physical Review D*, vol. 30, no. 5, pp. 1023–1033, 1984.
[28] N. F. Bell, V. Cirigliano, M. J. Ramsey-Musolf et al., "How magnetic is the dirac neutrino?" *Physical Review Letters*, vol. 95, no. 15, Article ID 151802, 4 pages, 2005.
[29] N. F. Bell, M. Gorchtein, M. J. Ramsey-Musolf, P. Vogel, and P. Wang, "Model independent bounds on magnetic moments of Majorana neutrinos," *Physics Letters B*, vol. 642, no. 4, pp. 377–383, 2006.
[30] H. Novales-Sanchez, A. Rosado, V. Santiago-Olan, and J. Toscano, "Effects of physics beyond the standard model on the neutrino charge radius: an effective Lagrangian approach," *Physical Review D*, vol. 78, no. 7, Article ID 073014, 8 pages, 2008.

[31] W. A. Bardeen, R. Gastmans, and B. Lautrup, "Static quantities in Weinberg's model of weak and electromagnetic interactions," *Nuclear Physics B*, vol. 46, no. 1, pp. 319–331, 1972.
[32] M. Dvornikov and A. Studenikin, "Electric charge and magnetic moment of a massive neutrino," *Physical Review D*, vol. 69, no. 7, Article ID 073001, 21 pages, 2004.
[33] M. Dvornikov and A. Studenikin, "Electromagnetic form factors of a massive neutrino," *Journal of Experimental and Theoretical Physics*, vol. 99, no. 2, pp. 254–269, 2004.
[34] S. L. Glashow, J. Iliopoulos, and L. Maiani, "Weak interactions with lepton-hadron symmetry," *Physical Review D*, vol. 2, no. 7, pp. 1285–1292, 1970.
[35] S. M. Barr, E. Freire, and A. Zee, "Mechanism for large neutrino magnetic moments," *Physical Review Letters*, vol. 65, no. 12, pp. 2626–2629, 1990.
[36] B. Pal, "Zee model on Majorana neutrino mass and magnetic moment," *Physical Review D*, vol. 44, no. 8, pp. 2261–2264, 1991.
[37] L. G. Cabral-Rosetti, J. Bernabeu, J. Vidal, and A. Zepeda, "Charge and magnetic moment of the neutrino in the background field method and in the linear R_xi^L gauge," *The European Physical Journal C*, vol. 12, pp. 633–642, 2000, http://arxiv.org/abs/hep-ph/9907249.
[38] The ALEPH Collaboration, The DELPHI Collaboration, The L3 Collaboration et al., "Precision electroweak measurements on the Z resonance," *Physics Reports*, vol. 427, no. 5-6, pp. 257–454, 2006.
[39] S. A. Cetin, G. W. S. Hou, V. E. Ozcan, A. N. Rozanov, and S. Sultansoy, "Status of the fourth generation—a brief summary of B3SM-III workshop in four parts," in *Proceedings of the B3SM-III*, Bogazici University, Istanbul, Turkey, October 2011, http://arxiv.org/abs/1112.2907.
[40] J. E. Kim, "Neutrino magnetic moment," *Physical Review D*, vol. 14, no. 11, pp. 3000–3002, 1976.
[41] M. Czakon, J. Gluza, and M. Zralek, "Neutrino magnetic moments in left-right symmetric models," *Physical Review D*, vol. 59, no. 1, Article ID 013010, 6 pages, 1999.
[42] M. A. B. Beg, W. J. Marciano, and M. Ruderman, "Properties of neutrinos in a class of gauge theories," *Physical Review D*, vol. 17, no. 5, pp. 1395–1401, 1978.
[43] A. G. Beda, V. B. Brudanin, E. V. Demidova et al., "First result for the neutrino magnetic moment from measurements with the GEMMA spectrometer," *Physics of Atomic Nuclei*, vol. 70, no. 11, pp. 1873–1884, 2007.
[44] R. N. Mohapatra, S. P. Ng, and H. Yu, "Reactor searches for neutrino magnetic moment as a probe of extra dimensions," *Physical Review D*, vol. 70, no. 5, Article ID 057301, 4 pages, 2004.
[45] J. F. Carlson and J. R. Oppenheimer, "The impacts of fast electrons and magnetic neutrons," *Physical Review*, vol. 41, no. 6, pp. 763–792, 1932.
[46] H. Bethe, "Ionization power of a neutrino with magnetic moment," *Mathematical Proceedings of the Cambridge Philosophical Society*, vol. 31, no. 1, pp. 108–115, 1935.
[47] G. Domogatsky and D. Nadezhin, "Modern theory of star evolution and experiments of F. Reines on anti ν e-scattering detection," *Yadernaya Fizika*, vol. 12, p. 1233, 1970.
[48] A. V. Kyuldjiev, "Searching for effects of neutrino magnetic moments at reactors and accelerators," *Nuclear Physics B*, vol. 243, no. 3, pp. 387–397, 1984.
[49] P. Vogel and J. Engel, "Neutrino electromagnetic form factors," *Physical Review D*, vol. 39, no. 11, pp. 3378–3383, 1989.
[50] W. Grimus and P. Stockinger, "Effects of neutrino oscillations and neutrino magnetic moments on elastic neutrino-electron scattering," *Physical Review D*, vol. 57, no. 3, pp. 1762–1768, 1998.
[51] H. T. Wong, H. B. Li, and S. T. Lin, "Enhanced sensitivities for the searches of neutrino magnetic moments through atomic ionization," *Physical Review Letters*, vol. 105, no. 6, Article ID 061801, 4 pages, 2010.
[52] M. Voloshin, "Neutrino scattering on atomic electrons in searches for the neutrino magnetic moment," *Physical Review Letters*, vol. 105, no. 20, Article ID 201801, 4 pages, 2010.
[53] K. A. Kouzakov and A. I. Studenikin, "Magnetic neutrino scattering on atomic electrons revisited," *Physics Letters B*, vol. 696, no. 3, pp. 252–256, 2011.
[54] K. A. Kouzakov and A. I. Studenikin, "Electromagnetic neutrino-atom collisions: the role of electron binding," *Nuclear Physics B*, vol. 217, no. 1, pp. 353–356, 2011.
[55] K. A. Kouzakov, A. I. Studenikin, and M. B. Voloshin, "Testing neutrino magnetic moment in ionization of atoms by neutrino impact," *JETP Letters*, vol. 93, pp. 699–703, 2011, http://arxiv.org/abs/1105.5543.
[56] K. A. Kouzakov, A. I. Studenikin, and M. B. Voloshin, "Neutrino-impact ionization of atoms in searches for neutrino magnetic moment," *Physical Review D*, vol. 83, no. 11, Article ID 113001, 11 pages, 2011.

[57] K. A. Kouzakov, A. I. Studenikin, and M. B. Voloshin, "On neutrino-atom scattering in searches for neutrino magnetic moments," in *Proceedings of the 24th International Conference on Neutrino Physics and Astrophysics*, Athens, Greece, June 2010, http://arxiv.org/abs/1102.0643.

[58] A. G. Beda, V. B. Brudanin, V. G. Egorov et al., "The results of search for the neutrino magnetic moment in GEMMA experiment," *Advances in High Energy Physics*, vol. 2012, Article ID 350150, 12 pages, 2012.

[59] J. F. Beacom and P. Vogel, "Neutrino magnetic moments, flavor mixing, and the super-kamiokande solar data," *Physical Review Letters*, vol. 83, no. 25, pp. 5222–5225, 1999.

[60] F. Reines, H. Gurr, and H. Sobel, "Detection of $\overline{\nu}_e$–e scattering," *Physical Review Letters*, vol. 37, no. 6, pp. 315–318, 1976.

[61] G. S. Vidyakin, V. N. Vyrodov, I.I. Gurevich et al., "Limitations on the magnetic moment and charge radius of the electron antineutrino," *JETP Letters*, vol. 55, no. 4, article 206, 1992.

[62] A. I. Derbin, A. V. Chernyi, L. A. Popeko, V. N. Muratova, G. A. Shishkina, and S. I. Bakhlanov, "Experiment on antineutrino scattering by electrons at a reactor of the Rovno nuclear power plant," *JETP Letters*, vol. 57, no. 12, article 768, 1993.

[63] H. T. Wong, H.B. Li, S.T. Lin et al., "Search of neutrino magnetic moments with a high-purity germanium detector at the Kuo-Sheng nuclear power station," *Physical Review D*, vol. 75, no. 1, Article ID 012001, 16 pages, 2007.

[64] C. Amsler, M. Avenier, G. Bagieu et al., "The MUNU experiment, general description," *Nuclear Instruments and Methods in Physics Research A*, vol. 396, no. 1-2, pp. 115–129, 1997.

[65] Z. Daraktchieva, C. Amsler, M. Avenier et al., "Final results on the neutrino magnetic moment from the MUNU experiment," *Physics Letters B*, vol. 615, no. 3-4, pp. 153–159, 2005.

[66] J. Beringer, J. F. Arguin, R. M. Barnett et al., "Review of particle physics," *Physical Review D*, vol. 86, no. 1, Article ID 010001, 1528 pages, 2012.

[67] L. B. Auerbach, R. L. Burman, D. O. Caldwell et al., "Measurement of electron-neutrino electron elastic scattering," *Physcial Review D*, vol. 63, no. 11, 11 pages, 2001.

[68] R. Schwienhorst, D. Ciampa, C. Erickson et al., "A new upper limit for the tau-neutrino magnetic moment," *Physics Letters B*, vol. 513, no. 1-2, pp. 23–29, 2001.

[69] D. W. Liu, Y. Ashie, S. Fukuda et al., "Limits on the neutrino magnetic moment using 1496 days of Super-Kamiokande-I solar neutrino data," *Physical Review Letters*, vol. 93, no. 2, Article ID 021802, 5 pages, 2004.

[70] C. Arpesella, H. O. Back, M. Balata et al., "Direct measurement of the 7Be solar neutrino flux with 192 days of Borexino data," *Physical Review Letters*, vol. 101, no. 9, Article ID 091302, 6 pages, 2008.

[71] D. Montanino, M. Picariello, and J. Pulido, "Probing neutrino magnetic moment and unparticle interactions with Borexino," *Physical Review D*, vol. 77, no. 9, Article ID 093011, 9 pages, 2008.

[72] W. Grimus, M. Maltoni, T. Schwetz, M. A. Tortola, and J. W. F. Valle, "Constraining Majorana neutrino electromagnetic properties from the LMA-MSW solution of the solar neutrino problem," *Nuclear Physics B*, vol. 648, no. 1-2, pp. 376–396, 2003.

[73] M. A. Tortola, "Constraining neutrino magnetic moment with solar and reactor neutrino data," in *Proceedings of the International Workshop on Astroparticle and High Energy Physics (AHEP-2003)*, Valencia, Spain, October 2003, http://arxiv.org/abs/hep-ph/0401135.

[74] J. Bernabeu, J. Papavassiliou, and M. Passera, "Dynamical zero in $\overline{\nu}_e$–e^- scattering and the neutrino magnetic moment," *Physics Letters B*, vol. 613, no. 3-4, pp. 162–169, 2005.

[75] J. Segura, J. Bernabeu, F. Botella, and J. Penarrocha, "Dynamical zeros in neutrino-electron elastic scattering at leading order," *Physical Review D*, vol. 49, no. 3, pp. 1633–1636, 1994.

[76] M. Gozdz, W. A. Kaminski, F. Simkovic, and A. Faessler, "Transition magnetic moments of Majorana neutrinos in supersymmetry without R-parity in light of neutrino oscillations," *Physical Review D*, vol. 74, no. 5, Article ID 055007, 11 pages, 2006.

[77] M. Gozdz, W. A. Kaminski, and F. Simkovic, "Majorana neutrino magnetic moments," *International Journal of Modern Physics E*, vol. 15, no. 2, article 441, 2006.

[78] M. Voloshin, "On compatibility of small mass with large magnetic moment of neutrino," *Soviet Journal of Nuclear Physics*, vol. 48, p. 512, 1988.

[79] P. B. Pal, "Particle physics confronts the solar neutrino problem," *International Journal of Modern Physics A*, vol. 7, no. 22, article 5387, 1992.

[80] N. F. Bell, "How magnetic is the neutrino?" *International Journal of Modern Physics A*, vol. 22, no. 27, article 4891, 2007, Festschrift in honour of B. H. J McKellar and G. C. Joshi.

[81] M. Leurer and N. Marcus, "A model for a large neutrino magnetic transition moment and naturally small mass," *Physics Letters B*, vol. 237, no. 1, pp. 81–87, 1990.

[82] K. Babu and R. N. Mohapatra, "Large transition magnetic moment of the neutrino from horizontal symmetry," *Physical Review D*, vol. 42, no. 11, pp. 3778–3793, 1990.
[83] H. Georgi and L. Randall, "Charge conjugation and neutrino magnetic moments," *Physics Letters B*, vol. 244, no. 2, pp. 196–202, 1990.
[84] G. Ecker, W. Grimus, and H. Neufeld, "A light Zeldovich-Konopinski-Mahmoud neutrino with a large magnetic moment," *Physics Letters B*, vol. 232, no. 2, pp. 217–221, 1989.
[85] D. Chang, W. Y. Keung, S. Lipovaca, and G. Senjanovic, "Neutrino magnetic moment and the dicyclic group ," *Physical Review Letters*, vol. 67, no. 8, pp. 953–956, 1991.
[86] R. Barbieri and R. N. Mohapatra, "A neutrino with a large magnetic moment and a naturally small mass ," *Physics Letters B*, vol. 218, no. 2, pp. 225–229, 1989.
[87] A. Zee, "A theory of lepton number violation and neutrino Majorana masses," *Physics Letters B*, vol. 93, no. 4, pp. 389–393, 1980.
[88] S. Lee, "Higher-order corrections to leptonic processes and the renormalization of Weinberg's theory of weak interactions in the unitary gauge," *Physical Review D*, vol. 6, no. 6, pp. 1701–1717, 1972.
[89] J. Lucio, A. Rosado, and A. Zepeda, "Neutrino charge in the linear R_ξ gauge," *Physical Review D*, vol. 29, no. 7, pp. 1539–1541, 1984.
[90] J. Lucio, A. Rosado, and A. Zepeda, "Characteristic size for the neutrino," *Physical Review D*, vol. 31, no. 5, pp. 1091–1096, 1985.
[91] G. Degrassi, A. Sirlin, and W. J. Marciano, "Effective electromagnetic form factor of the neutrino," *Physical Review D*, vol. 39, no. 1, pp. 287–294, 1989.
[92] J. Bernabeu, L. G. Cabral-Rosetti, J. Papavassiliou, and J. Vidal, "Charge radius of the neutrino," *Physial Review D*, vol. 62, no. 11, Article ID 113012, 19 pages, 2000.
[93] J. Bernabeu, J. Papavassiliou, and J. Vidal, "Observability of the Neutrino Charge Radius," *Physical Review Letters*, vol. 89, no. 10, Article ID 101802, 4 pages, 2002.
[94] J. Bernabeu, J. Papavassiliou, and J. Vidal, "The neutrino charge radius as a physical observable," *Nuclear Physics B*, vol. 680, no. 1–3, pp. 450–478, 2004.
[95] K. Fujikawa and R. Shrock, "Comment on 'observability of the neutrino charge Radius'," in press, http://arxiv.org/abs/hep-ph/0303188.
[96] K. Fujikawa and R. Shrock, "Neutrino electroweak radius," *Physical Review D*, vol. 69, no. 1, Article ID 013007, 8 pages, 2004.
[97] J. Papavassiliou, J. Bernabeu, D. Binosi, and J. Vidal, "The effective neutrino charge radius," *The European Physical Journal C*, vol. 33, supplement 1, pp. S865–s867, 2004, Proceedings of the EPS 2003, Aachen, Germany.
[98] J. Bernabeu, D. Binosi, and J. Papavassiliou, "The effective neutrino charge radius in the presence of fermion masses," *Nuclear Physics B*, vol. 716, no. 1-2, pp. 352–372, 2005.
[99] A. Grau and J. Grifols, "Neutrino charge radius and substructure," *Physics Letters B*, vol. 166, no. 2, pp. 233–237, 1986.
[100] K. Hagiwara, S. Matsumoto, D. Haidt, and C. Kim, "A novel approach to confront electroweak data and theory," *Zeitschrift Für Physik C*, vol. 64, no. 4, pp. 559–620, 1994.
[101] M. Deniz, S. T. Lin, V. Singh et al., "Measurement of $\overline{\nu}_e$-electron scattering cross section with a CsI(Tl) scintillating crystal array at the Kuo-Sheng nuclear power reactor," *Physical Review D*, vol. 81, no. 7, Article ID 072001, 17 pages, 2001.
[102] J. Grifols and E. Masso, "Bound on the neutrino charge radius from primordial nucleosynthesis," *Modern Physics Letters A*, vol. 2, no. 3, article 205, 1987.
[103] J. A. Grifols and E. Masso, "Charge radius of the neutrino: a limit from SN 1987 A," *Physical Review D*, vol. 40, no. 12, pp. 3819–3820, 1989.
[104] R. Allen, H. H. Chen, P. J. Doe et al., "Experimental bound on the charge radius of the electron neutrino," *Physical Review D*, vol. 43, no. 1, pp. R1–R3, 1991.
[105] A. M. Mourao, J. Pulido, and J. P. Ralston, "The neutrino electromagnetic moments and charge radius confront Kamiokande II and Homestake experimental results," *Physics Letters B*, vol. 285, no. 4, pp. 364–370, 1992.
[106] A. M. Mourao, J. Pulido, and J. P. Ralston, "Erratum: the neutrino electromagnetic moments and charge radius confront Kamiokande II and Homestake experimental results," *Physics Letters B*, vol. 288, no. 3-4, p. 421, 1992.
[107] J. Barranco, O. G. Miranda, and T. I. Rashba, "Improved limit on electron neutrino charge radius through a new evaluation of the weak mixing angle," *Physics Letters B*, vol. 662, no. 5, pp. 431–435, 2008.

[108] M. Hirsch, E. Nardi, and D. Restrepo, "Bounds on the tau and muon neutrino vector and axial vector charge radius," *Physical Review D*, vol. 67, no. 3, Article ID 033005, 10 pages, 2003.
[109] T. Goldman and G. J. Stephenson Jr., "Limits on the mass of the muon neutrino in the absence of muon-lepton-number conservation," *Physical Review D*, vol. 16, no. 7, pp. 2256–2259, 1977.
[110] G. T. Zatsepin and A. Y. Smirnov, "Neutrino decay in gauge theories," *Yadernaya Fizika*, vol. 28, p. 1569, 1978.
[111] D. A. Dicus, E. W. Kolb, and V. L. Teplitz, "Cosmological upper bound on heavy-neutrino lifetimes," *Physical Review Letters*, vol. 39, no. 4, pp. 168–171, 1977.
[112] K. Sato and M. Kobayashi, "Cosmological constraints on the mass and the number of heavy Lepton neutrinos," *Progress of Theoretical Physics*, vol. 58, no. 6, pp. 1775–1789, 1977.
[113] F. Stecker, "Have massive cosmological neutrinos already been detected?" *Physical Review Letters*, vol. 45, no. 17, pp. 1460–1462, 1980.
[114] R. Kimble, S. Bowyer, and P. Jakobsen, "Astrophysical constraints on the radiative lifetime of neutrinos with mass between 10 and 100 eV/c^2," *Physical Review Letters*, vol. 46, no. 2, pp. 80–83, 1981.
[115] A. Melott and D. Sciama, "Neutrino-lifetime constraints from neutral hydrogen in the galactic halo," *Physical Review Letters*, vol. 46, no. 21, pp. 1369–1372, 1981.
[116] A. De Rujula and S. Glashow, "Galactic neutrinos and uv astronomy," *Physical Review Letters*, vol. 45, no. 11, pp. 942–944, 1980.
[117] G. Raffelt, *Stars As Laboratories for Fundamental Physics: The Astrophysics of Neutrinos, Axions, and Other Weakly Interacting Particles*, University of Chicago Press, Chicago, Ill, USA, 1996.
[118] G. G. Raffelt, "Limits on neutrino electromagnetic properties—an update," *Physics Reports*, vol. 320, no. 1–6, pp. 319–327, 1999.
[119] S. Petcov, "Remarks on the Zee model of neutrino mixing ($\mu \to e + \gamma, v_{H \to v_L + \gamma}$, etc.)," *Physics Letters B*, vol. 115, no. 5, pp. 401–406, 1982.
[120] A. N. Ioannisian and G. G. Raffelt, "Cherenkov radiation by massless neutrinos in a magnetic field," *Physical Review D*, vol. 55, no. 11, pp. 7038–7043, 1997.
[121] A. Lobanov and A. Studenikin, *Physics Letters B*, vol. 564, no. 1-2, pp. 27–34, 2003.
[122] A. I. Studenikin, "Neutrinos in electromagnetic fields and moving media," *Physics of Atomic Nuclei*, vol. 67, no. 5, pp. 993–1002, 2004.
[123] A. I. Studenikin, "Neutrinos in matter and external fields," *Physics of Atomic Nuclei*, vol. 70, no. 7, pp. 1275–1287, 2007.
[124] A. Studenikin, "Neutrinos and electrons in background matter: a new approach," *Annales de la Fondation Louis de Broglie*, vol. 31, no. 2-3, pp. 286–316, 2006.
[125] D. V. Galtsov and N. S. Nikitina, "Photoneutrino processes in a strong field," *Soviet Physics, JETP*, vol. 35, p. 1047, 1972.
[126] A. Borisov, V. C. Zhukovsky, and A. Ternov, "Electromagnetic properties of the massive dirac neutrino in an external electromagnetic field," *Soviet Physics Journal*, vol. 31, no. 3, pp. 228–233, 1988.
[127] A. Borisov, V. C. Zhukovsky, and A. Ternov, "Electromagnetic properties of massive neutrinos," *Soviet Physics, Doklady*, vol. 34, p. 884, 1989.
[128] V. Skobelev, "Reaction whereby a photon decays into a neutrino-antineutrino pair and a neutrino decays into a photon-neutrino pair in a strong magnetic field," *Zhurnal Èksperimental'noi i Teoreticheskoi Fiziki*, vol. 71, pp. 1263–1267, 1976.
[129] V. Skobelev, "Interaction between a massive neutrino and a plane wave field," *Soviet Physics, JETP*, vol. 73, p. 40, 1991.
[130] M. Radomski, "Neutrino magnetic moment, plasmon Čerenkov radiation, and the solar-neutrino problem," *Physical Review D*, vol. 12, no. 8, pp. 2208–2211, 1975.
[131] W. Grimus and H. Neufeld, "Cherenkov radiation of neutrinos," *Physics Letters B*, vol. 315, no. 1-2, pp. 129–133, 1993.
[132] M. Sakuda, "Proposed method to measure the neutrino magnetic moment," *Physical Review Letters*, vol. 72, no. 6, pp. 804–807, 1994.
[133] M. Sakuda and Y. Kurihara, "Transition radiation of the neutrino magnetic moment," *Physical Review Letters*, vol. 74, no. 8, pp. 1284–1287, 1995.
[134] W. Grimus and H. Neufeld, "Transition radiation of ultrarelativistic neutral particles," *Physics Letters B*, vol. 344, no. 1-4, pp. 252–258, 1995.
[135] V. Oraevsky, V. Semikoz, and Y. Smorodinsky, "Polarization loss and induced electric charge of neutrinos in plasmas," *JETP Letters*, vol. 43, no. 12, article 709, 1986.

[136] J. C. D'Olivo, J. F. Nieves, and P. B. Pal, "Cherenkov radiation by massless neutrinos," *Physics Letters B*, vol. 365, no. 1–4, pp. 178–184, 1996.

[137] A. Borisov, V. C. Zhukovsky, A. Kurilin, and A. Ternov, "Radiative corrections to neutrino mass in external electromagnetic field," *Yadernaya Fizika*, vol. 41, p. 743, 1985.

[138] S. S.] Masood, A. Perez Martinez, H. Perez-Rojas, R. Gaitan, and S. Rodriguez-Romo, "Effective magnetic moment of neutrinos in strong magnetic fields," *Revista Mexicana de Fisica*, vol. 48, no. 006, p. 501, 2002.

[139] J. F. Nieves, "Electromagnetic vertex of neutrinos in an electron background and a magnetic field," *Physical Review D*, vol. 68, no. 11, Article ID 113003, 13 pages, 2003.

[140] S. Mohanty and M. K. Samal, "Čerenkov radiation by neutrinos in a supernova core," *Physical Review Letters*, vol. 77, no. 5, pp. 806–809, 1996.

[141] C. Giunti, C. Kim, and W. Lam, "Radiative decay and magnetic moment of neutrinos in matter," *Physical Review D*, vol. 43, no. 1, pp. 164–169, 1991.

[142] A. Gvozdev, N. Mikheev, and L. Vasilevskaya, "The magnetic catalysis of the radiative decay of a massive neutrino in the standard model with lepton mixing," *Physics Letters B*, vol. 289, no. 1-2, pp. 103–108, 1992.

[143] V. Skobelev, "Decay of massive neutrinos in a strong magnetic field," *Journal of Experimental and Theoretical Physics*, vol. 81, p. 1, 1995.

[144] V. C. Zhukovsky, P. Eminov, and A. Grigoruk, "Radiative decay of a massive neutrino in the Weinberg-Salam model with mixing in a constant uniform magnetic field," *Modern Physics Letters A*, vol. 11, no. 39-40, pp. 3119–3126, 1996.

[145] M. Kachelriess and G. Wunner, "Radiative neutrino decays in very strong magnetic fields," *Physics Letters B*, vol. 390, no. 1–4, pp. 263–267, 1997.

[146] A. Ternov and P. Eminov, "Radiative decay of the massive neutrino in magnetized plasma," *Journal of Physics G*, vol. 29, no. 2, article 357, 2003.

[147] A. Lobanov and A. Studenikin, "Neutrino self-polarization effect in matter," *Physics Letters B*, vol. 601, no. 3-4, pp. 171–175, 2004.

[148] A. Grigoriev, M. Dvornikov, and A. Studenikin, "Spin light of neutrino in gravitational fields," *International Journal of Modern Physics D*, vol. 14, no. 2, article 309, 2005.

[149] A. Studenikin and A. Ternov, "Neutrino quantum states and spin light in matter," *Physics Letters B*, vol. 608, no. 1-2, pp. 107–114, 2005.

[150] A. Grigorev, A. Studenikin, and A. Ternov, "Quantum theory of neutrino spin light in dense matter," *Physics Letters B*, vol. 622, no. 3-4, pp. 199–206, 2005.

[151] A. Lobanov, "High energy neutrino spin light," *Physics Letters B*, vol. 619, no. 1-2, pp. 136–144, 2005.

[152] J. Bernstein, M. Ruderman, and G. Feinberg, "Electromagnetic properties of the neutrino," *Physical Review*, vol. 132, no. 3, pp. 1227–1233, 1963.

[153] G. Raffelt, "New bound on neutrino dipole moments from globular-cluster stars," *Physical Review Letters*, vol. 64, no. 24, pp. 2856–2858, 1990.

[154] A. Heger, A. Friedland, M. Giannotti, and V. Cirigliano, "The impact of neutrino magnetic moments on the evolution of massive stars," *The Astrophysical Journal*, vol. 696, no. 1, article 608, 2009.

[155] A. Cisneros, "Effect of neutrino magnetic moment on solar neutrino observations," *Astrophysics and Space Science*, vol. 10, no. 1, pp. 87–92, 1971.

[156] M. B. Voloshin and M. I. Vysotsky, "Neutrino magnetic moment and time variation of solar neutrino flux," *Soviet Journal of Nuclear Physics*, vol. 44, article 544, 1986.

[157] L. B. Okun, M. B. Voloshin, and M. I. Vysotsky, "Neutrino electrodynamics and possible consequences for solar neutrinos," *Soviet Physics, JETP*, vol. 64, p. 446, 1986.

[158] E. K. Akhmedov, "Resonance enhancement of the neutrino spin precession in matter and the solar neutrino problem," *Soviet Journal of Nuclear Physics*, vol. 48, p. 382, 1988.

[159] C. S. Lim and W. J. Marciano, "Resonant spin-flavor precession of solar and supernova neutrinos," *Physical Review D*, vol. 37, no. 6, pp. 1368–1373, 1988.

[160] J. Vidal and J. Wudka, "Non-dynamical contributions to left-right transitions in the solar neutrino problem," *Physics Letters B*, vol. 249, no. 3-4, pp. 473–477, 1990.

[161] A. Smirnov, "The geometrical phase in neutrino spin precession and the solar neutrino problem," *Physics Letters B*, vol. 260, no. 1-2, pp. 161–164, 1991.

[162] E. K. Akhmedov, S. Petcov, and A. Smirnov, "Neutrinos with mixing in twisting magnetic fields," *Physical Review D*, vol. 48, no. 5, pp. 2167–2181, 1993.
[163] G. Likhachev and A. Studenikin, "Neutrino oscillations in the magnetic field of the sun, supernovae, and neutron stars," *Journal of Experimental and Theoretical Physics*, vol. 81, pp. 419–425, 1995.
[164] E. K. Akhmedov and M. Y. Khlopov, "Resonant amplification of neutrino oscillations in longitudinal magnetic field," *Modern Physics Letters A*, vol. 3, no. 5, pp. 451–457, 1988.
[165] E. K. Akhmedov and M. Khlopov, "Resonant Enhancement of Neutrino Oscillations in Longitudinal Magnetic Field," *Soviet Journal of Nuclear Physics*, vol. 47, p. 689, 1988.
[166] A. Egorov, A. Lobanov, and A. Studenikin, "Neutrino oscillations in electromagnetic fields," *Physics Letters B*, vol. 491, no. 1-2, pp. 137–142, 2000.
[167] A. Lobanov and A. Studenikin, "Neutrino oscillations in moving and polarized matter under the influence of electromagnetic field," *Physics Letters B*, vol. 515, no. 1-2, pp. 94–98, 2001.
[168] M. S. Dvornikov and A. I. Studenikin, "Neutrino oscillations in the field of a linearly polarized electromagnetic wave," *Physics of Atomic Nuclei*, vol. 64, no. 9, pp. 1624–1627, 2001.
[169] M. S. Dvornikov and A. I. Studenikin, "Parametric resonance in neutrino oscillations in periodically varying electromagnetic fields," *Physics of Atomic Nuclei*, vol. 67, no. 4, pp. 719–725, 2004.
[170] E. K. Akhmedov, "Antineutrinos from the sun," *Physics Letters B*, vol. 255, no. 1, pp. 84–88, 1991.
[171] X. Shi, D. Schramm, R. Rosner, and D. Dearborn, "On using a neutrino magnetic moment to attack the solar neutrino problem," *Comments on Nuclear and Particle Physics*, vol. 21, pp. 151–182, 1993.
[172] E. K. Akhmedov and J. Pulido, "Solar neutrino oscillations and bounds on neutrino magnetic moment and solar magnetic field," *Physics Letters B*, vol. 553, no. 1-2, pp. 7–17, 2003.
[173] O. G. Miranda, T. I. Rashba, A. I. Rez, and J. W. F. Valle, "Constraining the neutrino magnetic moment with anti-neutrinos from the sun," *Physical Review Letters*, vol. 93, no. 5, Article ID 051304, 4 pages, 2004.
[174] O. G. Miranda, T. I. Rashba, A. I. Rez, and J. W. F. Valle, "Enhanced solar anti-neutrino flux in random magnetic fields," *Physical Review D*, vol. 70, no. 11, Article ID 113002, 11 pages, 2004.
[175] A. Friedland, "Do solar neutrinos constrain the electromagnetic properties of the neutrino?" in press, http://arxiv.org/abs/hep-ph/0505165.
[176] G. Raffelt and T. Rashba, "Mimicking diffuse supernova antineutrinos with the sun as a source," *Physics of Atomic Nuclei*, vol. 73, no. 4, pp. 609–613, 2010.
[177] Y. Gando, S. Fukuda, Y. Fukuda et al., "Search for $\overline{v}_e$ from the sun at Super-Kamiokande-I," *Physical Review Letters*, vol. 90, no. 17, Article ID 171302, 5 pages, 2003.
[178] J. N. Bahcall, M. H. Pinsonneault, and S. Basu, "Solar models: current epoch and time dependences, neutrinos, and helioseismological properties," *The Astrophysical Journal*, vol. 555, no. 2, article 990, 2001.
[179] K. Eguchi, S. Enomoto, K. Furuno et al., "High sensitivity search for $\overline{v}_e$'s from the sun and other sources at KamLAND," *Physical Review Letters*, vol. 92, no. 7, Article ID 071301, 5 pages, 2004.
[180] G. Bellini, J. Benziger, S. Bonettii et al., "Study of solar and other unknown anti-neutrino fluxes with Borexino at LNGS," *Physics Letters B*, vol. 696, no. 3, pp. 191–196, 2011.
[181] A. Serenelli, S. Basu, J. W. Ferguson, and M. Asplund, "New solar composition: the problem with solar models revisited," *The Astrophysical Journal*, vol. 705, no. 2, article L123, 2009.
[182] B. C. Chauhan, J. Pulido, and E. Torrente-Lujan, "KamLAND, solar antineutrinos and the solar magnetic field," *Physical Review D*, vol. 68, no. 3, Article ID 033015, 6 pages, 2003.
[183] A. B. Balantekin and C. Volpe, "Does the neutrino magnetic moment have an impact on solar neutrino physics?" *Physical Review D*, vol. 72, no. 3, Article ID 033008, 7 pages, 2005.
[184] M. M. Guzzo, P. C. de Holanda, and O. L. G. Peres, "Random magnetic fields inducing solar neutrino spin-flavor precession in a three generation context," *Physical Review D*, vol. 72, no. 7, Article ID 073004, 10 pages, 2005.
[185] D. Yilmaz, "Joint analysis of solar neutrino and new KamLAND data in the RSFP framework," in press, http://arxiv.org/abs/0810.1037.
[186] J. Pulido, B. C. Chauhan, and R. S. Raghavan, "Solar neutrinos: spin flavour precession and LMA," in *Proceedings of the 12th Lomonosov Conference in Elementary Particle Physics*, Moscow, Russia, August 2005, http://arxiv.org/abs/hep-ph/0511341.
[187] S. Couvidat, S. Turck-Chièze, and A. G. Kosovichev, "Solar seismic models and the neutrino predictions," *Astrophysical Journal*, vol. 599, no. 2 I, pp. 1434–1448, 2003.
[188] A. Dar, Print-87-0178 IAS, Princeton, NJ, USA ,1987.
[189] S. Nussinov and Y. Rephaeli, "Magnetic moments of neutrinos: particle and astrophysical aspects," *Physical Review D*, vol. 36, no. 8, pp. 2278–2282, 1987.

[190] I. Goldman, Y. Aharonov, G. Alexander, and S. Nussinov, "Implications of the supernova SN1987A neutrino signals," *Physical Review Letters*, vol. 60, no. 18, pp. 1789–1792, 1988.
[191] J. M. Lattimer and J. Cooperstein, "Limits on the neutrino magnetic moment from SN1987A," *Physical Review Letters*, vol. 61, no. 1, pp. 23–26, 1988.
[192] Y. Pehlivan, A. B. Balantekin, T. Kajino, and T. Yoshida, "Invariants of collective neutrino oscillations," *Physical Review D*, vol. 84, no. 6, Article ID 065008, 19 pages, 2011.

Quantization of Free Scalar Fields in the Presence of Natural Cutoffs

K. Nozari, F. Moafi, and F. Rezaee Balef

Department of Physics, Islamic Azad University, Sari Branch, Sari, Iran

Correspondence should be addressed to K. Nozari, knozari@umz.ac.ir

Academic Editor: Joseph Formaggio

We construct a quantum theory of free scalar fields in (1+1)-dimensions based on the deformed Heisenberg algebra $[\hat{x}, \hat{p}] = i\hbar(1 - \beta p + 2\beta^2 p^2)$, that admits the existence of both a minimal measurable length and a maximal momentum, where β is a deformation parameter. We consider both canonical and path integral formalisms of the scenario. Finally a higher dimensional extension is easily performed in the path integral formalism.

1. Introduction

Physics in extremely high energy scales is particularly of interest to particle physicists. It is now a well-known issue that gravity induces uncertainty in measurement of physical quantities. By now, string theory is one of the most successful theoretical frameworks which overcomes the difficulty of ultraviolet divergences in quantum theory of gravity. Incorporation of gravity in quantum field theory leads naturally to an effective cutoff (a minimal measurable length) in the ultraviolet regime. Therefore, if we construct a field theory which captures some stringy nature and/or includes stringy corrections, then it would play a crucial role in investigation of physics at high energy scales towards the Planck scale. Some of the stringy corrections appear as higher order polynomials of momentum leading to modified dispersion relations (see e.g., [1] and references therein). One of the stringy corrections is deforming the Heisenberg uncertainty principle to a generalized uncertainty principle (GUP). Some approaches to quantum gravity such as string theory [2–9], loop quantum gravity [10], and quantum geometry [11–13] all indicated the existence of a minimal measurable length of the order of the Planck length, $l_{\text{pl}} \sim 10^{-35}$ m (see also [14–17]). Moreover some Gedanken experiments in the spirit of black hole physics have also supported the idea of existence of a minimal measurable length [18]. So, existence of a minimal observable length is a common feature of all promising quantum gravity candidates. The existence of a minimal measurable length modifies the Heisenberg uncertainty principle (HUP) to the so-called generalized (gravitational) uncertainty principle (GUP). On the other hand, Doubly Special Relativity Theories (for review see, for instance, [19–24]), suggest that existence of a minimal measurable length would restrict a test particle's momentum to take any arbitrary values. In other words, a test particle's momentum measurement cannot be arbitrarily imprecise leading nontrivially to an upper bound $P_{\max}$ on test particles' momentum. This means that there is a maximal particles' momentum because of fundamental structure of spacetime at the Planck scale [25–28]. Quantum field theory in the presence of minimal length as a natural ultraviolet cutoff has been studied by some authors [29–35]. Specially, Matsuo and Shibusa have studied quantization of fields based on a generalized uncertainty principle that admits just a minimal measurable length [34]. Also Shibusa has extended this study to the supersymmetric field theory based on the GUP with minimal length [35]. Since position and momentum are dual to each other, it is natural to argue that existence of a minimal measurable length naturally leads to the existence of a maximal momentum. This issue has not been considered in the mentioned studies of scalar field theory. It is obvious that existence of a cutoff on particles' momentum affects considerably the formulation of the quantum field theory with just a minimal length cutoff. Based on this argument, our central task in this study is to construct a field theory for free scalar fields in the presence of quantum gravity effects encoded in a GUP that admits existence of a minimal measurable length and a maximal particles' momentum. Following our recent work on Hilbert space representation of quantum mechanics in this case [36], we reformulate the main structure of a free scalar field theory in this setup. Our primary input is the following GUP [37–39]:

$$[\hat{x}, \hat{p}] = i\hbar\left(1 - \beta p + 2\beta^2 p^2\right). \tag{1.1}$$

We investigate the quantization of free scalar field based on the deformed algebra (1.1) in the canonical formalism in $(1+1)$-dimensions and in the path integral formalism as well. In the process of quantization using the latter formalism, we use the Bjorken-Johnson-Low prescription [40, 41] which states that T-product of fields must take the same value in the two formalisms. We compare our results with Matsuo and Shibusa work [34] in each step.

2. A Brief about GUP

2.1. GUP with a Minimal Length

In ordinary quantum mechanics, the standard Heisenberg uncertainty principle (HUP) is given by

$$\Delta x \Delta p \geq \frac{\hbar}{2}. \tag{2.1}$$

There is no trace of gravity in this relation. Today we know that HUP breaks down for energies close to the Planck scale where the corresponding Schwarzschild radius becomes comparable with the Compton wavelength with both becoming approximately of the order of the Planck length. As we have said in the Introduction, existence of a minimal measurable distance in quantum gravity is inevitable. This is encoded in the following Generalized (Gravitational) uncertainty principle (GUP)

$$\Delta x \Delta p \geq \frac{\hbar}{2} + \frac{\beta_0 l_{\text{pl}}^2 (\Delta p)^2}{\hbar}. \tag{2.2}$$

The additional term, $\beta_0 l_{\text{pl}}^2 (\Delta p)^2/\hbar$, has its origin on the very nature of space time at the Planck energy scale (see also [42–44] for alternative generalizations). It was shown in [17] that the simplest GUP relation which implies the appearance of a nonzero minimal uncertainty Δx_0 in position has the form

$$\Delta x \Delta p \geq \frac{\hbar}{2\left(1 + \beta (\Delta p)^2 + \beta \langle p \rangle^2\right)}, \tag{2.3}$$

where β is the GUP parameter defined as $\beta = \beta_0/(M_{\text{pl}}c)^2 = \beta_0 l_{\text{pl}}^2/\hbar^2$, and $M_{\text{pl}}c^2 \approx 10^{19}$ GeV is the 4-dimensional fundamental scale.

At energies much below the Planck energy, the extra term in the right hand side of (2.2) would be irrelevant, which means $\beta \longrightarrow 0$ and the standard HUP relation is recovered. Instead, approaching the Planck energy scale, this term becomes relevant and, as has been said, it is related to the minimal measurable length. From a string theory viewpoint since a string cannot probe distances smaller than its length, existence of a minimal observable length is reliable. Since for any pair of observable A and B (which are represented as symmetric operators on a domain of $\mathbf{A}^2$ and $\mathbf{B}^2$) one has [17]

$$\Delta \mathbf{A} \Delta \mathbf{B} \geq \frac{\hbar}{2} |\langle [\mathbf{A}, \mathbf{B}] \rangle|, \tag{2.4}$$

one find the following algebraic structure

$$[x, p] = i\hbar\left(1 + \beta p^2\right). \tag{2.5}$$

Following [17], we define position and momentum operators for the GUP case as

$$\begin{aligned} X &= x, \\ P &= p\left(1 + \beta p^2\right), \end{aligned} \tag{2.6}$$

where x and p ensure the Jacobi identities, namely,

$$[x_i, p_j] = i\hbar\delta_{ij}, \qquad [x_i, x_j] = 0, \qquad [p_i, p_j] = 0. \tag{2.7}$$

Now it is easy to show that X and P satisfy the generalized uncertainty principle. We interpret p as the momentum operator at low energies which has the standard representation in position space, that is, $p_j = (\hbar/i)(\partial/\partial x_j)$, and P as the momentum operator at high energies, which has the generalized representation in position space, that is, $P_j = (\hbar/i)(\partial/\partial x_j)[1+\beta((\hbar/i)(\partial/\partial x_j))^2]$.

2.2. *GUP with Minimal Length and Maximal Momentum*

Magueijo and Smolin have shown that in the context of the Doubly Special Relativity a test particle's momentum cannot be arbitrarily imprecise and therefore there is an upper bound for momentum fluctuation [25–27]. Then it has been shown by Cortes and Gamboa [28] that this may lead to a maximal measurable momentum for a test particle. In this framework, the GUP that predicts both a minimal observable length and a maximal momentum can be written as follows [37–39] (we note that maximal momentum arises also by considering the fact that the modification of the standard dispersion relation is such that the value of momentum saturates as one varies energy. See [45] for more details.):

$$\Delta x \Delta p \geq \frac{\hbar}{2}\left(1 - 2\beta\langle p\rangle + 4\beta^2\left\langle p^2\right\rangle\right), \tag{2.8}$$

or

$$\Delta x \Delta p \geq \frac{\hbar}{2}\left[1 - \beta(\Delta p) + 2\beta^2(\Delta p)^2\right]. \tag{2.9}$$

In this framework the following algebraic structure can be deduced (see [37–39]):

$$[x,p] = i\hbar\left(1 - \beta p + 2\beta^2 p^2\right), \tag{2.10}$$

where β is the GUP parameter in the presence of both minimal length and maximal momentum. Similar to the minimal length case, we can define [36]

$$\begin{gathered} X = x, \\ P = p\left(1 - \beta p + 2\beta^2 p^2\right), \end{gathered} \tag{2.11}$$

where as before x and p satisfy the canonical commutation relations via the Jacobi identity, and X and P satisfy the generalized commutation relations in the presence of minimal length and maximal momentum as

$$[X,P] = i\hbar\left(1 - \beta p + 2\beta^2 p^2\right). \tag{2.12}$$

3. Variational Matrix Product State Approach

We start with the following deformed Heisenberg algebra:

$$[\hat{x},\hat{p}] = i\hbar\left(1 - \beta p + 2\beta^2 p^2\right), \tag{3.1}$$

which has a representation in momentum space as

$$\hat{p} = p,$$
$$\hat{x} = i\hbar\left(1 - \beta p + 2\beta^2 p^2\right)\frac{\partial}{\partial p}. \tag{3.2}$$

In general, states which have minimal uncertainty obey the following equation:

$$\left(\hat{x} - \langle\hat{x}\rangle + \frac{\langle[\Delta\hat{x}, \Delta\hat{p}]\rangle}{2(\Delta\hat{p})^2}\Delta\hat{p}\right)|\psi\rangle = 0, \tag{3.3}$$

where $\Delta\hat{A} \equiv \hat{A} - \langle\hat{A}\rangle$ and $\Delta\hat{A} \equiv \sqrt{\langle(\Delta\hat{A})^2\rangle}$. Among these states, we will focus on the states which are subjected to the following conditions:

$$(\Delta x)_{\min} = \hbar\sqrt{\beta},$$
$$(\Delta p)_{\min} = \frac{1}{\sqrt{\beta}}, \tag{3.4}$$

and for simplicity we set

$$\langle\hat{p}\rangle = 0. \tag{3.5}$$

One can easily find the wavefunction of maximal localization states in momentum space as [36]

$$\psi_\zeta(p) = \frac{N_1}{\sqrt{2\pi\hbar}}\frac{1}{\sqrt{1 - \beta p + 2\beta^2 p^2}}\exp\left(\frac{\zeta}{i\hbar\sqrt{\beta}}\tan^{-1}\left(\sqrt{\beta}p\right)\right). \tag{3.6}$$

Here $\zeta \equiv \langle\hat{x}\rangle$ and N_1 is a normalization factor. These maximal localization states are identified as one particle state in the theory based on GUP [36]. We will use these states as the basic ingredients in constructing quantum fields in forthcoming sections of this paper. Note that these states are not eigenstates of the position operator $\hat{x}$ because of limitation on resolution of space points encoded in the existence of minimal length. Nevertheless, momentum space representation is still applicable. So, we introduce a first quantized Heisenberg space as the space equipped with the complete basis $\{|p\rangle\}$ which has the completeness relation as

$$\mathbf{1} = \int D_p |p\rangle\langle p|. \tag{3.7}$$

We determine the measure D_p so that the operator $\hat{x} = i\hbar(1-\beta p+2\beta^2p^2)(\partial/\partial p)$ is Hermitian (i.e., $\langle \chi \mid \hat{x}\psi\rangle = \langle \hat{x}\chi \mid \psi\rangle$)

$$D_p \equiv N_2 \frac{dp}{1-\beta p+2\beta^2p^2}, \tag{3.8}$$

where N_2 is a normalization factor independent of p. We set $N_2 = 1$ by recalling the basis as $\sqrt{N_2}|p\rangle \rightarrow |p\rangle$. We note that this measure is not singular since the denominator of equation (3.8) has no real root. This measure leads to the orthogonality condition as

$$\langle p \mid p'\rangle = \left(1-\beta p+2\beta^2p^2\right)\delta(p-p'). \tag{3.9}$$

We denote by $|\psi_\zeta\rangle$ the maximal localization states and then we can rewrite the wave function of one particle state (3.6) in the following form:

$$\psi_\zeta(p) = \langle p \mid \psi_\zeta\rangle. \tag{3.10}$$

So we find

$$i\hbar\left(1-\beta p+2\beta^2p^2\right)\frac{\partial\psi_\zeta(p)}{\partial p} = \zeta\psi_\zeta(p), \tag{3.11}$$

or

$$\frac{\zeta}{i\hbar(1-\beta p+2\beta^2p^2)}\partial p = \frac{\partial\psi_\zeta(p)}{\psi_\zeta(p)}. \tag{3.12}$$

Integrating this equation gives

$$\ln\psi_\zeta(p) = \frac{\zeta}{i\hbar}\int_{-p_{\text{pl}}}^{+p_{\text{pl}}}\frac{dp}{(1-\beta p+2\beta^2p^2)} = \frac{8\zeta}{7i\hbar}\int_{-p_{\text{pl}}}^{+p_{\text{pl}}}\frac{dp}{\left((4\beta p-1)/\sqrt{7}\right)^2+1}, \tag{3.13}$$

where the limits of integral are coming from existence of maximal momentum which is assumed to be the Planck momentum, p_{pl}. So we find

$$\psi_\zeta(p) = \psi_\zeta(0)\exp\left[\frac{-i2\zeta}{\beta\hbar\sqrt{7}}\left\{\tan^{-1}\left(\frac{1}{\sqrt{7}}\right)+\tan^{-1}\left(\frac{4\beta p-1}{\sqrt{7}}\right)\right\}\right]. \tag{3.14}$$

Then, by normalization $\langle \psi \mid \psi \rangle = 1$ we have

$$\begin{aligned} 1 &= \int_{-p_{\mathrm{pl}}}^{+p_{\mathrm{pl}}} \frac{dp}{(1 - \beta p + 2\beta^2 p^2)} \psi_\zeta^*(p) \psi_\zeta(p) \\ &= \psi_\zeta^*(0)\psi_\zeta(0) \int_{-p_{\mathrm{pl}}}^{+p_{\mathrm{pl}}} \frac{dp}{(1 - \beta p + 2\beta^2 p^2)}. \end{aligned} \tag{3.15}$$

So we have

$$\psi_\zeta(0) = \sqrt{\frac{\beta\sqrt{7}}{2}} \left[\tan^{-1}\left(\frac{4\beta p_{\mathrm{pl}} - 1}{\sqrt{7}} \right) + \tan^{-1}\left(\frac{4\beta p_{\mathrm{pl}} + 1}{\sqrt{7}} \right) \right]^{-1/2}. \tag{3.16}$$

Thus the formal position eigenvectors in momentum space are given by

$$\begin{aligned} \psi_\zeta(p) &= \sqrt{\frac{\beta\sqrt{7}}{2}} \left[\tan^{-1}\left(\frac{4\beta p_{\mathrm{pl}} - 1}{\sqrt{7}} \right) + \tan^{-1}\left(\frac{4\beta p_{\mathrm{pl}} + 1}{\sqrt{7}} \right) \right]^{-1/2} \\ &\times \exp\left[\frac{-i2\zeta}{\beta\hbar\sqrt{7}} \left\{ \tan^{-1}\left(\frac{1}{\sqrt{7}} \right) + \tan^{-1}\left(\frac{4\beta p - 1}{\sqrt{7}} \right) \right\} \right]. \end{aligned} \tag{3.17}$$

This is the generalized, momentum space eigenstate of the position operator in the presence of both a minimal length and a maximal momentum. Now we calculate the scaler product of the formal position eigenstate as follows:

$$\begin{aligned} \langle \psi_{\zeta'} \mid \psi_\zeta \rangle &= \int_{-p_{\mathrm{pl}}}^{+p_{\mathrm{pl}}} \frac{dp}{(1 - \beta p + 2\beta^2 p^2)} \psi_{\zeta'}^*(p) \psi_\zeta(p) \\ &= \frac{\beta\sqrt{7}}{2} \left[\tan^{-1}\left(\frac{4\beta p_{\mathrm{pl}} - 1}{\sqrt{7}} \right) + \tan^{-1}\left(\frac{4\beta p_{\mathrm{pl}} + 1}{\sqrt{7}} \right) \right]^{-1} \int_{-p_{\mathrm{pl}}}^{+p_{\mathrm{pl}}} \frac{dp}{(1 - \beta p + 2\beta^2 p^2)} \\ &\times \exp\left(\frac{2(\zeta - \zeta')}{i\hbar\beta\sqrt{7}} \left\{ \tan^{-1}\left(\frac{1}{\sqrt{7}} \right) + \tan^{-1}\left(\frac{4\beta p - 1}{\sqrt{7}} \right) \right\} \right) \\ &= i\lambda \left[\exp\left(\frac{2(\zeta - \zeta')}{i\hbar\beta\sqrt{7}} \left\{ \tan^{-1}\left(\frac{1}{\sqrt{7}} \right) + \tan^{-1}\left(\frac{4\beta p_{\mathrm{pl}} - 1}{\sqrt{7}} \right) \right\} \right) \right. \\ &\quad \left. - \exp\left(\frac{2(\zeta - \zeta')}{i\hbar\beta\sqrt{7}} \left\{ \tan^{-1}\left(\frac{1}{\sqrt{7}} \right) - \tan^{-1}\left(\frac{4\beta p_{\mathrm{pl}} + 1}{\sqrt{7}} \right) \right\} \right) \right], \end{aligned} \tag{3.18}$$

where by definition

$$\lambda \equiv \frac{\hbar\beta\sqrt{7}}{2(\zeta-\zeta')}\left[\tan^{-1}\left(\frac{4\beta p_{\rm pl}-1}{\sqrt{7}}\right)+\tan^{-1}\left(\frac{4\beta p_{\rm pl}+1}{\sqrt{7}}\right)\right]^{-1}. \tag{3.19}$$

Furthermore, we find a completeness of basis $\{|\psi_\zeta\rangle\}$ on this Hilbert space as

$$\begin{aligned} 1 &= \int_{-p_{\rm pl}}^{+p_{\rm pl}} \frac{d\zeta}{|N_1|^2}\left(1-\beta p+2\beta^2p^2\right)|\psi_\zeta\rangle\langle\psi_\zeta| \\ &= \int_{-p_{\rm pl}}^{+p_{\rm pl}} \frac{d\zeta}{|N_1|^2}|\psi_\zeta\rangle\langle\psi_\zeta|\left(1-\beta p+2\beta^2p^2\right). \end{aligned} \tag{3.20}$$

In what follows we set $N_1 = 1$ by rescaling the basis as $(N_1)^{-1}|\psi_\zeta\rangle \to |\psi_\zeta\rangle$. Here we have two bases $\{|p\rangle\}$ and $\{|\psi_\zeta\rangle\}$. Thus we can perform a Fourier transformation which interchanges the two representations. We write the p-space wavefunction and ζ-space wavefunction for a state $|\phi\rangle$ in this Hilbert space as follows (See [36] for more details):

$$\begin{aligned} \phi(p) &\equiv \langle p \mid \phi\rangle, \\ \phi(\zeta) &\equiv \langle \psi_\zeta \mid \phi\rangle. \end{aligned} \tag{3.21}$$

The Fourier transformation in this Hilbert space is

$$\begin{aligned} \phi(\zeta) &= \int_{-p_{\rm pl}}^{+p_{\rm pl}} \frac{dp}{(1-\beta p+2\beta^2p^2)}\psi_\zeta^*(p)\phi(p) \\ &= \int_{-p_{\rm pl}}^{+p_{\rm pl}} \frac{|N_1|}{\sqrt{2\pi\hbar}}\frac{dp}{(1-\beta p+2\beta^2p^2)^{3/2}}\phi(p)\exp\left(\frac{i\zeta}{\hbar\sqrt{\beta}}\tan^{-1}\left(\sqrt{\beta}p\right)\right), \end{aligned} \tag{3.22}$$

$$\begin{aligned} \phi(p) &= \int_{-p_{\rm pl}}^{+p_{\rm pl}} d\zeta\left(1-\beta p+2\beta^2p^2\right)\psi_\zeta(p)\phi(\zeta) \\ &= \int_{-p_{\rm pl}}^{+p_{\rm pl}} \frac{(1-\beta p+2\beta^2p^2)^{1/2}}{\sqrt{2\pi\hbar}}\exp\left(\frac{-i\zeta}{\hbar\sqrt{\beta}}\tan^{-1}\left(\sqrt{\beta}p\right)\right)\phi(\zeta)d\zeta. \end{aligned} \tag{3.23}$$

Here there are some points which are crucial to be mentioned. The transformation (3.23) is an inverse transformation to (3.22) for arbitrary square integrable functions as in the case of usual Fourier transformation. However the transformation (3.22) is not an inverse transformation to (3.23) for arbitrary square integrable functions. This can be seen easily by inserting (3.23) into (3.22), then we find

$$\begin{aligned}
&\int_{-p_{\rm pl}}^{+p_{\rm pl}} \frac{dp}{(1-\beta p+2\beta^2p^2)}\varphi_\zeta^*(p)\phi(p) \\
&= \int_{-p_{\rm pl}}^{+p_{\rm pl}} \frac{dp}{(1-\beta p+2\beta^2p^2)}\varphi_\zeta^*(p)\int_{-p_{\rm pl}}^{+p_{\rm pl}}\left(1-\beta p+2\beta^2p^2\right)\varphi_{\zeta'}(p)\phi(\zeta')d\zeta' \\
&= \int_{-p_{\rm pl}}^{+p_{\rm pl}} \frac{dp}{(1-\beta p+2\beta^2p^2)}\varphi_\zeta^*(p)\varphi_{\zeta'}(p)\int_{-p_{\rm pl}}^{+p_{\rm pl}}\left(1-\beta p+2\beta^2p^2\right)\phi(\zeta')d\zeta' \\
&= \int_{-p_{\rm pl}}^{+p_{\rm pl}} dp\frac{\beta\sqrt{7}}{2}\left[\tan^{-1}\left(\frac{4\beta p_{\rm pl}-1}{\sqrt{7}}\right)+\tan^{-1}\left(\frac{4\beta p_{\rm pl}+1}{\sqrt{7}}\right)\right]^{-1} \\
&\quad\times\exp\left(\frac{2(\zeta-\zeta')}{i\hbar\beta\sqrt{7}}\left\{\tan^{-1}\left(\frac{1}{\sqrt{7}}\right)+\tan^{-1}\left(\frac{4\beta p-1}{\sqrt{7}}\right)\right\}\right)\phi(\zeta')d\zeta' \\
&= \int_{-p_{\rm pl}}^{+p_{\rm pl}} i\lambda\left[e^{N(A+C)}-e^{(-N)(B-C)}\right]d\zeta'\left(1-\beta p+2\beta^2p^2\right)\phi(\zeta'),
\end{aligned} \tag{3.24}$$

where by definition

$$\begin{aligned}
\lambda &\equiv \frac{\hbar\beta\sqrt{7}}{2(\zeta-\zeta')}\left[\tan^{-1}\left(\frac{4\beta p_{\rm pl}-1}{\sqrt{7}}\right)+\tan^{-1}\left(\frac{4\beta p_{\rm pl}+1}{\sqrt{7}}\right)\right]^{-1}, \\
A &\equiv \tan^{-1}\left(\frac{4\beta p_{\rm pl}-1}{\sqrt{7}}\right), \\
B &\equiv \tan^{-1}\left(\frac{4\beta p_{\rm pl}+1}{\sqrt{7}}\right), \\
C &\equiv \tan^{-1}\left(\frac{1}{7}\right), \\
N &\equiv \frac{2(\zeta-\zeta')}{i\hbar\beta\sqrt{7}}.
\end{aligned} \tag{3.25}$$

This is not equal to the original function for arbitrary function $\phi(\zeta')$. We restrict the function $\phi(\zeta')$ to liner combination of $\varphi_{\zeta'}^*(p)$'s as

$$\phi(\zeta') = \int_{-p_{\rm pl}}^{+p_{\rm pl}} \frac{dp}{(1-\beta p+2\beta^2p^2)}a(p)\varphi_{\zeta'}^*(p). \tag{3.26}$$

For these functions, the transformation (3.22) is as follows:

$$\phi(\zeta') = \int_{-p_{\rm pl}}^{+p_{\rm pl}} \frac{dp}{(1-\beta p+2\beta^2 p^2)} a(p)\varphi_{\zeta'}^{*}(p). \tag{3.27}$$

Now for these functions the transformation (3.22) is truly the inverse transformation to (3.23). That is,

$$\begin{aligned}\int_{-p_{\rm pl}}^{+p_{\rm pl}} \frac{dp}{(1-\beta p+2\beta^2 p^2)} \varphi_{\zeta}^{*}(p)\phi(p) &= \int_{-p_{\rm pl}}^{+p_{\rm pl}} d\zeta'\left(1-\beta p+2\beta^2 p^2\right)\varphi_{\zeta'}(p)\phi(\zeta') \\ &= \int_{-p_{\rm pl}}^{+p_{\rm pl}} \frac{dp\,dp'\,d\zeta'}{(1-\beta p+2\beta^2 p^2)} a(p')\varphi_{\zeta}^{*}(p)\varphi_{\zeta'}(p)\varphi_{\zeta'}^{*}(p') \\ &= \int_{-p_{\rm pl}}^{+p_{\rm pl}} \frac{dp'}{(1-\beta p+2\beta^2 p^2)} a(p')\varphi_{\zeta}^{*}(p') = \phi(\zeta).\end{aligned} \tag{3.28}$$

Since

$$\begin{aligned}\int_{-p_{\rm pl}}^{+p_{\rm pl}} dp\,d\zeta'\varphi_{\zeta}^{*}(p)\varphi_{\zeta'}(p) &= \int_{-p_{\rm pl}}^{+p_{\rm pl}} \frac{\left(1-\beta p'+2\beta^2 p'^2\right)}{\left(1-\beta p'+2\beta^2 p'^2\right)} dp\varphi_{\zeta}^{*}(p)\varphi'_{\zeta'}(p)d\zeta', \\ \int_{-p_{\rm pl}}^{+p_{\rm pl}} \left(1-\beta p'+2\beta^2 p'^2\right)\langle\varphi_{\zeta}\mid\varphi_{\zeta'}\rangle &= \int_{-p_{\rm pl}}^{+p_{\rm pl}} \left(1-\beta p'+2\beta^2 p'^2\right)\delta(\zeta-\zeta')d\zeta' = 1.\end{aligned} \tag{3.29}$$

This restriction is nothing but the condition that the state $|\phi\rangle$ is included in our Hilbert space. Thus the Fourier transformation, (3.22) and (3.23), is well defined in this Hilbert space. It should be noted that the wave function (3.6) reduces to the plane wave solution in the limit of $\beta \to 0$

$$\varphi_{\zeta}(p) \longrightarrow \frac{1}{\sqrt{2\pi\hbar}} \exp\left(\frac{\zeta p}{i\hbar}\right). \tag{3.30}$$

This is because we have set the normalization constant as $N_1 = 1$ by rescaling $|\varphi_{\zeta}\rangle$. In [36] the normalization constant is fixed to $N_1 = 2\sqrt{\hbar\sqrt{\beta}}$ so that the norm of the wavefunction is equal to one. In this case, the wavefunction vanishes in the mentioned limit because of the extra dependence on β in the normalization factor.

4. A (1+1)-Dimensional Quantum Field Theory with GUP

In this section we construct a quantum field theory with a GUP (that admits both a minimal length and a maximal momentum) in $(1+1)$-dimensional spacetime in the canonical formalism. First of all, we define a classical field such that it solves the following Klein-Gordon equation (see e.g., [46]):

$$-\hbar^2 \frac{\partial^2 \Phi}{\partial t^2} = -\hbar^2 c^2 \nabla^2 \Phi + m^2 c^4 \Phi,$$
$$0 = \left[\left(\hbar \frac{\partial}{\partial t} \right)^2 + Z(p) + m^2 \right] \Phi(p,t). \tag{4.1}$$

In this equation we leave the function of momentum $Z(p)$ indefinite, where $Z(p)$ is an arbitrary even function whose explicit form does not have any influence on the following arguments [34]. This function determines the dispersion relation and depends on what kind of theory we want. Quantization of classical field is in order to provide the set of operator fields and Hamiltonian that lead the same Klein-Gordon equation by the Heisenberg equation. We define a field $\widehat{\Phi}(p,t)$ (see for instance [34])

$$\widehat{\Phi}(p,t) = \frac{\hbar}{\sqrt{E(p)}} \widehat{\phi}(p) \exp\left(\frac{t}{i\hbar} E(p) \right) + \frac{\hbar}{\sqrt{E(-p)}} \widehat{\phi}^{\dagger}(-p) \exp\left(\frac{-t}{i\hbar} E(-p) \right), \tag{4.2}$$

where $E(p) \equiv \sqrt{Z(p) + m^2}$. This field satisfies the Klein-Gordon equation (4.1). Its Fourier pair $\widehat{\phi}(\zeta, t)$ is written in terms of our one-particle states in space momentum

$$\widehat{\Phi}(\zeta,t) \equiv \int_{-p_{\mathrm{pl}}}^{+p_{\mathrm{pl}}} \frac{dp}{(1 - \beta p + 2\beta^2 p^2)} \left\{ \psi_{\zeta}^{*}(p) \widehat{\phi}(p) \frac{\hbar}{\sqrt{2E(p)}} \exp\left(\frac{1}{i\hbar} tE(p) \right) \right.$$
$$\left. + \psi_{\zeta}(p) \widehat{\phi}^{\dagger}(p) \frac{\hbar}{\sqrt{2E(p)}} \exp\left(\frac{-1}{i\hbar} tE(p) \right) \right\}. \tag{4.3}$$

In the above expression, the operators $\widehat{\phi}(p)$ and $\widehat{\phi}^{\dagger}(p)$ are, respectively, the annihilation and creation operators for one particle state with momentum p. They have the following commutation relation in momentum space:

$$\left[\widehat{\phi}(p), \widehat{\phi}^{\dagger}(p') \right] = \left(1 - \beta p + 2\beta^2 \mathrm{p}^2 \right) \delta(p - p'). \tag{4.4}$$

We construct a natural Hamiltonian in terms of the Fock space picture

$$\mathcal{H} \equiv \int_{-p_{\mathrm{pl}}}^{+p_{\mathrm{pl}}} \frac{dp}{(1 - \beta p + 2\beta^2 p^2)} E(p) \widehat{\phi}^{\dagger}(p) \widehat{\phi}(p). \tag{4.5}$$

This Hamiltonian produces the following Heisenberg equation of motion:

$$\frac{1}{i\hbar}\left[\hat{\Phi}(p,t),\mathscr{H}\right]=\frac{\partial}{\partial t}\hat{\Phi}(p,t), \tag{4.6}$$

which is consistent with the Klein-Gordon equation (4.1). We can define the conjugate momentum $\hat{\Pi}(p,t)\equiv(\partial/\partial t)\hat{\Phi}(p,t)$. Therefore, we can write the Hamiltonian in momentum space as

$$\mathscr{H}\equiv\int_{-p_{\rm pl}}^{+p_{\rm pl}}\frac{dp}{(1-\beta p+2\beta^2p^2)}\frac{1}{2}\left[\hat{\Pi}(-p,t)\hat{\Pi}(p,t)+\left(\frac{E(p)}{\hbar}\right)^2\hat{\Phi}(-p,t)\times\hat{\Phi}(p,t)\right]+\text{const.}. \tag{4.7}$$

Now we compute the following commutation relation using the commutation relation (4.4):

$$\left[\hat{\Phi}(p,t),\hat{\Pi}(p',t)\right]=i\hbar\left(1-\beta p+2\beta^2p^2\right)\delta(p+p'). \tag{4.8}$$

We have also

$$\begin{aligned}\left[\hat{\Phi}(\zeta,t),\hat{\Pi}(\zeta',t)\right]&=i\hbar\langle\varphi_\zeta\mid\varphi_{\zeta'}\rangle=\frac{-\sqrt{7}\beta\hbar^2}{2(\zeta-\zeta')}\left[\tan^{-1}\left(\frac{4\beta p_{\rm pl}-1}{\sqrt{7}}\right)+\tan^{-1}\left(\frac{4\beta p_{\rm pl}+1}{\sqrt{7}}\right)\right]^{-1}\\&\times\left\{\exp\left[\frac{2(\zeta-\zeta')}{i\hbar\beta\sqrt{7}}\left(\tan^{-1}\left(\frac{1}{\sqrt{7}}\right)+\tan^{-1}\left(\frac{4\beta p_{\rm pl}-1}{\sqrt{7}}\right)\right)\right]\right.\\&\left.-\exp\left[\frac{-2(\zeta-\zeta')}{i\hbar\beta\sqrt{7}}\left(\tan^{-1}\left(\frac{1}{\sqrt{7}}\right)+\tan^{-1}\left(\frac{4\beta p_{\rm pl}+1}{\sqrt{7}}\right)\right)\right]\right\}.\end{aligned} \tag{4.9}$$

5. Higher Dimensional Algebra

The generalization of the Heisenberg algebra to higher dimensions where rotation symmetry is preserved and there are both a minimal length and a maximal momentum is (see [36])

$$[\hat{x}_i,\hat{p}_j]=i\hbar\delta_{ij}\left(1-\beta p+2\beta^2p^2\right), \tag{5.1}$$

where in three dimensions $\vec{p}=p_x\vec{i}+p_y\vec{j}+p_z\vec{k}$, that is, $\vec{i}$, $\vec{j}$, and $\vec{k}$ are unit vectors of cartesian coordinates and $(\vec{p})^2=\sum_{i=1}^{d}(\hat{p}_i)^2$. So, we can generalize our operators acting on momentum space in higher dimensional theory as

$$\begin{aligned}\mathbf{p}_i\varphi(p)&=p_i\varphi(p),\\\mathbf{X}_j\varphi(p)&=i\hbar\left(1-\beta\mathbf{p}+2\beta^2\mathbf{p}^2\right)\frac{\partial}{\partial p_j}\varphi(p).\end{aligned} \tag{5.2}$$

Then, it is easy to show that

$$[\mathbf{X}_i, \mathbf{X}_j] = i\beta\hbar\left(\frac{4\beta P - 1}{P}\right)(\mathbf{P}_i\mathbf{X}_j - \mathbf{P}_j\mathbf{X}_i). \tag{5.3}$$

We see that in the presence of minimal length and maximal momentum, the very spacetime manifold is noncommutative in nature. Note that it was not the case for Matsuo and Shibusa analysis in the presence of just the minimal length [34]. So incorporation of natural cutoffs as minimal length and maximal momentum brings the spacetime manifold to be noncommutative in quantum gravity era. The generator of rotation in our framework are [36]

$$\begin{aligned} \mathbf{L}_{ij} &= \frac{1}{(1 - \beta\mathbf{p} + 2\beta^2\mathbf{p}^2)}(\mathbf{X}_i\mathbf{P}_j - \mathbf{X}_j\mathbf{P}_i), \\ [\mathbf{X}_i, \mathbf{X}_j] &= -i\beta\hbar\left(4\beta - \frac{1}{\mathbf{p}}\right)\left(1 - \beta\mathbf{p} + 2\beta^2\mathbf{p}^2\right)\mathbf{L}_{ij}. \end{aligned} \tag{5.4}$$

As Kempf et al. [17] case, from the positivity of the norm, that is,

$$\left\| \left(X - \langle X \rangle + \frac{\langle [X, P] \rangle}{2(\Delta P)^2}(P - \langle P \rangle) \right) |\varphi\rangle \right\| \geq 0, \tag{5.5}$$

on the boundary of the physically allowed region, we have

$$\left(X - \langle X \rangle + \frac{\langle [X, P] \rangle}{2(\Delta P)^2}(P - \langle P \rangle) \right) |\varphi\rangle = 0. \tag{5.6}$$

Using (5.2), the differential equation in momentum space corresponding to (5.6) is in the following form:

$$\left[i\hbar\left(1 - \beta p + 2\beta^2 p^2\right)\frac{\partial}{\partial p_i} - \left\langle \widehat{X}^i \right\rangle + i\hbar \times \frac{1 + 2\beta^2(\Delta p)^2 + 2\beta^2\langle p \rangle^2 - \beta\langle p \rangle}{2(\Delta p)^2}(p_i - \langle p_i \rangle) \right] |\varphi\rangle = 0. \tag{5.7}$$

By introducing the path integral measure and also Lagrangian through introduction of two parameters ℓ and m defined as follows:

$$1 = \int_{-p_{\rm pl}}^{+p_{\rm pl}} D\Phi(p,t) \exp\left(-\frac{1}{2}\int_{-p_{\rm pl}}^{+p_{\rm pl}} \frac{dtdp}{(1 - \beta p + 2\beta^2 p^2)^{\ell}} \Phi(p,t)\Phi(-p,t) \right) \tag{5.8}$$

$$\mathcal{L} = -\frac{1}{2}\int_{-p_{\rm pl}}^{+p_{\rm pl}} dp\left(1 - \beta p + 2\beta^2 p^2\right)^m \Phi(-p,t)\left[\partial_t^2 + \left(\frac{E(p)}{\hbar}\right)^2 \right]\Phi(p,t), \tag{5.9}$$

the action $S \equiv \int dt\mathcal{L}$ of the model can be written as follows:

$$S = -\frac{1}{2}\int_{-p_{pl}}^{+p_{pl}} dpdp'dqdq'\left(1-\beta p+2\beta^2p^2\right)^m \delta(p+p')\delta(q+q') \times \Phi(p,q)\left[-\left(\frac{q}{\hbar}\right)^2+\left(\frac{E(p)}{\hbar}\right)^2\right]\Phi(p',q'). \tag{5.10}$$

Now using assumptions (5.8) and (5.10), we can compute the T^*-product in this framework as follows:

$$\left\langle T^*\widehat{\Phi}(p,q)\widehat{\Phi}(p',q')\right\rangle = \frac{\hbar^3}{i}\left(1-\beta p+2\beta^2p^2\right)^{-m-\ell}\delta(p+p')\delta(q+q') \times \frac{1}{-q^2+E^2(p)-i\varepsilon}. \tag{5.11}$$

Using the Bjorken-Johnson-Low prescription, from behavior of

$$\lim_{q\to\infty}\left\langle T^*\widehat{\Phi}(p,q)\widehat{\Phi}(p',q')\right\rangle, \quad \lim_{q\to\infty} q\left\langle T^*\widehat{\Phi}(p,q)\widehat{\Phi}(p',q')\right\rangle, \tag{5.12}$$

we find

$$T^*\widehat{\Phi}(p,q)\widehat{\Phi}(p',q') = T\widehat{\Phi}(p,q)\widehat{\Phi}(p',q'), \qquad \left[\widehat{\Phi}(p,t),\widehat{\Phi}(p',t)\right] = 0. \tag{5.13}$$

Also through the behavior of $\lim_{q\to\infty}q^2\langle T^*\widehat{\Phi}(p,q)\widehat{\Phi}(p',q')\rangle$, we can obtain

$$\left[\widehat{\Phi}(p,t),\dot{\widehat{\Phi}}(p',t)\right] = i\hbar\left(1-\beta p+2\beta^2p^2\right)^{-m-\ell}\delta(p+p'). \tag{5.14}$$

If we define conjugate momentum $\Pi(p,t)$ in Lagrangian Formalism by using a free parameter r as

$$\Pi(p,t) \equiv \left(1-\beta p+2\beta^2p^2\right)^r \frac{\delta S}{\delta\dot{\Phi}(-p,t)}, \tag{5.15}$$

the commutation relation (5.14) takes the following form:

$$\left[\widehat{\Phi}(p,t),\widehat{\Pi}(p',t)\right] = i\hbar\left(1-\beta p+2\beta^2p^2\right)^{-2m-\ell-r}\delta(p+p'). \tag{5.16}$$

The right hand side of the above equation is a symplectic form [34]. Since the Legender transformation from Lagrangian to Hamiltonian is (see [34])

$$\frac{1}{i\hbar}\int \mathcal{H} dt \equiv (\text{symplectic form})^{-1}\sum_i \Pi_i d\Phi_i - \frac{1}{i\hbar}\int dt\mathcal{L}, \tag{5.17}$$

one can obtain a Hamiltonian as

$$\mathcal{H} = \int_{-p_{\rm pl}}^{+p_{\rm pl}} dp\left(1-\beta p+2\beta^2p^2\right)^{m+\ell}\left[\left\{1-\frac{1}{2}\left(1-\beta p+2\beta^2p^2\right)^{-2m-2\ell-2r}\right\}\hat{\Pi}(-p,t)\hat{\Pi}(p,t) + \frac{1}{2}\left(1-\beta p+2\beta^2p^2\right)^{-\ell}\left(\frac{E(p)}{\hbar}\right)^2\hat{\Phi}(-p,t)\hat{\Phi}(p,t)\right]. \tag{5.18}$$

The condition that the coefficient of $\Pi(-p,t)\Pi(p,t)$ must be homogeneous fixes the free parameter r so that

$$-\ell - 2m - 2r = 0. \tag{5.19}$$

So we find

$$r = -\frac{\ell}{2} - m. \tag{5.20}$$

Then the Hamiltonian becomes

$$\mathcal{H} = \int_{-p_{\rm pl}}^{+p_{\rm pl}} dp\left(1-\beta p+2\beta^2p^2\right)^{m+\ell} \times\frac{1}{2}\left[\hat{\Pi}(-p,t)\hat{\Pi}(p,t) + \left(1-\beta p+2\beta^2p^2\right)^{-\ell}\left(\frac{E(p)}{\hbar}\right)^2\hat{\Phi}(-p,t)\hat{\Phi}(p,t)\right]. \tag{5.21}$$

By comparing this with (4.7), we obtain the parameters m and ℓ as the form

$$\ell = 0, \quad m = -1. \tag{5.22}$$

We see explicitly that the two formalisms provide consistent structure of the deformed quantum field theory. Summarizing the results, we have

$$\begin{aligned}\mathcal{H} &= \int_{-p_{\rm pl}}^{+p_{\rm pl}} dp\left(1-\beta p+2\beta^2p^2\right)^{-1} \\ &\quad\times\frac{1}{2}\left[\hat{\Pi}(-p,t)\hat{\Pi}(p,t)+\left(\frac{E(p)}{\hbar}\right)^2\hat{\Phi}(-p,t)\hat{\Phi}(p,t)\right]+\text{const.} \\ &= \int_{-p_{\rm pl}}^{+p_{\rm pl}} dp\left(1-\beta p+2\beta^2p^2\right)^{-1}E(p)\left(\hat{\phi}^\dagger(p)\hat{\phi}(p)+\text{const.}\right).\end{aligned} \tag{5.23}$$

Now the annihilation and creation operators for one particle states in the presence of both minimal length and maximal momentum have the following commutation relation

$$\left[\hat{\phi}(p),\hat{\phi}^\dagger(p')\right]=\left(1-\beta p+2\beta^2p^2\right)\delta(p-p'), \tag{5.24}$$

and the canonical commutation relation is

$$\left[\hat{\Phi}(p,t),\hat{\Pi}(p',t)\right]=i\hbar\left(1-\beta p+2\beta^2p^2\right)\delta(p+p'). \tag{5.25}$$

Also Heisenberg equation of motion is given as

$$\frac{1}{i\hbar}\left[\hat{\Phi}(p,t),\mathcal{H}\right]=\frac{\partial}{\partial t}\hat{\Phi}(p,t)=\hat{\Pi}(p,t). \tag{5.26}$$

Now we calculate propagators from these quantities in the presence of minimal length and maximal momentum as follows:

$$\begin{aligned}\left\langle T\hat{\Phi}(p,q)\hat{\Phi}(p',q')\right\rangle &\equiv \frac{1}{2\pi\hbar}\int_{-p_{\rm pl}}^{+p_{\rm pl}} dtdt'\exp\left(-\frac{qt+q't'}{i\hbar}\right)\left\langle T\hat{\Phi}(p,t)\hat{\Phi}(p',t')\right\rangle \\ &= \frac{\hbar^3}{i}\left(1-\beta p+2\beta^2p^2\right)\delta(p+p')\delta(q+q')\frac{1}{-q^2+E^2(p)-i\varepsilon}.\end{aligned} \tag{5.27}$$

In this case the Lagrangian is given by

$$\mathcal{L}=-\frac{1}{2}\int_{-p_{\rm pl}}^{+p_{\rm pl}} dp\left(1-\beta p+2\beta^2p^2\right)^{-1}\Phi(-p,t)\left[\partial_t^2+\left(\frac{E(p)}{\hbar}\right)^2\right]\Phi(p,t). \tag{5.28}$$

Also

$$\mathbf{1}=\int_{-p_{\rm pl}}^{+p_{\rm pl}} D\Phi(p,t)\exp\left(-\frac{1}{2}\int_{-p_{\rm pl}}^{+p_{\rm pl}} dtdp\Phi(p,t)\Phi(-p,t)\right). \tag{5.29}$$

Lagrangian and path integral measure in $(d+1)$-dimensional spacetime and in the presence of both minimal length and maximal momentum can be written a follows:

$$\mathcal{L} = -\frac{1}{2}\int_{-p_{\text{pl}}}^{+p_{\text{pl}}} d^d p\left(1-\beta\vec{p}+2\beta^2\vec{p}^2\right)^{-1}\Phi(-\vec{p},t)\left[\partial_t^2+\left(\frac{E(\vec{p})}{\hbar}\right)^2\right]\Phi(\vec{p},t),$$
$$1 = \int_{-p_{\text{pl}}}^{+p_{\text{pl}}} D\Phi(\vec{p},t)\exp\left(-\frac{1}{2}\int_{-p_{\text{pl}}}^{+p_{\text{pl}}} dt d^d p\Phi(\vec{p},t)\Phi(-\vec{p},t)\right). \tag{5.30}$$

So we have provided the basic ingredients of a scalar field theory in the context of minimal length and maximal momentum hypothesis. We note that incorporation of maximal momentum as a new natural cutoff in quantum field theory restricts further the form of commutation relations and propagators.

6. Summary

Incorporation of quantum gravity effects in quantum mechanics and also quantum field theory leads to the existence of natural cutoffs on position and momentum measurements of test particles. These natural cutoffs regularize the underlying field theory in a phenomenologically viable manner. Quantum field theory of scalar fields in the presence of minimal length has been studied in literature (see [34]). Here, following our recent work on Hilbert space representation of quantum mechanics with minimal length and maximal momentum [36], we generalized the work reported in [34] to the case that there is a maximal momentum for test particles too. We have formulated a quantum theory of free scalar fields in $(1+1)$-dimensions based on a deformed Heisenberg algebra $[\hat{x},\hat{p}] = i\hbar(1-\beta p+2\beta^2 p^2)$, where β is a deformation parameter. The generalized uncertainty principle based on this algebra requires the existence of a minimal measurable length and also maximal measurable momentum of the order of Planck momentum for any test particle. In this framework we have considered both canonical and path integral quantization of free scalar field in the presence of these natural cutoffs. Finally we have extended our analysis to a higher dimensional spacetime in the path integral formalism. We have shown that in the presence of minimal length and maximal momentum, the underlying spacetime manifold becomes noncommutative in nature. We note that it was not the case for Matsuo and Shibusa analysis in the presence of just the minimal length as has been reported in [34]. Finally we note that this study can be simply generalized to the supersymmetric quantum field theory.

Acknowledgment

K. Nozari work is financially supported by the Research Deputy of the Islamic Azad University, Sari Branch, Sari, Iran.

References

[1] G. Amelino-Camelia, M. Arzano, Y. Ling, and G. Mandanici, "Black-hole thermodynamics with modified dispersion relations and generalized uncertainty principles," *Classical and Quantum Gravity*, vol. 23, no. 7, pp. 2585–2606, 2006.

[2] G. Veneziano, "A stringy nature needs just two constants," *Europhysics Letters*, vol. 2, p. 199, 1986.
[3] D. Amati, M. Cialfaloni, and G. Veneziano, "Superstring collisions at planckian energies," *Physics Letters B*, vol. 197, pp. 81–88, 1987.
[4] D. Amati, M. Cialfaloni, and G. Veneziano, "Can spacetime be probed below the string size?" *Physics Letters B*, vol. 216, pp. 41–47, 1989.
[5] D. J. Gross and P. F. Mende, "The high-energy behavior of string scattering amplitudes," *Physics Letters B*, vol. 197, pp. 129–134, 1987.
[6] K. Konishi, G. Paffuti, and P. Provero, "Minimum physical length and the generalized uncertainty principle in string theory," *Physics Letters B*, vol. 234, no. 3, pp. 276–284, 1990.
[7] R. Guida, K. Konishi, and P. Provero, "On the short distance behavior of string theories," *Modern Physics Letters A*, vol. 6, p. 1487, 1991.
[8] M. Kato, "Particle theories with minimum observable length and open string theory," *Physics Letters B*, vol. 245, pp. 43–47, 1990.
[9] L. J. Garay, "Quantum gravity and minimum length," *International Journal of Modern Physics A*, vol. 10, p. 145, 1995.
[10] S. Capozziello, G. Lambiase, and G. Scarpetta, "Generalized uncertainty principle from quantum geometry," *International Journal of Theoretical Physics*, vol. 39, no. 1, pp. 15–22, 2000.
[11] M. Maggiore, "A generalized uncertainty principle in quantum gravity ," *Physics Letters B*, vol. 304, pp. 65–69, 1993.
[12] M. Maggiore, "The algebraic structure of the generalized uncertainty principle," *Physics Letters B*, vol. 319, pp. 83–86, 1993.
[13] M. Maggiore, "Quantum groups, gravity, and the generalized uncertainty principle," *Physical Review D*, vol. 49, no. 10, pp. 5182–5187, 1994.
[14] S. Hossenfelder, "Can we measure structures to a precision better than the Planck length?" *Classical and Quantum Gravity*, vol. 29, Article ID 115011, 2012.
[15] P. Pedram, "New approach to nonperturbative quantum mechanics with minimal length uncertainty," *Physical Review D*, vol. 85, Article ID 024016, 2012.
[16] R. J. Adler, "Six easy roads to the Planck scale," *American Journal of Physics*, vol. 78, p. 925, 2010.
[17] A. Kempf, G. Mangano, and R. B. Mann, "Hilbert space representation of the minimal length uncertainty relation," *Physical Review D*, vol. 52, no. 2, pp. 1108–1118, 1995.
[18] F. Scardigli, "Generalized uncertainty principle in quantum gravity from micro-black hole gedanken experiment," *Physics Letters B*, vol. 452, pp. 39–44, 1999.
[19] G. Amelino-Camelia, "Relativity in spacetimes with short-distance structure governed by an observer-independent (planckian) length scale," *International Journal of Modern Physics D*, vol. 11, p. 35, 2000.
[20] G. Amelino-Camelia, "Relativity: special treatment," *Nature*, vol. 418, pp. 34–35, 2002.
[21] G. Amelino-Camelia, "Doubly-special relativity: first results and key open problems," *International Journal of Modern Physics D*, vol. 11, p. 1643, 2002.
[22] J. Kowalski-Glikman, "Introduction to doubly special relativity," *Lecture Notes in Physics*, vol. 669, pp. 131–159, 2005.
[23] G. Amelino-Camelia, J. Kowalski-Glikman, G. Mandanici, and A. Procaccini, "Phenomenology of doubly special relativity," *International Journal of Modern Physics A*, vol. 20, p. 6007, 2005.
[24] K. Imilkowska and J. Kowalski-Glikman, "Doubly special relativity as a limit of gravity," *Lecture Notes in Physics*, vol. 702, p. 279, 2006.
[25] J. Magueijo and L. Smolin, "Lorentz invariance with an invariant energy scale," *Physical Review Letters*, vol. 88, Article ID 1904303, 2002.
[26] J. Magueijo and L. Smolin, "Generalized Lorentz invariance with an invariant energy scale," *Physical Review D*, vol. 67, Article ID 044017, 2003.
[27] J. Magueijo and L. Smolin, "String theories with deformed energy-momentum relations, and a possible nontachyonic bosonic string," *Physical Review D*, vol. 71, Article ID 026010, 2005.
[28] J. L. Cortes and J. Gamboa, "Quantum uncertainty in doubly special relativity," *Physical Review D*, vol. 71, Article ID 065015, 2005.
[29] A. Kempf, "Quantum field theory with nonzero minimal uncertainties in positions and momenta," *Journal of Mathematical Physics*, vol. 38, Article ID 124014, p. 1374, 1997.

[30] A. Kempf and G. Mangano, "Minimal length uncertainty relation and ultraviolet regularization," *Physical Review D*, vol. 55, Article ID 124014, p. 7909, 1997.

[31] A. Kempf, "Fields with finite information density ," *Physical Review D*, vol. 69, Article ID 124014, 2004.

[32] S. Hossenfelder, "Interpretation of quantum field theories with a minimal length scale," *Physical Review D*, vol. 73, Article ID 105013, 2006.

[33] S. Hossenfelder, "A note on quantum field theories with a minimal length scale," *Classical and Quantum Gravity*, vol. 25, Article ID 038003, 2008.

[34] T. Matsuo and Y. Shibusa, "Quantization of fields based on generalized uncertainty principle," *Modern Physics Letters A*, vol. 21, p. 1285, 2006.

[35] Y. Shibusa, "Supersymmetric field theory based on generalized uncertainty principle," *International Journal of Modern Physics A*, vol. 22, p. 5279, 2007.

[36] K. Nozari and A. Etemadi, "Minimal length, maximal momentum, and Hilbert space representation of quantum mechanics," *Physical Review D*, vol. 85, Article ID 104029, 2012.

[37] A. F. Ali, S. Das, and E. C. Vagenas, "Discreteness of space from the generalized uncertaintyprinciple," *Physics Letters B*, vol. 678, pp. 497–499, 2009.

[38] S. Das, E. C. Vagenas, and A. Farag Ali, "Discreteness of space from GUP II: relativistic wave equations," *Physics Letters B*, vol. 692, no. 5, p. 342, 2010.

[39] P. Pedram, K. Nozari, and S. H. Taheri, "The effects of minimal length and maximal momentum on the transition rate of ultra cold neutrons in gravitational field," *Journal of High Energy Physics*, vol. 3, article 93, 2011.

[40] J. D. Bjorken, "Applications of the chiral U(6)⊗U(6) algebra of current densities," *Physical Review*, vol. 148, p. 1467, 1966.

[41] K. Johnson and F. E. Low, "Current algebras in a simple model," *Progress of Theoretical Physics, Supplement*, no. 37-38, pp. 74–93, 1966.

[42] D. V. Ahluwalia, "Quantum measurement, gravitation, and locality," *Physics Letters B*, vol. 339, pp. 301–303, 1994.

[43] S. Doplicher, K. Fredenhagen, and J. E. Roberts, "Spacetime quantization induced by classical gravity," *Physics Letters B*, vol. 331, pp. 39–44, 1994.

[44] G. Amelino-Camelia, "Gravity-wave interferometers as probes of a low-energy effective quantum gravity," *Physical Review D*, vol. 62, Article ID 024015, 2000.

[45] G. Amelino-Camelia, D. Benedetti, F. D'Andrea, and A. Procaccini, "Comparison of relativity theories with observer-independent scales of both velocity and length/mass," *Classical and Quantum Gravity*, vol. 20, no. 24, pp. 5353–5370, 2003.

[46] D. McMahon, *Quantum Field Theory Demystified*, McGraw-Hill, 2008.

3

Leptogenesis in the Universe

Chee Sheng Fong,[1] Enrico Nardi,[1] and Antonio Riotto[2]

[1] *INFN Laboratori Nazionali di Frascati, Via Enrico Fermi 40, 00044 Frascati, Italy*
[2] *Department of Theoretical Physics, University of Geneva and Center for Astroparticle Physics (CAP), 24 quai E. Ansermet, 1211 Geneva 4, Switzerland*

Correspondence should be addressed to Enrico Nardi, enrico.nardi@lnf.infn.it

Academic Editor: Gian Luigi Fogli

Leptogenesis is a class of scenarios in which the cosmic baryon asymmetry originates from an initial lepton asymmetry generated in the decays of heavy sterile neutrinos in the early Universe. We explain why leptogenesis is an appealing mechanism for baryogenesis. We review its motivations and the basic ingredients and describe subclasses of effects, like those of lepton flavours, spectator processes, scatterings, finite temperature corrections, the role of the heavier sterile neutrinos, and quantum corrections. We then address leptogenesis in supersymmetric scenarios, as well as some other popular variations of the basic leptogenesis framework.

1. The Baryon Asymmetry of the Universe

1.1. Observations

Up to date no traces of cosmological antimatter have been observed. The presence of a small amount of antiprotons and positrons in cosmic rays can be consistently explained by their secondary origin in cosmic particles collisions or in highly energetic astrophysical processes, but no antinuclei, even as light as antideuterium or as tightly bounded as anti-α particles, have ever been detected.

The absence of annihilation radiation $p\overline{p} \rightarrow \ldots\pi^0 \rightarrow \ldots 2\gamma$ excludes significant matter-antimatter admixtures in objects up to the size of galactic clusters ~20 Mpc [1]. Observational limits on anomalous contributions to the cosmic diffuse γ-ray background and the absence of distortions in the cosmic microwave background (CMB) imply that little antimatter is to be found within ~1 Gpc and that within our horizon an equal amount of matter and antimatter can be excluded [2]. Of course, at larger superhorizon scales the vanishing of the average asymmetry cannot be ruled out, and this would indeed be the case if the fundamental Lagrangian is C and CP symmetric and charge invariance is broken spontaneously [3].

Quantitatively, the value of baryon asymmetry of the Universe is inferred from observations in two independent ways. The first way is by confronting the abundances of the light elements, D, ^{3}He, ^{4}He, and ^{7}Li, with the predictions of Big Bang nucleosynthesis (BBN) [4–9]. The crucial time for primordial nucleosynthesis is when the thermal bath temperature falls below $T \lesssim 1\,\text{MeV}$. With the assumption of only three light neutrinos, these predictions depend on a single parameter, that is, the difference between the number of baryons and antibaryons normalized to the number of photons:

$$\eta \equiv \left.\frac{n_B - n_{\overline{B}}}{n_\gamma}\right|_0 , \tag{1.1}$$

where the subscript 0 means "at present time." By using only the abundance of deuterium, that is particularly sensitive to η, [4] quotes:

$$10^{10}\,\eta = 5.7 \pm 0.6 \quad (95\%\,\text{c.l.}). \tag{1.2}$$

In this same range there is also an acceptable agreement among the various abundances, once theoretical uncertainties as well as statistical and systematic errors are accounted for [6].

The second way is from measurements of the CMB anisotropies (for pedagogical reviews, see [10, 11]). The crucial time for CMB is that of recombination, when the temperature dropped down to $T \lesssim 1\,\text{eV}$ and neutral hydrogen can be formed. CMB observations measure the relative baryon contribution to the energy density of the Universe multiplied by the square of the (reduced) Hubble constant $h \equiv H_0/(100\,\text{km sec}^{-1}\,\text{Mpc}^{-1})$:

$$\Omega_B h^2 \equiv h^2 \frac{\rho_B}{\rho_{\text{crit}}}, \tag{1.3}$$

that is related to η through $10^{10}\,\eta = 274\,\Omega_B h^2$. The physical effect of the baryons at the onset of matter domination, which occurs quite close to the recombination epoch, is to provide extra gravity which enhances the compression into potential wells. The consequence is enhancement of the compressional phases which translates into enhancement of the odd peaks in the spectrum. Thus, a measurement of the odd/even peak disparity constrains the baryon energy density. A fit to the most recent observations (WMAP7 data only, assuming a ΛCDM model with a scale-free power spectrum for the primordial density fluctuations) gives at 68% c.l. [12]

$$10^2\,\Omega_B h^2 = 2.258^{+0.057}_{-0.056}. \tag{1.4}$$

There is a third way to express the baryon asymmetry of the Universe, that is, by normalizing the baryon asymmetry to the entropy density $s = g_*(2\pi^2/45)T^3$, where g_* is the number of degrees of freedom in the plasma and T is the temperature:

$$Y_{\Delta B} \equiv \left.\frac{n_B - n_{\overline{B}}}{s}\right|_0 . \tag{1.5}$$

The relation with the previous definitions is given by the conversion factor $s_0/n_{\gamma 0} = 7.04$. $Y_{\Delta B}$ is a convenient quantity in theoretical studies of the generation of the baryon asymmetry from very early times, because it is conserved throughout the thermal evolution of the Universe. In terms of $Y_{\Delta B}$ the BBN results (1.2) and the CMB measurement (1.4) (at 95% c.l.) read

$$Y_{\Delta B}^{\rm BBN} = (8.10 \pm 0.85) \times 10^{-11}, \qquad Y_{\Delta B}^{\rm CMB} = (8.79 \pm 0.44) \times 10^{-11}. \tag{1.6}$$

The impressive consistency between the determinations of the baryon density of the Universe from BBN and CMB that, besides being completely independent, also refer to epochs with a six orders of magnitude difference in temperature, provides a striking confirmation of the hot Big Bang cosmology.

1.2. Theory

From the theoretical point of view, the question is where the Universe baryon asymmetry comes from. The inflationary cosmological model excludes the possibility of a fine tuned initial condition, and since we do not know any other way to construct a consistent cosmology without inflation, this is a strong veto.

The alternative possibility is that the Universe baryon asymmetry is generated dynamically, a scenario that is known as *baryogenesis*. This requires that baryon number (B) is not conserved. More precisely, as Sakharov pointed out [13], the ingredients required for baryogenesis are three.

(1) B violation is required to evolve from an initial state with $Y_{\Delta B} = 0$ to a state with $Y_{\Delta B} \neq 0$.

(2) C and CP violation: if either C or CP was conserved, then processes involving baryons would proceed at the same rate as the C- or CP-conjugate processes involving antibaryons, with the overall effect that no baryon asymmetry is generated.

(3) Out of equilibrium dynamics: equilibrium distribution functions $n_{\rm eq}$ are determined solely by the particle energy E, chemical potential μ, and by its mass which, because of the CPT theorem, is the same for particles and antiparticles. When charges (such as B) are not conserved, the corresponding chemical potentials vanish, and thus $n_B = \int (d^3p/(2\pi^3))n_{\rm eq} = n_{\overline{B}}$.

Although these ingredients are all present in the Standard Model (SM), so far all attempts to reproduce quantitatively the observed baryon asymmetry have failed.

(1) In the SM B is violated via the triangle anomaly. Although at zero temperature B violating processes are too suppressed to have any observable effect [14], at high temperatures they occur with unsuppressed rates [15]. The first condition is then quantitatively realized in the early Universe.

(2) SM weak interactions violate C maximally. However, the amount of CP violation from the Kobayashi-Maskawa complex phase [16], as quantified by means of the Jarlskog invariant [17], is only of order 10^{-20}, and this renders impossible generating $Y_{\Delta B} \sim 10^{-10}$ [18–20].

(3) Departures from thermal equilibrium occur in the SM at the electroweak phase transition (EWPT) [21, 22]. However, the experimental lower bound on the Higgs mass implies that this transition is not sufficiently first order as required for successful baryogenesis [23].

This shows that baryogenesis requires new physics that extends the SM in at least two ways. It must introduce new sources of CP violation and it must either provide a departure from thermal equilibrium in addition to the EWPT or modify the EWPT itself. In the past thirty years or so, several new physics mechanisms for baryogenesis have been put forth. Some among the most studied are *GUT baryogenesis* [24–33], *electroweak baryogenesis* [21, 34, 35], the *affleck-Dine mechanism* [36, 37], and *spontaneous Baryogenesis* [38, 39]. However, soon after the discovery of neutrino masses, because of its connections with the seesaw model [40–44] and its deep interrelations with neutrino physics in general, the mechanism of baryogenesis via *Leptogenesis* acquired a continuously increasing popularity. Leptogenesis was first proposed by Fukugita and Yanagida in [45]. Its simplest and theoretically best motivated realization is precisely within the seesaw mechanism. To implement the seesaw, new Majorana $SU(2)_L$ singlet neutrinos with a large mass scale M are added to the SM particle spectrum. The complex Yukawa couplings of these new particles provide new sources of CP violation, departure from thermal equilibrium can occur if their lifetime is not much shorter than the age of the Universe when $T \sim M$, and their Majorana masses imply that lepton number is not conserved. A lepton asymmetry can then be generated dynamically, and SM sphalerons will partially convert it into a baryon asymmetry [46]. A particularly interesting possibility is "thermal leptogenesis" where the heavy Majorana neutrinos are produced by scatterings in the thermal bath starting from a vanishing initial abundance, so that their number density can be calculated solely in terms of the seesaw parameters and of the reheat temperature of the Universe.

This paper is organized as follows. In Section 2 the basis of leptogenesis is reviewed in the simple scenario of the one-flavour regime, while the role of flavour effects is described in Section 3. Theoretical improvements of the basic pictures, like spectator effects, scatterings and CP violation in scatterings, thermal corrections, the possible role of the heavier singlet neutrinos, and quantum effects, are reviewed in Section 4. Leptogenesis in the supersymmetric seesaw is reviewed in Section 5, while in Section 6 we mention possible leptogenesis realizations that go beyond the type-I seesaw. Finally, in Section 7 we draw the conclusions.

2. N_1 Leptogenesis in the Single Flavour Regime

The aim of this section is to give a pedagogical introduction to leptogenesis [45] and establish the notations. We will consider the classic example of leptogenesis from the lightest right-handed (RH) neutrino N_1 (the so-called N_1 leptogenesis) in the type-I seesaw model [40, 41, 43, 44] in the single flavour regime. First in Section 2.1 we introduce the type-I seesaw Lagrangian and the relevant parameters. In Section 2.2, we will review the CP violation in RH neutrino decays induced at 1-loop level. Then in Section 2.3, we will write down the classical Boltzmann equations taking into account only decays and inverse decays of N_1 and give a simple but rather accurate analytical estimate of the solution. In Section 2.4 we will relate the lepton asymmetry generated to the baryon asymmetry of the Universe. Finally in Section 2.5, we will discuss the lower bound on N_1 mass and the upper bound on light neutrino mass scale from successful leptogenesis.

2.1. Type-I Seesaw, Neutrino Masses, and Leptogenesis

With m ($m \geq 2$) (neutrino oscillation data and leptogenesis both require $m \geq 2$.) singlet RH neutrinos N_{R_i} ($i = 1, m$), we can add the following Standard Model (SM) gauge invariant terms to the SM Lagrangian:

$$\mathcal{L}_I = \mathcal{L}_{\rm SM} + i\overline{N_{R_i}}\partial\!\!\!/ N_{R_i} - \left(\frac{1}{2}M_i\overline{N_{R_i}^c}N_{R_i} + \epsilon_{ab}Y_{\alpha i}\overline{N_{R_i}}\ell_\alpha^a H^b + \text{h.c.}\right), \tag{2.1}$$

where M_i are the Majorana masses of the RH neutrinos, $\ell_\alpha = (\nu_{\alpha L}, \alpha_L^-)$ with $\alpha = e, \mu, \tau$ and $H = (H^+, H^0)$ are, respectively, the left-handed (LH) lepton and Higgs $SU(2)_L$ doublets and $\epsilon_{ab} = -\epsilon_{ba}$ with $\epsilon_{12} = 1$. Without loss of generality, we have chosen the basis where the Majorana mass term is diagonal. The physical mass eigenstates of the RH neutrinos are the Majorana neutrinos $N_i = N_{R_i} + N_{R_i}^c$. Since N_i are gauge singlets, the scale of M_i is naturally much larger than the electroweak (EW) scale $M_i \gg \langle\Phi\rangle \equiv v = 174\,\text{GeV}$. Hence after EW symmetry breaking, the light neutrino mass matrix is given by the famous seesaw relation [40, 41, 43, 44]:

$$m_\nu \simeq -v^2 Y \frac{1}{M} Y^T. \tag{2.2}$$

Assuming $Y \sim \mathcal{O}(1)$ and $m_\nu \simeq \sqrt{\Delta m_{\rm atm}^2} \simeq 0.05\,\text{eV}$, we have $M \sim 10^{15}\,\text{GeV}$ not far below the GUT scale.

Besides giving a natural explanation of the light neutrino masses, there is another bonus: the *three* Sakharov's conditions [13] for leptogenesis are implicit in (2.1) with the *lepton number violation* provided by M_i, the CP *violation* from the complexity of $Y_{i\alpha}$, and the *departure from thermal equilibrium condition* given by an additional requirement that N_i decay rate Γ_{N_i} is not very fast compared to the Hubble expansion rate of the Universe $H(T)$ at temperature $T = M_i$ with

$$\Gamma_{N_i} = \frac{(Y^\dagger Y)_{ii} M_i}{8\pi}, \qquad H(T) = \frac{2}{3}\sqrt{\frac{g_*\pi^3}{5}}\frac{T^2}{M_{\rm pl}}, \tag{2.3}$$

where $M_{\rm pl} = 1.22 \times 10^{19}\,\text{GeV}$ is the Planck mass, g_* (=106.75 for the SM excluding RH neutrinos) is the total number of relativistic degrees of freedom contributing to the energy density of the Universe.

To quantify the departure from thermal equilibrium, we define the *decay parameter* as follows:

$$K_i \equiv \frac{\Gamma_{N_i}}{H(M_i)} = \frac{\widetilde{m}_i}{m_*}, \tag{2.4}$$

where $\widetilde{m}_i$ is the *effective neutrino mass* defined as [47]

$$\widetilde{m}_i \equiv \frac{(Y^\dagger Y)_{ii} v^2}{M_i}, \tag{2.5}$$

with $m_* \equiv (16\pi^2 v^2/3M_{\rm pl})\sqrt{g_*\pi/5} \simeq 1 \times 10^{-3}$ eV. The regimes where $K_i \ll 1$, $K_i \approx 1$ and $K_i \gg 1$ are, respectively, known as weak, intermediate, and strong washout regimes.

2.2. CP Asymmetry

The CP asymmetry in the decays of RH neutrinos N_i can be defined as

$$\epsilon_{i\alpha} = \frac{\gamma(N_i \longrightarrow \ell_\alpha H) - \gamma\left(N_i \longrightarrow \overline{\ell_\alpha} H^*\right)}{\sum_\alpha \gamma(N_i \longrightarrow \ell_\alpha H) + \gamma\left(N_i \longrightarrow \overline{\ell_\alpha} H^*\right)} \equiv \frac{\Delta\gamma^\alpha_{N_i}}{\gamma_{N_i}}, \tag{2.6}$$

where $\gamma(i \rightarrow f)$ is the thermally averaged decay rate defined as (here the Pauli-blocking and Bose-enhancement statistical factors have been ignored and we also assume Maxwell-Boltzmann distribution for the particle i, that is, $f_i = e^{-E_i/T}$; see [48, 49] for detailed studies of their effects)

$$\gamma(i \rightarrow f) \equiv \int \frac{d^3 p_i}{(2\pi)^3 2E_i} \frac{d^3 p_f}{(2\pi)^3 2E_f} (2\pi)^4 \delta^{(4)}(p_i - p_f) |\mathcal{A}(i \longrightarrow f)|^2 e^{-E_i/T}, \tag{2.7}$$

where $\mathcal{A}(i \rightarrow f)$ is the decay amplitude. Ignoring all thermal effects [48, 49], (2.6) simplifies to

$$\epsilon_{i\alpha} = \frac{|\mathcal{A}_0(N_i \longrightarrow \ell_\alpha H)|^2 - \left|\mathcal{A}_0\left(N_i \longrightarrow \overline{\ell_\alpha} H^*\right)\right|^2}{\sum_\alpha |\mathcal{A}_0(N_i \longrightarrow \ell_\alpha H)|^2 + \left|\mathcal{A}_0\left(N_i \longrightarrow \overline{\ell_\alpha} H^*\right)\right|^2}, \tag{2.8}$$

where $\mathcal{A}_0(i \rightarrow f)$ denotes the decay amplitude at zero temperature. Equation (2.8) vanishes at tree level but is induced at 1-loop level through the interference between tree and 1-loop diagrams shown in Figure 1. There are two types of contributions from the 1-loop diagrams: the self-energy or wave diagram (middle) [50] and the vertex diagram (right) [45]. At leading order, we obtain the CP asymmetry [51]:

$$\begin{aligned} \epsilon_{i\alpha} &= \frac{1}{8\pi} \frac{1}{(Y^\dagger Y)_{ii}} \sum_{j \neq i} \mathrm{Im}\left[\left(Y^\dagger Y\right)_{ji} Y_{\alpha i} Y^*_{\alpha j}\right] g\left(\frac{M_j^2}{M_i^2}\right) \\ &+ \frac{1}{8\pi} \frac{1}{(Y^\dagger Y)_{ii}} \sum_{j \neq i} \mathrm{Im}\left[\left(Y^\dagger Y\right)_{ij} Y_{\alpha i} Y^*_{\alpha j}\right] \frac{M_i^2}{M_i^2 - M_j^2}, \end{aligned} \tag{2.9}$$

where the loop function is

$$g(x) = \sqrt{x}\left[\frac{1}{1-x} + 1 - (1+x)\ln\left(\frac{1+x}{x}\right)\right]. \tag{2.10}$$

The first term in (2.9) comes from L-violating wave and vertex diagrams, while the second term is from the L-conserving wave diagram. The terms of the form $(M_i^2 - M_j^2)^{-1}$ in (2.9)

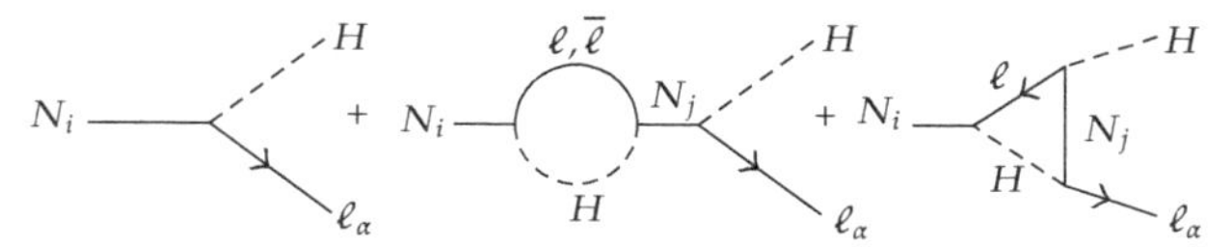

Figure 1: The CP asymmetry in type-I seesaw leptogenesis results from the interference between tree and 1-loop wave and vertex diagrams. For the 1-loop wave diagram, there is an additional contribution from L-conserving diagram to the CP asymmetry which vanishes when summing over lepton flavours.

are from the wave diagram contributions which can resonantly enhance the CP asymmetry if $M_i \approx M_j$ (resonant leptogenesis scenario, see Section 6.1). Notice that the resonant term becomes singular in the degenerate limit $M_i = M_j$. This singularity can be regulated by using, e.g., an effective field-theoretical approach based on resummation [52]. Let us also note that at least two RH neutrinos are needed otherwise the CP asymmetry vanishes because the Yukawa couplings combination becomes real.

In the one flavour regime, we sum over the flavour index α in (2.9) and obtain

$$\epsilon_i \equiv \sum_\alpha \epsilon_{i\alpha} = \frac{1}{8\pi} \frac{1}{(Y^\dagger Y)_{ii}} \sum_{j \neq i} \text{Im}\left[\left(Y^\dagger Y\right)_{ji}^2\right] g\left(\frac{M_j^2}{M_i^2}\right), \tag{2.11}$$

where the second term in (2.9) vanishes because the combination of the Yukawa couplings is real.

2.3. *Classical Boltzmann Equations*

We work in the one-flavour regime and consider only the decays and inverse decays of N_1. If leptogenesis occurs at $T \gtrsim 10^{12}$ GeV, then the charged lepton Yukawa interactions are out of equilibrium, and this defines the one-flavour regime. The assumption that only the dynamics of N_1 is relevant can be realized if, for example, the reheating temperature after inflation is $T_{\text{RH}} \ll M_{2,3}$ such that $N_{2,3}$ are not produced. In order to scale out the effect of the expansion of the Universe, we will introduce the *abundances*, that is, the ratios of the particle densities $n_i = \int d^3p f_i$ to the entropy density $s = (2\pi^2/45) g_* T^3$:

$$Y_i \equiv \frac{n_i}{s}. \tag{2.12}$$

The evolution of the N_1 density and the lepton asymmetry $Y_{\Delta L} = 2Y_{\Delta \ell} \equiv 2(Y_\ell - Y_{\bar{\ell}})$ (the factor of 2 comes from the $SU(2)_L$ degrees of freedoms) can be described by the following classical Boltzmann equations (BE) [53]:

$$\frac{dY_{N_1}}{dz} = -D_1\left(Y_{N_1} - Y_{N_1}^{\text{eq}}\right), \tag{2.13}$$

$$\frac{dY_{\Delta L}}{dz} = \epsilon_1 D_1\left(Y_{N_1} - Y_{N_1}^{\text{eq}}\right) - W_1 Y_{\Delta L}, \tag{2.14}$$

where $z \equiv M_1/T$ and the decay and washout terms are, respectively, given by

$$D_1(z) = \frac{\gamma_{N_1} z}{sH(M_1)} = K_1 z \frac{\mathcal{K}_1(z)}{\mathcal{K}_2(z)}, \qquad W_1(z) = \frac{1}{2} D_1(z) \frac{Y_{N_1}^{\text{eq}}(z)}{Y_\ell^{\text{eq}}}, \tag{2.15}$$

with $\mathcal{K}_n$ the nth order modified Bessel function of second kind. $Y_{N_1}^{\text{eq}}$ and Y_ℓ^{eq} read (to write down a simple analytic expression for the equilibrium density of N_1, we assume Maxwell-Boltzmann distribution. However, following [54], the normalization factor Y_ℓ^{eq} is obtained from a Fermi-Dirac distribution)

$$Y_N^{\text{eq}}(z) = \frac{45}{2\pi^4 g_*} z^2 \mathcal{K}_2(z), \qquad Y_\ell^{\text{eq}} = \frac{15}{4\pi^2 g_*}. \tag{2.16}$$

From (2.13) and (2.14), the solution for $Y_{\Delta L}$ can be written down as follows:

$$Y_{\Delta L}(z) = Y_{\Delta L}(z_i) e^{-\int_{z_i}^{z} dz' W_1(z')} - \int_{z_i}^{z} dz'^{\epsilon_1}(z') \frac{dY_{N_1}}{dz'} e^{-\int_{z'}^{z} dz'' W_1(z'')}, \tag{2.17}$$

where z_i is some initial temperature when N_1 leptogenesis begins, and we have assumed that any preexisting lepton asymmetry vanishes $Y_{\Delta L}^0(z_i) = 0$. Notice that ignoring thermal effects, the CP asymmetry is independent of the temperature $\epsilon_1(z) = \epsilon_1$ (c.f. (2.11)).

2.3.1. *Weak Washout Regime*

In the weak washout regime ($K_1 \ll 1$), the initial condition on the N_1 density $Y_{N_1}(z_i)$ is important. If we assume thermal initial abundance of N_1, that is, $Y_{N_1}(z_i) = Y_{N_1}^{\text{eq}}(0)$, we can ignore the washout when N_1 starts decaying at $z \gg 1$ and we have

$$Y_{\Delta L}^t(\infty) \simeq -\epsilon_1 \int_0^\infty dz' \frac{dY_{N_1}^{\text{eq}}}{dz'} = \epsilon_1 Y_{N_1}^{\text{eq}}(0). \tag{2.18}$$

On the other hand, if we have zero initial N_1 abundance, that is, $Y_{N_1}(z_i) = 0$, we have to consider the opposite sign contributions to lepton asymmetry from the inverse decays when N_1 is being populated ($Y_{N_1} < Y_{N_1}^{\text{eq}}$) and from the period when N_1 starts decaying ($Y_{N_1} > Y_{N_1}^{\text{eq}}$). Taking this into account the term which survives the partial cancellations is given by [55] (this differs from the efficiency in [55] by the factor $12/\pi^2$, which is due to the different normalization Y_ℓ^{eq} (2.16))

$$Y_{\Delta L}^0(\infty) \simeq \frac{27}{16} \epsilon_1 K_1^2 \, Y_{N_1}^{\text{eq}}(0). \tag{2.19}$$

2.3.2. *Strong Washout Regime*

In the strong washout regime ($K_1 \gg 1$) any lepton asymmetry generated during the N_1 creation phase is efficiently washed out. Here we adopt the *strong washout balance approximation* [56] which states that in the strong washout regime, the lepton asymmetry at each instant takes the value that enforces a balance between the production and the destruction rates of the asymmetry. Equating the decay and washout terms in (2.14), we have

$$Y_{\Delta L}(z) \approx -\frac{1}{W(z)}\epsilon_1 \frac{dY_{N_1}}{dz} \simeq -\frac{1}{W(z)}\epsilon_1 \frac{dY_{N_1}^{\rm eq}}{dz} = \frac{2}{zK_1}\epsilon_1 Y_\ell^{\rm eq}, \tag{2.20}$$

where in the second approximation, we assume $Y_{N_1} \simeq Y_{N_1}^{\rm eq}$. The approximation no longer holds when $Y_{\Delta L}$ freezes, and this happens when the washout decouples at z_f, that is, $W(z_f) < 1$. Hence, the final lepton asymmetry is given by (compare this to a more precise analytical approximation in [55])

$$Y_{\Delta L}(\infty) = \frac{2}{z_f K_1}\epsilon_1 Y_\ell^{\rm eq} = \frac{\pi^2}{6 z_f K_1}\epsilon_1 Y_{N_1}^{\rm eq}(0). \tag{2.21}$$

The freeze out temperature z_f depends mildly on K_1. For K_1 = 10–100 we have, for example, z_f = 7– 10. We also see that independently of initial conditions, in the strong regime $Y_{\Delta L}(\infty)$ goes as K_1^{-1}.

2.4. Baryon Asymmetry from EW Sphaleron

The final lepton asymmetry $Y_{\Delta L}(\infty)$ can be conveniently parametrized as follows:

$$Y_{\Delta L}(\infty) = \epsilon_1 \eta_1 Y_{N_1}^{\rm eq}(0), \tag{2.22}$$

where η_1 is known as the *efficiency factor*. In the weak washout regime ($K_1 \ll 1$) from (2.18) we have $\eta_1 = 1$ ($= (27/16)K_1^2 < 1$) for thermal (zero) initial N_1 abundance. In the strong washout regime ($K_1 \gg 1$), from (2.21), we have $\eta_1 = (\pi^2/6 z_f K_1) < 1$.

If leptogenesis ends before EW sphaleron processes become active ($T \gtrsim 10^{12}$ GeV), the $B - L$ asymmetry $Y_{\Delta_{B-L}}$ is simply given by

$$Y_{\Delta_{B-L}} = -Y_{\Delta L}. \tag{2.23}$$

At the later stage, the $B - L$ asymmetry is partially transferred to a B asymmetry by the EW sphaleron processes through the relation [57]

$$Y_{\Delta B}(\infty) = \frac{28}{79} Y_{\Delta_{B-L}}(\infty), \tag{2.24}$$

that holds if sphalerons decouple before EWPT. This relation will change if the EW sphaleron processes decouple after the EWPT [57, 58] or if threshold effects for heavy particles like the top quark and Higgs are taken into account [58, 59].

2.5. Davidson-Ibarra Bound

Assuming a hierarchical spectrum of the RH neutrinos ($M_1 \ll M_2, M_3$), and that the dominant lepton asymmetry is from the N_1 decays, from (2.11) the CP asymmetry from N_1 decays can be written as

$$\epsilon_1 = -\frac{3}{16\pi}\frac{1}{(Y^\dagger Y)_{11}}\sum_{j\neq 1}\text{Im}\left[\left(Y^\dagger Y\right)_{j1}^2\right]\frac{M_1}{M_j}. \tag{2.25}$$

Assuming three generations of RH neutrinos ($n = 3$) and using the Casas-Ibarra parametrization [60] for the Yukawa couplings

$$Y_{\alpha i} = \frac{1}{v}\left(\sqrt{D_{m_N}}R\sqrt{D_{m_\nu}}U_\nu^\dagger\right)_{\alpha i}, \tag{2.26}$$

where $D_{m_N} = \text{diag}(M_1, M_2, M_3)$, $D_{m_\nu} = \text{diag}(m_{\nu_1}, m_{\nu_2}, m_{\nu_3})$ and R any complex orthogonal matrix satisfying $R^T R = RR^T = 1$, (2.25) becomes

$$\epsilon_1 = -\frac{3}{16\pi}\frac{M_1}{v^2}\frac{\sum_i m_{\nu_i}\text{Im}(R_{1i}^2)}{\sum_i m_{\nu_i}|R_{1i}|^2}. \tag{2.27}$$

Using the orthogonality condition $\sum_i R_{1i}^2 = 1$, we then obtain the Davidson-Ibarra (DI) bound [61]

$$|\epsilon_1| \le \epsilon^{\text{DI}} = \frac{3}{16\pi}\frac{M_1}{v^2}(m_{\nu_3} - m_{\nu_1}) = \frac{3}{16\pi}\frac{M_1}{v^2}\frac{\Delta m_{\text{atm}}^2}{m_{\nu_1} + m_{\nu_3}}, \tag{2.28}$$

where m_{ν_3} (m_{ν_1}) is the heaviest (lightest) light neutrino mass. Applying the DI bound on (2.22)–(2.24), and requiring that $Y_{\Delta B}(\infty) \ge Y_B^{\text{CMB}} \simeq 10^{-10}$, we obtain

$$M_1\left(\frac{0.1\,\text{eV}}{m_{\nu_1} + m_{\nu_3}}\right)\eta_1^{\max}(M_1) \gtrsim 10^9\,\text{GeV}, \tag{2.29}$$

where the $\eta_1^{\max}(M_1)$ is the efficiency factor maximized with respect to K_1 (2.4) for a particular value of M_1. This allows us to make a plot of region which satisfies (2.29) on the (M_1, m_{ν_1}) plane and hence obtain bounds on M_1 and m_{ν_1}. Many careful numerical studies have been carried out, and it was found that successful leptogenesis with a hierarchical spectrum of the RH neutrinos requires $M_1 \gtrsim 10^9$ GeV [61–63] and $m_{\nu_1} \lesssim 0.1$ eV [55, 64–66]. This bound implies that the RH neutrinos must be produced at temperatures $T \gtrsim 10^9$ GeV which in turn implies the reheating temperature after inflation has to be $T_{\text{RH}} \gtrsim 10^9$ GeV in order to have sufficient RH neutrinos in the thermal bath. To conclude this section, let us note that the DI bound (2.28) holds if and only if all the following conditions apply.

(1) N_1 dominates the contribution to leptogenesis.

(2) The mass spectrum of RH neutrinos is hierarchical $M_1 \ll M_2, M_3$.

(3) Leptogenesis occurs in the unflavoured regime $T \gtrsim 10^{12}$ GeV.

As we will see in the following sections, violation of one or more of the previous conditions allows us to lower somewhat the scale of leptogenesis.

3. Lepton Flavour Effects

3.1. When Are Lepton Flavour Effects Relevant?

The first leptogenesis calculations were performed in the single lepton flavour regime. In short, this amounts to assuming that the leptons and antileptons which couple to the lightest RH neutrino N_1 maintain their coherence as flavour superpositions throughout the leptogenesis era, that is $\ell_1 = \sum_\alpha c_{\alpha 1}\ell_\alpha$ and $\overline{\ell_1'} = \sum_\alpha c_{\alpha 1}^{'*}\overline{\ell}_\alpha$. Note that at the tree level the coefficients c and c'^* are simply the Yukawa couplings: $c_{\alpha 1} = Y_{\alpha 1}$ and $c_{\alpha 1}^{'*} = Y_{\alpha 1}^*$. However it should be kept in mind that since CP is violated by loops, beyond the tree level approximation the antilepton state $\overline{\ell_1'}$ is not the CP conjugate of the ℓ_1, that is, $c'_{\alpha 1} \neq c_{\alpha 1}$.

The single flavour regime is realized only at very high temperatures ($T \gtrsim 10^{12}$ GeV) when both ℓ_1 and ℓ_1' remain coherent flavour superpositions and thus are the correct states to describe the dynamics of leptogenesis. However, at lower temperatures scatterings induced by the charged lepton Yukawa couplings occur at a sufficiently fast pace to distinguish the different lepton flavours, ℓ_1 and ℓ_1' decohere in their flavour components, and the dynamics of leptogenesis must then be described in terms of the flavour eigenstates ℓ_α. Of course, there is great interest to extend the validity of quantitative leptogenesis studies also at lower scale $T \lesssim 10^{12}$ GeV, and this requires accounting for flavour effects. The role of lepton flavour in leptogenesis was first discussed in [67]; however the authors did not highlight in what the results were significantly different from the single flavour approximation. Therefore, until the importance of flavour effects was fully clarified in [68–70], they had been included in leptogenesis studies only in a few cases [71–75]. Nowadays lepton flavour effects have been investigated in full detail [76–89] and are a mandatory ingredient of any reliable analysis of leptogenesis.

The specific temperature when leptogenesis becomes sensitive to lepton flavour dynamics can be estimated by requiring that the rates of processes Γ_α ($\alpha = e, \mu, \tau$) that are induced by the charged lepton Yukawa couplings h_α become faster than the Universe expansion rate $H(T)$. An approximate relation gives [90, 91]

$$\Gamma_\alpha(T) \simeq 10^{-2}\, h_\alpha^2 T, \tag{3.1}$$

which implies that (in supersymmetric case, since $h_\alpha = m_\alpha/(v_u \cos\beta)$, we have $T \lesssim T_\alpha\ (1 + \tan^2\beta)$)

$$\Gamma_\alpha(T) > H(T) \quad \text{when } T \lesssim T_\alpha, \tag{3.2}$$

where $T_e \simeq 4\times 10^4$ GeV, $T_\mu \simeq 2\times 10^9$ GeV, and $T_\tau \simeq 5\times 10^{11}$ GeV. Notice that to fully distinguish the three flavours it is sufficient that the τ and μ Yukawa reactions attain thermal equilibrium. It has been pointed out that besides being faster than the expansion of the Universe, the charged lepton Yukawa interactions should also be faster than the N_1 interactions [69, 83, 84]. In general whenever $\Gamma_\tau(M_1) > H(M_1)$ we also have $\Gamma_\tau(M_1) > \Gamma_{N_1}(M_1)$. However, there exists parameter space where $\Gamma_\tau(M_1) > H(M_1)$ but $\Gamma_\tau(M_1) < \Gamma_{N_1}(M_1)$. This scenario was studied in [83].

3.2. The Effects on CP Asymmetry and Washout

The CP violation in N_i decays can manifest itself in two ways [69]

(i) The leptons and antileptons are produced at different rates:

$$\gamma_i \neq \overline{\gamma}_i, \tag{3.3}$$

where $\gamma_i \equiv \gamma(N_i \to \ell_i H)$ and $\overline{\gamma}_i \equiv \gamma(N_i \to \overline{\ell}'_i H^*)$.

(ii) The leptons and antileptons produced are not CP conjugate states:

$$\mathrm{CP}\left(\overline{\ell}'_i\right) = \ell'_i \neq \ell_i, \tag{3.4}$$

that is, due to loops effects they are slightly misaligned in flavour space.

We can rewrite the CP asymmetry for N_i decays from (2.6) as follows:

$$\epsilon_{i\alpha} = \frac{P_{i\alpha}\gamma_i - \overline{P}_{i\alpha}\overline{\gamma}_i}{\gamma_i + \overline{\gamma}_i} = \frac{P_{i\alpha} + \overline{P}_{i\alpha}}{2}\epsilon_i + \frac{P_{i\alpha} - \overline{P}_{i\alpha}}{2} \simeq P^0_{i\alpha}\epsilon_i + \frac{\Delta P_{i\alpha}}{2}, \tag{3.5}$$

where terms of order $\mathcal{O}(\epsilon_i \, \Delta P_{i\alpha})$ and higher have been neglected. $P_{i\alpha}$ is the projector from state ℓ_i into flavour state ℓ_α and $\Delta P_{i\alpha} = P_{i\alpha} - \overline{P}_{i\alpha}$. At tree level, clearly, $P_{i\alpha} = \overline{P}_{i\alpha} \equiv P^0_{i\alpha}$ where the tree level flavour projector is given by

$$P^0_{i\alpha} = \frac{Y_{\alpha i} Y^*_{\alpha i}}{(Y^\dagger Y)_{ii}}. \tag{3.6}$$

From (3.5), we can identify the two types of CP violation, the first term being of type (i) equation (3.3) while the second being of type (ii) equation (3.4). Since $\sum_\alpha P_{i\alpha} = \sum_\alpha \overline{P}_{i\alpha} = 1$, when summing over flavour indices α, the second term vanishes $\sum_\alpha \Delta P_{i\alpha} = 0$. Note that the lepton-flavour-violating but L-conserving terms in the second line of (2.9) are part of type (ii). In fact, they come from $d = 6$ L-conserving operators which have nothing to do with the unique $d = 5$ L-violating operator (the Weinberg operator [92]) responsible for neutrino masses. However, in some cases they can still dominate the CP asymmetries but, as we will see in Section 3.4, lepton flavour equilibration effects [93] then impose important constraints on their overall effects. Note also that due to flavour misalignment, the CP asymmetry in a particular flavour direction $\epsilon_{i\alpha}$ can be much larger and even of opposite sign from the total CP asymmetry ϵ_i. In fact the relevance of CP violation of type (ii) in the flavour regimes is what allows to evade the DI bound (2.28). As regards the washout of the lepton asymmetry of flavour α, it is proportional to

$$W_{i\alpha} \propto P_{i\alpha}\gamma_i + \overline{P}_{i\alpha}\overline{\gamma}_i \simeq P^0_{i\alpha} W_i, \tag{3.7}$$

which results in a reduction of washout by a factor of $P^0_{i\alpha} \leq 1$ compared to unflavoured case. As we will see next, the new CP-violating sources from flavour effects and the reduction in the washout could result in great enhancement of the final lepton asymmetry, and, as was first pointed out in [69], leptogenesis with a vanishing total CP asymmetry $\epsilon_i = 0$ also becomes possible.

3.3. Classical Flavoured Boltzmann Equations

Here again we only consider leptogenesis from the decays and inverse decays of N_1. In this approximation, the BE for Y_{N_1} is still given by (2.13) while the BE for $Y_{\Delta L_\alpha}$ the lepton asymmetry in the flavour α is given by (to study the transition between different flavour regimes (from one to two or from two to three flavours), a density matrix formalism has to be used [68, 84, 94]).

$$\frac{dY_{\Delta L_\alpha}}{dz} = \epsilon_{1\alpha} D_1 \left(Y_{N_1} - Y^{\rm eq}_{N_1} \right) - P^0_{1\alpha} W_1 Y_{\Delta L_\alpha}. \tag{3.8}$$

Notice that as long as L violation from sphalerons is neglected (see Section 4) the BEs for $Y_{\Delta L_\alpha}$ are independent of each other, and hence the solutions for the weak and strong washout regimes are given, respectively, by (2.19) and (2.21), after replacing $\epsilon_1 \rightarrow \epsilon_{1\alpha}$ and $K_1 \rightarrow K_{1\alpha} \equiv P^0_{1\alpha} K_1$.

As an example let us assume that leptogenesis occurs around $T \sim 10^{10}$ GeV, that is, in the two-flavour regime. Due to the fast τ Yukawa interactions $\ell_1(\ell'_1)$ gets projected onto $\ell_\tau(\ell'_\tau)$ and a coherent mixture of $e + \mu$ eigenstate $\ell_{e+\mu}(\ell'_{e+\mu})$. For illustrative purpose, here we consider a scenario in which lepton flavour effects are most prominent. We take both $K_{1\tau}, K_{1e+\mu} \gg 1$, so that both $Y_{\Delta L_\tau}$ and $Y_{\Delta L_{e+\mu}}$ are in the strong regime. From (2.21) we can write down the solution:

$$\begin{aligned} Y_{\Delta L}(\infty) &= Y_{\Delta L_\tau}(\infty) + Y_{\Delta L_{e+\mu}}(\infty) \\ &= \frac{\pi^2}{6 z_f K_1} Y^{\rm eq}_{N_1}(0) \left(\frac{\epsilon_{1\tau}}{P^0_{1\tau}} + \frac{\epsilon_{1e+\mu}}{P^0_{1e+\mu}} \right) \\ &\simeq \frac{\pi^2}{3 z_f K_1} \epsilon_1 Y^{\rm eq}_{N_1}(0) + \frac{\pi^2}{12 z_f K_1} Y^{\rm eq}_{N_1}(0) \left(\frac{\Delta P_{1\tau}}{P^0_{1\tau}} + \frac{\Delta P_{1e+\mu}}{P^0_{1e+\mu}} \right), \end{aligned} \tag{3.9}$$

where in the last line we have used (3.5). If $P^0_{1\tau} \simeq P^0_{1e+\mu}$, then since $\Delta P_{1\tau} + \Delta P_{1e+\mu} = 0$, the second term approximately cancels, and (3.9) reduces to

$$Y_{\Delta L}(\infty) \simeq \frac{\pi^2}{3 z_f K_1} \epsilon_1 Y^{\rm eq}_{N_1}(0). \tag{3.10}$$

We see that the final asymmetry is enhanced by a factor of 2 compared to the unflavoured case. If there exists some hierarchy between the flavour projectors, then the second term in (3.9) plays an important role and can further enhance the asymmetry. For example, we can have $P^0_{1\tau} > P^0_{1e+\mu}$ while $\Delta P_{1\tau} \ll \Delta P_{1e+\mu}$. In this case, the second term can dominate over the first term. Finally from (3.9) we also notice that leptogenesis with $\epsilon_1 = 0$, the so-called *purely flavoured leptogenesis* (PFL) (this can also refer to the case where the total CP asymmetry is negligible $\epsilon_1 \approx 0$), can indeed proceed [69, 95–98]. In this scenario some symmetry has to be imposed to realize the condition $\epsilon_1 = 0$, as, for example, an approximate global lepton number $\mathrm{U}(1)_L$. In the limit of exact $\mathrm{U}(1)_L$ the active neutrinos will be exactly massless. Instead of the seesaw mechanism, the small neutrino masses are explained by $\mathrm{U}(1)_L$ which is slightly broken by a small parameter μ (the "inverse seesaw") [99] which is technically natural since the Lagrangian exhibits an enhanced symmetry when $\mu \to 0$ [100]. In the next section, we will discuss another aspect of flavour effects which are in particular crucial for PFL.

3.4. Lepton Flavour Equilibration

Another important effect is lepton flavour equilibration (LFE) [93]. LFE processes violate lepton flavour but conserve total lepton number, for example, $\ell_\alpha H \to \ell_\beta H$, and can proceed, for example, via off-shell exchange of $N_{2,3}$. In thermal equilibrium, LFE processes can quickly equilibrate the asymmetries generated in different flavours. In practice this would be equivalent to a situation where all the flavour projector equations (3.6) are equal, in which case the flavoured BE equation (3.8) can be summed up into a single BE:

$$\frac{dY_{\Delta L}}{dz} = \epsilon_1 D_1 \left(Y_{N_1} - Y^{\mathrm{eq}}_{N_1} \right) - P^0_{1\alpha} W_1 Y_{\Delta L}, \tag{3.11}$$

where $P^0_{1\alpha} = 1/2\ (1/3)$ in the two- (three-) flavour regime. In this case the BE is just like the unflavoured case but with a reduced washout which, in the strong washout regime, would result in enhancement of a factor of 2 (3) in the two- (three-) flavour regime (c.f. (3.10)). Clearly, LFE can make PFL with $\epsilon_1 = 0$ impotent [56, 93]. Since LFE $N_{2,3}$ processes scale as T^3 while the Universe expansion scales as T^2, in spite of the fact that PFL evades the DI bound, they eventually prevent the possibility of lowering too much the leptogenesis scale. A generic study in PFL scenario taking into account LFE effects concluded that successful leptogenesis still requires $M_1 \gtrsim 10^8$ GeV [97]. A more accurate study in the same direction recently carried out in [98] showed that in fact the leptogenesis scale can be lowered down to $M_1 \sim 10^6$ GeV.

4. Beyond the Basic Boltzmann Equations

Within factors of a few, the amount of baryon asymmetry that is generated via leptogenesis in N_1 decays is determined essentially by the size of the (flavoured) CP asymmetries and by the rates of the (flavoured) washout reactions. However, to obtain more precise results (say, within an $\mathcal{O}(1)$ uncertainty) several additional effects must be taken into account, and the formalism must be extended well beyond the basic BE discussed in the previous sections. In the following we review some of the most important sources of corrections, namely, spectator processes (Section 4.1), scatterings with top quarks and gauge bosons (Section 4.2), thermal effects (Section 4.3), contributions from heavier RH neutrinos (Section 4.4), and we also discuss the role of quantum corrections evaluated in the quantum BE approach (Section 4.5). Throughout this paper we use integrated BE; that is, we assume kinetic equilibrium for all particle species, and thus we use particles densities instead than particles distribution functions. Corrections arising from using nonintegrated BE have been studied for example in [101–104], and are generally subleading.

4.1. Spectator Processes

Reactions that without involving violation of $B - L$ can still affect the final amount of baryon asymmetry are classified as *"spectator processes"* [105, 106]. The basic way through which they act is that of redistributing the asymmetry generated in the lepton doublets among the other particle species. Since the density asymmetries of the lepton doublets are what weights the rates of the washout processes, it can be expected that spectator processes would render the washouts less effective and increase the efficiency of leptogenesis. However, in most cases this is not true: proper inclusion of spectator processes implies accounting for all the particle asymmetries and in particular also for the density asymmetry of the Higgs $Y_{\Delta H}$ [106]. This was omitted in Section 2 but in fact has to be added to the density asymmetry of the leptons $Y_{\Delta \ell}$ in weighting, for example, washouts from inverse decays. Equation (2.14) would then become

$$\frac{dY_{\Delta L}}{dz} = \epsilon_1 D_1 \left(Y_{N_1} - Y_{N_1}^{\text{eq}}\right) - 2\,(Y_{\Delta \ell} + Y_{\Delta H}) W_1, \tag{4.1}$$

where the factor of two in front of the washout term counts the leptons and Higgs gauge multiplicity. Clearly, in some regimes in which $Y_{\Delta \ell}$ and $Y_{\Delta H}$ are not sufficiently diluted by interacting with other particles, this can have the effect of enhancing the washout rates and suppressing the efficiency.

In the study of spectator processes it is fundamental to specify the range of temperature in which leptogenesis occurs. This is because at each specific temperature T, particle reactions must be treated in a different way depending on if their characteristic time scale τ (given by inverse of their thermally averaged rates) is [89, 107]

(1) much shorter than the age of the Universe: $\tau \ll t_U(T)$;

(2) much larger than the age of the Universe: $\tau \gg t_U(T)$;

(3) comparable with the Universe age: $\tau \sim t_U(T)$.

Spectator processes belong to the first type of reactions which occur very frequently during one expansion time. Their effects can be accounted for by imposing on the thermodynamic system the chemical equilibrium condition appropriate for each specific reaction, that is, $\sum_I \mu_I = \sum_F \mu_F$, where μ_I denotes the chemical potential of an initial state particle and μ_F that of a final state particle (the relation between chemical potentials and particle density asymmetries is given in (5.3)). The numerical values of the parameters that are responsible for these reactions only determine the precise temperature T when chemical equilibrium is attained but, apart from this, have no other relevance and do not appear explicitly in the formulation of the problem. Reactions of type (2) cannot have any effect on the system, since they basically do not occur. All physical processes are blind to the corresponding parameters, that can be set to zero in the effective Lagrangian. In most cases this results in exact global symmetries corresponding to conserved charges, and these conservation laws impose constraints on the particle chemical potentials. Reactions of type (3) in general violate some symmetry and thus spoil the corresponding conservation conditions, but are not fast enough to enforce chemical equilibrium. These are the only reactions that need to be studied by means of BE, and for which the precise value of the parameters that control their rates is of utmost importance.

A simple case to illustrate how to include spectator processes is the one-flavour regime at particularly high temperatures (say, $T \gtrsim 10^{13}$ GeV). The Universe expansion is fast implying that except for processes induced by the large Yukawa coupling of the top and for gauge interactions, all other $B - L$-conserving reactions fall in class (ii). Then there are several conserved quantities as, for example, the total number density asymmetries of the RH leptons as well as those of all the quarks except the top. Since electroweak sphalerons are also out of equilibrium, B is conserved too (and vanishing, if we assume that there is no preexisting asymmetry). $B = 0$ then translates in the condition:

$$2Y_{\Delta Q_3} + Y_{\Delta t} = 0, \tag{4.2}$$

where $Y_{\Delta Q_3}$ is the density asymmetries of one degree of freedom of the top $SU(2)_L$ doublet and color triplet which, being gauge interactions in equilibrium, is the same for all the six gauge components and $Y_{\Delta t}$ is the density asymmetry of the $SU(2)_L$ singlet top. Hypercharge is always conserved, yielding

$$Y_{\Delta Q_3} + 2Y_{\Delta t} - Y_{\Delta \ell} + Y_{\Delta H} = 0. \tag{4.3}$$

Finally, in terms of density asymmetries chemical equilibrium for the top-Yukawa-related reactions $\mu_{Q_3} + \mu_H = \mu_t$ translates into

$$Y_{\Delta Q_3} + \frac{1}{2} Y_{\Delta H} = Y_{\Delta t}. \tag{4.4}$$

We have three conditions for four density asymmetries, which allows to express the Higgs density asymmetry in terms of the density asymmetry of the leptons as $Y_{\Delta H} = (2/3) Y_{\Delta \ell}$. Moreover, given that only the LH lepton degrees of freedom are populated, we have $Y_{\Delta L} = 2Y_{\Delta \ell}$ so that the coefficient weighting W_1 in (4.1) becomes 2 $(Y_{\Delta \ell} + Y_{\Delta H}) = (5/3) Y_{\Delta L}$ and the washout is accordingly stronger.

With decreasing temperatures, more reactions attain chemical equilibrium, and accounting for spectator processes becomes accordingly more complicated. When the temperature drops below $T \sim 10^{12}$ GeV, EW sphalerons are in equilibrium, and baryon number is no more conserved. Then the condition (4.2) is no more satisfied, and, more importantly, the BE equation (4.1) is no more valid since sphalerons violate also lepton number within equilibrium rates. However, sphalerons conserve $B - L$, which is then violated only by slow reactions of type (3), and we should then write down a BE for this quantity. Better said, since at $T \lesssim 10^{12}$ GeV all the third generation Yukawa reactions, including the ones of the τ-lepton, are in equilibrium, the dynamical regime is that of two flavours in which the relevant quasiconserved charges are $\Delta_\tau = B/3 - L_\tau$ and $\Delta_{e\mu} = B/3 - L_{e\mu}$. The fact that only two charges are relevant is because there is always a direction in e-μ space which remains decoupled from N_1. The corresponding third charge $\Delta'_{e\mu}$ is then exactly conserved, its value can be set to zero, and the corresponding BE dropped. In this regime, the BE corresponding to (4.1) becomes

$$-\frac{dY_{\Delta_\alpha}}{dz} = \epsilon_\alpha D_1 \left(Y_{N_1} - Y_{N_1}^{\rm eq} \right) - 2 \left(Y_{\Delta \ell_\alpha} + Y_{\Delta H} \right) W_1 \quad (\alpha = \tau, e\mu). \tag{4.5}$$

To rewrite these equations in a solvable closed form, $Y_{\Delta\ell_\tau}$, $Y_{\Delta\ell_{e\mu}}$, and $Y_{\Delta H}$ must be expressed in terms of the two charge densities Y_{Δ_τ} and $Y_{\Delta_{e\mu}}$. This can be done by imposing the hypercharge conservation condition (4.3) and the chemical equilibrium conditions that, in addition to (4.4), are appropriate for the temperature regime we are considering. They are [106] (1) QCD sphaleron equilibrium; (2) EW sphaleron equilibrium; (3) b-quark and τ-lepton Yukawa equilibrium. The "rotation" from the particle density asymmetries $Y_{\Delta\ell_\alpha}$ and $Y_{\Delta H}$ to the charge densities Y_{Δ_α} can be expressed in terms of the A matrix introduced in [67] $Y_{\Delta\ell_\alpha} = A^{\ell}_{\alpha\beta} Y_{\Delta_\beta} (\alpha, \beta = \tau, e\mu)$ and C-vector $Y_{\Delta H} = C^H_\alpha Y_{\Delta_\alpha}$ introduced in [69]. For the present case, with the ordering $(e\mu, \tau)$ they are [69]

$$A^{\ell} = \frac{1}{460} \begin{pmatrix} 196 & -24 \\ -9 & 156 \end{pmatrix}, \qquad C^H = \frac{1}{230}(41, 56). \tag{4.6}$$

It is important to stress that in each temperature regime there are always enough constraints (conservation laws and chemical equilibrium conditions) to allow to express all the relevant particle density asymmetries in terms of the quasiconserved charges Y_{Δ_α}. This is because each time a conservation law has to be dropped (like B conservation above), it gets replaced by a chemical equilibrium condition (like EW sphalerons equilibrium), and each time the chemical potential of a new particle species becomes relevant, it is precisely because a new reaction involving that particle attains chemical equilibrium, enforcing the corresponding condition. As regards the quantitative corrections ascribable to spectator processes, several numerical studies have confirmed that they generally remain below order one. Thus, differently from flavour effects, for order of magnitude estimates they can be neglected.

4.2. Scatterings and CP Violation in Scatterings

Scattering processes are relevant for the production of the N_1 population, because decay and inverse decay rates are suppressed by a time dilation factor $\propto M_1/T$. The $N_1 = \overline{N}_1$ particles can be produced by scatterings with the top quark in s-channel H-exchange $qt^c \to N\ell_\alpha$ and $\overline{q}\overline{t}^c \to N\overline{\ell}_\alpha$, by t-channel H-exchange in $q\overline{\ell}_\alpha \to N\overline{t}^c$, $\overline{q}\ell_\alpha \to Nt^c$ and by u-channel H-exchange in $\ell_\alpha \overline{t}^c \to Nq$, $t^c\overline{\ell}_\alpha \to N\overline{q}$; see the diagrams (a) in Figure 2. Several scattering channels with gauge bosons also contribute to the production of N_1; the corresponding diagrams are (b) and (c) in the same figure.

When the effects of scatterings in populating the N_1 degree of freedom are included, for consistency CP violation in scatterings must also be included. In doing so some care has to be put in treating properly also all the processes of higher order in the couplings ($h_t^2\lambda^4$, $g^2\lambda^4$, where g is a gauge coupling) with an on-shell intermediate state N_1 subtracted out. This can be done by following the procedure adopted in [108], and we refer to that paper for details.

In the first approximation, the CP asymmetry in scattering processes is the same as in decays and inverse decays [70, 109]. This result was first found in [75, 110, 111] for

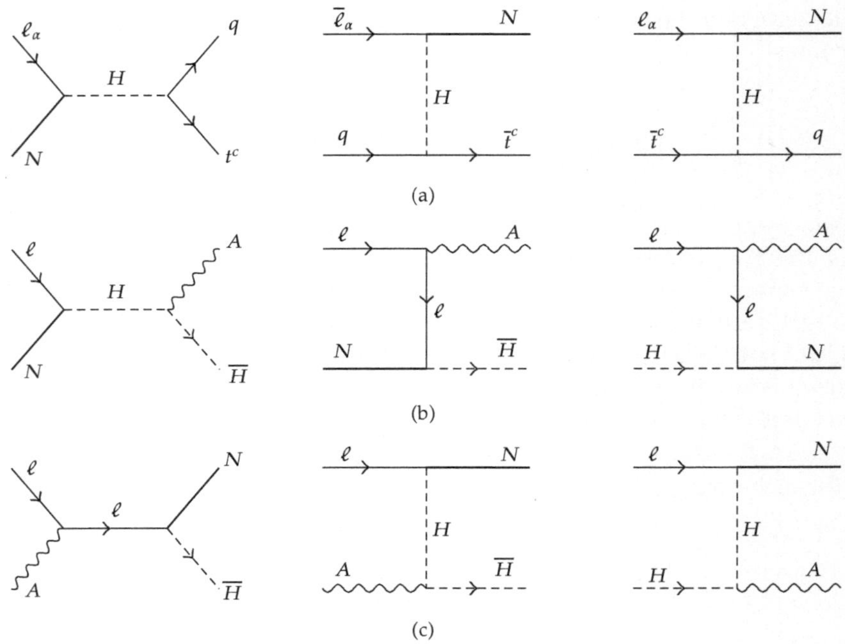

Figure 2: Diagrams for various $2 \leftrightarrow 2$ scattering processes: (a) scatterings with the top quarks; (b) and (c) scatterings with the gauge bosons ($A = B, W_i$ with $i = 1, 2, 3$).

the case of resonant leptogenesis and was later derived in [70] for the case of hierarchical N_j. A full calculation of the CP asymmetry in scatterings involving the top quark was carried out in [108], and the validity of approximating it with the CP asymmetry in decays was analyzed, finding that the approximation is generally good for sufficiently strong RH neutrino hierarchies, for examlpe, $M_2/M_1 \gg 10$. Corrections up to several tens of percent can appear around temperatures of order $T \sim M_2/10$ and can be numerically relevant in case of milder hierarchies.

Regarding the scattering processes with gauge bosons such as $N\ell_\alpha \rightarrow A\overline{H}$ $NH \rightarrow A\overline{\ell}_\alpha$ or $NA \rightarrow \ell_\alpha H$, their effects in leptogenesis were estimated in [108] under the assumption that it can also be factorized in terms of the decay CP asymmetry. However, with respect to scatterings involving the top quark, there is a significant difference that now box diagrams in which the gauge boson is attached to a lepton or Higgs in the loop of the vertex-type diagrams are also present, leading to more complicated expressions that were explicitly calculated in [112]. There it was shown that the presence of box diagrams implies that for scatterings with gauge bosons the CP asymmetry is different from the decay CP asymmetry even for hierarchical RH neutrinos. Still, this difference remains within a factor of two [112] so that related effects are in any case not very large. In general, it turns out that CP asymmetry in scatterings is more relevant at high temperatures ($T > M_1$) when the scattering rates are larger than the decay rate. Hence, it can be of some relevance to the final value of the baryon asymmetry when some of the lepton flavours are weakly washed out, and some memory of the asymmetries generated at high temperature is preserved in the final result.

4.3. *Thermal Corrections*

At the high temperatures at which leptogenesis occurs, the light particles involved in the leptogenesis processes are in equilibrium with the hot plasma. Thermal effects give corrections to several ingredients in the analysis: (i) coupling constants, (ii) particle propagators (leptons, quarks, gauge bosons, and the Higgs), and (iii) CP-violating asymmetries, which we briefly discuss later. A detailed study of thermal corrections can be found in [49].

4.3.1. Coupling Constants

Renormalization of gauge and Yukawa couplings in a thermal plasma is studied in [113]. In practice, it is a good approximation to use the zero-temperature renormalization group equations for the couplings, with a renormalization scale $\Lambda \sim 2\pi T$ [49]. The value $\Lambda > T$ is related to the fact that the average energy of the colliding particles in the plasma is larger than the temperature.

The renormalization effects for the neutrino couplings are also well known [114, 115]. In the nonsupersymmetric case, to a good approximation these effects can be described by a simple rescaling of the low energy neutrino mass matrix $m(\mu) = r \cdot m$, where $1.2 \lesssim r \lesssim 1.3$ for 10^8 GeV $\lesssim \mu \lesssim 10^{16}$ GeV [49], and therefore can be accounted for by increasing the values of the neutrino mass parameters (e.g., $\tilde{m}$) as measured at low energy by $\approx$ 20%–30% (depending on the leptogenesis scale). In the supersymmetric case one expects a milder enhancement, but uncertainties related with the precise value of the top-Yukawa coupling can be rather large (see Figure 3 in [49]).

4.3.2. Decays and Scatterings

In the thermal plasma, any particle with sizable couplings to the background acquires a thermal mass which is proportional to the plasma temperature. Consequently, decay and scattering rates get modified. Particle thermal masses have been thoroughly studied in both the SM and the supersymmetric SM [91, 116–120]. The singlet neutrinos have no gauge interactions, their Yukawa couplings are generally small, and, during the relevant era, their bare masses are of the order of the temperature or larger. Consequently, to a good approximation, corrections to their masses can be neglected. We thus need to account for the thermal masses of the leptons and Higgs doublets and, when scatterings are included, also of the third generation quarks and of the gauge bosons (and of their superpartners in the supersymmetric case). For a qualitative discussion, it is enough to keep in mind that, within the leptogenesis temperature range, we have $m_H(T) \gtrsim m_{q_3,t}(T) \gg m_\ell(T)$. The most important effects are related to four classes of leptogenesis processes.

(i) *Decays and inverse decays*: since thermal corrections to the Higgs mass are particularly large ($m_H(T) \approx 0.4\,T$), decays and inverse decays become kinematically forbidden in the temperature range in which $m_H(T) - m_\ell(T) < M_{N_1} < m_H(T) + m_\ell(T)$. For lower temperatures, the usual processes $N_1 \leftrightarrow \ell H$ can occur. For higher temperatures, the Higgs is heavy enough that it can decay: $H \leftrightarrow \ell N_1$. A rough estimate of the kinematically forbidden region yields $2 \lesssim T/M_1 \lesssim 5$. The important point is that these corrections are effective only at $T > M_1$. In the parameter region $\tilde{m} > 10^{-3}$ eV, that is favoured by the measurements of the neutrino mass-squared differences, the N_1 number density and its L-violating reactions attain thermal equilibrium at $T \approx M_1$ and erase quite efficiently any memory of the specific conditions at higher temperatures. Consequently, in the strong washout regime, these corrections have practically no effect on the final value of the baryon asymmetry.

(ii) $\Delta L = 1$ *scatterings with top quark*: a comparison between the corrected and uncorrected rates of the top-quark scattering $\gamma_{H_s}^{\text{top}} \equiv \gamma(q_3\bar{t} \leftrightarrow \ell N_1)$ with the Higgs exchanged in the s-channel and of the sum of the t- and u-channel scatterings $\gamma_{H_{t+u}}^{\text{top}} \equiv \gamma(q_3 N_1 \leftrightarrow \ell t) + \gamma(\bar{t}N_1 \leftrightarrow \ell\bar{q}_3)$ shows that the only corrections appearing at low temperatures, and thus more relevant, are for $\gamma_{H_{t+u}}^{\text{top}}$ (see Figure 7.1 in [109]). They reduce the scattering rates and suppress the related washouts. This peculiar situation arises from the fact that in the zero temperature limit there is a large enhancement $\sim \ln(M_{N_1}/m_H)$ from the quasimassless Higgs exchanged in the t- and u-channels, which disappears when the Higgs thermal mass $m_H(T) \sim T \sim M_{N_1}$ is included.

(iii) $\Delta L = 1$ *scatterings with the gauge bosons*: here the inclusion of thermal masses is required to avoid IR divergences that would arise when massless ℓ (and H) states are exchanged in the t- and u-channels. A naive use of some cutoff for the phase space integrals to control the IR divergences can yield incorrect estimates of the gauge bosons scattering rates [49] and would be particularly problematic at low temperatures, where gauge bosons scatterings dominate over top-quark scatterings.

4.3.3. CP Asymmetries

CP asymmetries arise from the interference of tree level and one-loop amplitudes when the couplings involved have complex phases and the loop diagrams have an absorptive part. This last requirement is satisfied whenever the loop diagram can be cut in such a way that the particles in the cut lines can be produced on shell. For the CP asymmetry in decay (at zero temperature) this is guaranteed by the fact that the Higgs and the lepton final states coincide with the states circulating in the loops. However, in the hot plasma in which N_1 decays occur, the Higgs and the lepton doublets are in thermal equilibrium and their interactions with the background introduce in the CP asymmetries a dependence on the temperature $\epsilon \rightarrow \epsilon(T)$ that arises from various effects

(i) Absorption and reemission of the loop particles by the medium require the use of finite temperature propagators.

(ii) Stimulation of decays into bosons and blocking of decays into fermions in the dense background require proper modification of the final states density distributions.

(iii) Thermal motion of the decaying N's with respect to the background breaks the Lorentz symmetry and affects the evaluation of the CP asymmetries.

(iv) Thermal masses should be included in the finite temperature resummed propagators, and they also modify the fermion and boson dispersion relations. Their inclusion yields the most significant modifications to the zero temperature results for the CP asymmetries.

The first three effects were investigated in [48] while the effects of thermal masses were included in [49]. In principle, at finite temperature, there are additional effects related to new cuts that involve the heavy $N_{2,3}$ neutrino lines. These new cuts appear because the heavy particles in the loops may absorb energy from the plasma and go onshell. However, for hierarchical spectrum, $M_{2,3} \gg M_1$, the related effects are Boltzmann suppressed by $\exp(-M_{2,3}/T)$ that at $T \sim M_1$ is a tiny factor. For a nonhierarchical spectrum, the effect of these new cuts can however be sizable. A detailed study can be found in [121].

4.3.4. Propagators and Statistical Distributions

Particle propagators at finite temperature are computed in the real-time formalism of thermal field theory [122, 123]. In this formalism, ghost fields dual to each of the physical fields have to be introduced, and consequently the thermal propagators have 2×2 matrix structures. For the one-loop computations of the absorptive parts of the Feynman diagrams, the relevant propagator components are just those of the physical lepton and Higgs fields. The usual zero temperature propagators $-iS_\ell^0(p, m_\ell) = (\not{p} - m_\ell + i0^+)^{-1}$ and $-iD_H^0(p, m_H) = (p^2 - m_H^2 + i0^+)^{-1}$ acquire an additive term that is proportional to the particle density distribution $n_{\ell,H} = [\exp(E_{\ell,H}/T) \pm 1]^{-1}$:

$$\begin{aligned} \delta S_\ell(T) &= -2\pi n_\ell (\not{p} - m_\ell)\delta\left(p^2 - m_\ell^2\right), \\ \delta D_H(T) &= +2\pi n_H \delta\left(p^2 - m_H^2\right). \end{aligned} \tag{4.7}$$

For the fermionic thermal propagators, there are other higher order corrections (see [49]). Unlike the case of bosons, the interactions of the fermions with the thermal bath lead to two different types of excitations with different dispersion relations, that are generally referred to as "particles" and "holes" [49]. The contributions of these two fermionic modes were studied in [124–126] where it was argued that in the strong washout regime they could give nonnegligible effects [126]. The leading effects in (i) are proportional to the factor $-n_\ell + n_H - 2n_\ell n_H$ that vanishes when the final states thermal masses are neglected, because the Bose-Einstein and Fermi-Dirac statistical distributions depend on the same argument, $E_\ell = E_H = M_1/2$. As a consequence, the thermal corrections to the fermion and boson propagators (n_ℓ and n_H) and the product of the two thermal corrections ($n_\ell n_H$) cancel each other. This was interpreted as a complete compensation between stimulated emission and Pauli blocking. As regards the effects in (ii), they lead to overall factors that cancel between numerator and denominator in the expression for the CP asymmetry. (A similar cancellation holds also in the supersymmetric case. However, because of the presence of the superpartners $\tilde{\ell}, \widetilde{H}$ both as final states and in the loops, the cancellation is more subtle and it involves a compensation between the two types of corrections (i) and (ii). We refer to [48] for details.) More recently, on the basis of a first principle derivation of the CP asymmetry within a quantum BE approach (see Section 4.5) it has been claimed that the statistical factor induced by thermal loops is instead $-n_\ell + n_H$, which does not vanish even in the massless approximation. This would result in a further enhancement in the CP asymmetry from the thermal effects [127].

4.3.5. Particle Motion

Given that the decaying particle N_1 is moving with respect to the background (with velocity $\vec{\beta}$) the fermionic decay products are preferentially emitted in the direction antiparallel to the plasma velocity (for which the thermal distribution is less occupied), while the bosonic ones are emitted preferentially in the forward direction (for which stimulated emission is more effective). Particle motion then induces an angular dependence in the decay distribution at order $\mathcal{O}(\beta)$. In the total decay rate the $\mathcal{O}(\beta)$ anisotropy effect is integrated out, and only $\mathcal{O}(\beta^2)$ effects remain [48]. Therefore, while accounting for thermal motion does modify the zero temperature results, these corrections are numerically small [48, 49] and generally negligible.

4.3.6. Thermal Masses

When the finite values of the light particle thermal masses are taken into account, the arguments of the Bose-Einstein and Fermi-Dirac statistical distributions are different. It is a good approximation [49] to use for the particle energies $E_{\ell,H} = M_1/2 \mp (m_H^2 - m_\ell^2)/2M_1$. Since now $E_\ell \neq E_H$, the prefactor $-n_\ell + n_H - 2n_\ell n_H$ that multiplies the thermal corrections does not vanish anymore, and sizable corrections become possible. The most relevant effect is that the CP asymmetry vanishes when, as the temperature increases, the sum of the light particles thermal masses approaches M_1 [49]. This is not surprising, since the particles in the final state coincide with the particles in the loop, and therefore when the decay becomes kinematically forbidden, also the particles in the loop cannot go on the mass shell. When the temperature is large enough that $m_H(T) > m_\ell(T) + M_1$, the Higgs can decay, and then there is a new source of lepton number asymmetry associated with $H \to \ell N_1$. The CP asymmetry in Higgs decays ϵ_H can be up to one order of magnitude larger than the CP asymmetry in N_1 decays [49]. While this could represent a dramatic enhancement of the CP asymmetry, ϵ_H is nonvanishing only at temperatures $T \gtrsim T_H \sim 5M_1$, when the kinematic condition for its decays is satisfied. Therefore, in the strong washout regime, no trace of this effect survives. On the other hand, rather large couplings are required in order that Higgs decays can occur before the phase space closes: the decay rate can attain thermal equilibrium only when $\tilde{m} \gtrsim (T_H/M_1)^2 m_* \gg m_*$, and therefore, in the weak washout regime ($\tilde{m} \lesssim m_*$), these decays always remain strongly out of equilibrium. This means that only a small fraction of the Higgs particles have actually time to decay, and the lepton asymmetry generated in this way is accordingly suppressed.

In summary, while the corrections to the CP asymmetries can be significant at $T \gtrsim M_1$ (and quite large at $T \gg M_1$ for Higgs decays), in the low temperature regime, where the precise value of ϵ plays a fundamental role in determining the final value of the baryon asymmetry, there are almost no effects, and the zero temperature results still give a reliable approximation.

4.4. Decays of the Heavier Right-Handed Neutrinos

In leptogenesis studies, the effects of $N_{2,3}$ are often neglected, which in many cases is not a good approximation. This is obvious, for example, when N_1 dynamics is irrelevant for leptogenesis: $\epsilon_1 \ll 10^{-6}$ cannot provide enough CP asymmetry to account for baryogenesis, and $\tilde{m}_1 \ll m_*$ implies that N_1 washout effects are negligible. It is then clear that any asymmetry generated in $N_{2,3}$ decays can survive and becomes crucial for the success of leptogenesis. Another case in which it is intuitively clear that $N_{2,3}$ effects can be important is when the RH neutrino spectrum is compact, which means that $M_{2,3}$ have values within a factor of a few from M_1. Then N_1 and $N_{2,3}$ contributions to leptogenesis can be equally important and must be summed up. A model with compact RH neutrino spectrum in which $N_{2,3}$ dynamics is of crucial importance was recently discussed in [128].

It is less obvious that $N_{2,3}$ effects can also be important for a hierarchical RH spectrum and when N_1 is strongly coupled. This can happen because decoherence effects related to N_1-interactions can project the asymmetry generated in $N_{2,3}$ decays onto a flavour direction that remains protected against N_1 washouts [67, 129–131]. Let us illustrate this with an example. Let us assume that a sizable asymmetry is generated in N_2 decays, while N_1 leptogenesis is inefficient and fails, that is:

$$\tilde{m}_2 \not\gg m_*, \qquad \tilde{m}_1 \gg m_*. \tag{4.8}$$

Assuming also a strong hierarchy and that leptogenesis occurs thermally guarantees that [131]

$$n_{N_1}(T \sim M_2) \approx 0, \qquad n_{N_2}(T \sim M_1) \approx 0. \tag{4.9}$$

Thus, the dynamics of N_2 and N_1 are decoupled: there are neither N_1-related washout effects during N_2 leptogenesis nor N_2-related washout effects during N_1 leptogenesis. The N_2 decays into a combination of lepton doublets that we denote by ℓ_2:

$$|\ell_2\rangle = \left(Y^\dagger Y\right)_{22}^{-1/2} \sum_\alpha Y_{\alpha 2} |\ell_\alpha\rangle. \tag{4.10}$$

The second condition in (4.8) implies that already at $T \gtrsim M_1$ the N_1-Yukawa interactions are sufficiently fast to quickly destroy the coherence of ℓ_2. Then a statistical mixture of ℓ_1 and of the state orthogonal to ℓ_1 builds up, and it can be described by a suitable diagonal density matrix. Let us consider the simple case where both N_2 and N_1 decay at $T \gtrsim 10^{12}$ GeV, so that flavour effects are irrelevant. A convenient choice for an orthogonal basis for the lepton doublets is $(\ell_1, \ell_0, \ell_0')$ where, without loss of generality, ℓ_0' satisfies $\langle \ell_0' \mid \ell_2 \rangle = 0$. Then the asymmetry ΔY_{ℓ_2} produced in N_2 decays decomposes into two components:

$$\Delta Y_{\ell_0} = c^2 \Delta Y_{\ell_2}, \qquad \Delta Y_{\ell_1} = s^2 \Delta Y_{\ell_2}, \tag{4.11}$$

where $c^2 \equiv |\langle \ell_0 \mid \ell_2 \rangle|^2$ and $s^2 = 1 - c^2$. The crucial point here is that we expect, in general, $c^2 \neq 0$ and, since $\langle \ell_0 \mid \ell_1 \rangle = 0$, ΔY_{ℓ_0} is protected against N_1 washout. Consequently, a finite part of the asymmetry ΔY_{ℓ_2} from N_2 decays survives through N_1 leptogenesis. A more detailed analysis [131] finds that ΔY_{ℓ_1} is not entirely washed out, resulting in the final lepton asymmetry $Y_{\Delta L} = (3/2)\Delta Y_{\ell_0} = (3/2)c^2 \Delta Y_{\ell_2}$.

For 10^9 GeV $\lesssim M_1 \lesssim 10^{12}$ GeV, flavour issues modify the quantitative details, but the qualitative picture, and in particular the survival of a finite part of ΔY_{ℓ_2}, still holds. In contrast, for $M_1 \lesssim 10^9$ GeV, the full flavour basis $(\ell_e, \ell_\mu, \ell_\tau)$ is resolved, and thus there are no directions in flavour space where an asymmetry is protected, so that Y_{ℓ_2} can be erased entirely.

A dedicated study in which the various flavour regimes for $N_{1,2,3}$ decays are considered can be found in [132].

In conclusion $N_{2,3}$ leptogenesis cannot be ignored, unless one of the following conditions holds.

(1) The reheat temperature is below M_2.

(2) The asymmetries and/or the washout factors vanish, $\epsilon_{N_2}\eta_2 \approx 0$ and $\epsilon_{N_3}\eta_3 \approx 0$.

(3) N_1-related washout is still significant at $T \lesssim 10^9$ GeV.

4.5. Quantum Boltzmann Equations

So far we have analyzed the leptogenesis dynamics by adopting the classical BE of motion. An interesting question which has attracted some attention recently [85, 94, 121, 127, 133–139] is under which circumstances the classical BE can be safely applied to get reliable results and, conversely, when a more rigorous quantum approach is needed. Quantum BEs are obtained starting from the nonequilibrium quantum field theory based on the closed time-path (CTP) formulation [140]. Both, CP violation from wave function and vertex corrections are incorporated. Unitarity issues are resolved, and an accurate account of all quantum-statistical effects on the asymmetry is made. Moreover, the formulation in terms of Green functions bears the potential of incorporating corrections from thermal field theory within the CTP formalism.

In the CTP formalism, particle number densities are replaced by Green's functions obeying a set of equations which, under some assumptions, can be reduced to a set of kinetic equations describing the evolution of the lepton asymmetry and the RH neutrinos. These kinetic equations are nonMarkovian and present memory effects. In other words, differently from the classical approach where every scattering in the plasma is independent of the previous one, the particle abundances at a given time depend upon the history of the system. The more familiar energy-conserving delta functions are replaced by retarded time integrals of time-dependent kernels, and cosine functions whose arguments are the energy involved in the various processes. Therefore, the nonMarkovian kinetic equations include the contribution of coherent processes throughout the history of the kernels and the relaxation times are expected to be typically longer than the one dictated by the classical approach.

If the time range of the kernels is shorter than the relaxation time of the particles abundances, the solutions to the quantum and the classical BE differ only by terms of the order of the ratio of the timescale of the kernel to the relaxation timescale of the distribution. In thermal leptogenesis this is typically the case. However, there are situations where this does not happen. For instance, in the case of resonant leptogenesis, two RH (s)neutrinos N_1 and N_2 are almost degenerate in mass, and the CP asymmetry from the decay of the first RH neutrino N_1 is resonantly enhanced if the mass difference $\Delta M = (M_2 - M_1)$ is of the order of the decay rate of the second RH neutrino Γ_{N_2}. The typical timescale to build up coherently the CP asymmetry is of the order of $1/\Delta M$, which can be larger than the timescale $\sim 1/\Gamma_{N_1}$ for the change of the abundance of the N_1's.

Since we need the time evolution of the particle asymmetries with definite initial conditions and not simply the transition amplitude of particle reactions, the ordinary equilibrium quantum field theory at finite temperature is not the appropriate tool. The most appropriate extension of the field theory to deal with nonequilibrium phenomena amounts to generalizing the time contour of integration to a closed time path. More precisely, the time integration contour is deformed to run from t_0 to $+\infty$ and back to t_0. The CTP formalism is a powerful Green's function formulation for describing nonequilibrium phenomena in field theory. It allows to describe phase-transition phenomena and to obtain a self-consistent set of quantum BE. The formalism yields various quantum averages of operators evaluated in the instate without specifying the out-state. On the contrary, the ordinary quantum field theory yields quantum averages of the operators evaluated with an instate at one end and an outstate at the other.

For example, because of the time-contour deformation, the partition function in the in-in formalism for a complex scalar field is defined to be

$$\begin{aligned} Z\left[J, J^\dagger\right] &= \mathrm{Tr}\left[\mathcal{T}\left(\exp\left[i\int_C \left(J(x)\phi(x) + J^\dagger(x)\phi^\dagger(x)\right)d^4x\right]\right)\rho\right] \\ &= \mathrm{Tr}\left[\mathcal{T}_+\left(\exp\left[i\int \left(J_+(x)\phi_+(x) + J_+^\dagger(x)\phi_+^\dagger(x)\right)d^4x\right]\right)\right. \\ &\qquad \left.\times\mathcal{T}_-\left(\exp\left[-i\int\left(J_-(x)\phi_-(x) + J_-^\dagger(x)\phi_-^\dagger(x)\right)d^4x\right]\right)\rho\right], \end{aligned} \tag{4.12}$$

where C in the integral denotes that the time integration contour runs from t_0 to plus infinity and then back to t_0 again. The symbol ρ represents the initial density matrix, and the fields are in the Heisenberg picture and defined on this closed time contour (plus and minus subscripts refer to the positive and negative directional branches of the time path, resp.). The time-ordering operator along the path is the standard one ($\mathcal{T}_+$) on the positive branch, and the antitime-ordering ($\mathcal{T}_-$) on the negative branch. As with the Euclidean-time formulation, scalar (fermionic) fields ϕ are still periodic (antiperiodic) in time, but with $\phi(t,\vec{x}) = \phi(t-i\beta,\vec{x})$, $\beta = 1/T$. The temperature T appears due to boundary condition, and time is now explicitly present in the integration contour.

We must now identify field variables with arguments on the positive or negative directional branches of the time path. This doubling of field variables leads to six different real-time propagators on the contour. These six propagators are not independent, but using all of them simplifies the notation. For a generic charged scalar field ϕ they are defined as

$$\begin{aligned} G_\phi^>(x,y) &= -G_\phi^{-+}(x,y) = -i\left\langle \phi(x)\phi^\dagger(y)\right\rangle, \\ G_\phi^<(x,y) &= -G_\phi^{+-}(x,y) = -i\left\langle \phi^\dagger(y)\phi(x)\right\rangle, \\ G_\phi^t(x,y) &= G_\phi^{++}(x,y) = \theta(x,y)G_\phi^>(x,y) + \theta(y,x)G_\phi^<(x,y), \\ G_\phi^{\bar{t}}(x,y) &= G_\phi^{--}(x,y) = \theta(y,x)G_\phi^>(x,y) + \theta(x,y)G_\phi^<(x,y), \\ G_\phi^r(x,y) &= G_\phi^t - G_\phi^< = G_\phi^> - G_\phi^{\bar{t}}, \qquad G_\phi^a(x,y) = G_\phi^t - G_\phi^> = G_\phi^< - G_\phi^{\bar{t}}, \end{aligned} \tag{4.13}$$

where the last two Green's functions are the retarded and advanced Green's functions, respectively, and $\theta(x,y) \equiv \theta(t_x - t_y)$ is the step function.

For a generic fermion field ψ the six different propagators are analogously defined as

$$\begin{aligned}
G_{\psi}^{>}(x,y) &= -G_{\psi}^{-+}(x,y) = -i\langle\psi(x)\overline{\psi}(y)\rangle, \\
G_{\psi}^{<}(x,y) &= -G_{\psi}^{+-}(x,y) = +i\langle\overline{\psi}(y)\psi(x)\rangle, \\
G_{\psi}^{t}(x,y) &= G_{\psi}^{++}(x,y) = \theta(x,y)G_{\psi}^{>}(x,y) + \theta(y,x)G_{\psi}^{<}(x,y), \\
G_{\psi}^{\bar{t}}(x,y) &= G_{\psi}^{--}(x,y) = \theta(y,x)G_{\psi}^{>}(x,y) + \theta(x,y)G_{\psi}^{<}(x,y), \\
G_{\psi}^{r}(x,y) &= G_{\psi}^{t} - G_{\psi}^{<} = G_{\psi}^{>} - G_{\psi}^{\bar{t}}, \qquad G_{\psi}^{a}(x,y) = G_{\psi}^{t} - G_{\psi}^{>} = G_{\psi}^{<} - G_{\psi}^{\bar{t}}.
\end{aligned} \tag{4.14}$$

From the definitions of the Green's functions, one can see that the hermiticity properties

$$\left(i\gamma^0 G_{\psi}(x,y)\right)^{\dagger} = i\gamma^0 G_{\psi}(y,x), \qquad (iG_{\phi}(x,y))^{\dagger} = iG_{\phi}(y,x) \tag{4.15}$$

are satisfied. For interacting systems, whether in equilibrium or not, one must define and calculate self-energy functions. Again, there are six of them: Σ^t, $\Sigma^{\bar{t}}$, $\Sigma^<$, $\Sigma^>$, Σ^r, and Σ^a. The same relationships exist among them as for the Green's functions in (4.13) and (4.14), such as

$$\Sigma^r = \Sigma^t - \Sigma^< = \Sigma^> - \Sigma^{\bar{t}}, \qquad \Sigma^a = \Sigma^t - \Sigma^> = \Sigma^< - \Sigma^{\bar{t}}. \tag{4.16}$$

The self-energies are incorporated into the Green's functions through the use of Dyson's equations. A useful notation may be introduced which expresses four of the six Green's functions as the elements of two-by-two matrices:

$$\tilde{G} = \begin{pmatrix} G^t & \pm G^< \\ G^> & -G^{\bar{t}} \end{pmatrix}, \qquad \tilde{\Sigma} = \begin{pmatrix} \Sigma^t & \pm\Sigma^< \\ \Sigma^> & -\Sigma^{\bar{t}} \end{pmatrix}, \tag{4.17}$$

where the upper signs refer to the bosonic case and the lower signs to the fermionic case. For systems either in equilibrium or in nonequilibrium, Dyson's equation is most easily expressed by using the matrix notation:

$$\tilde{G}(x,y) = \tilde{G}^0(x,y) + \int d^4z_1 \int d^4z_2 \tilde{G}^0(x,z_1)\tilde{\Sigma}(z_1,z_2)\tilde{G}(z_2,y), \tag{4.18}$$

where the superscript "0" on the Green's functions means to use those for noninteracting system. It is useful to notice that Dyson's equation can be written in an alternative form, instead of (4.18), with $\tilde{G}^0$ on the right in the interaction terms:

$$\tilde{G}(x,y) = \tilde{G}^0(x,y) + \int d^4z_3 \int d^4z_4 \tilde{G}(x,z_3)\tilde{\Sigma}(z_3,z_4)\tilde{G}^0(z_4,y). \tag{4.19}$$

Equations (4.18) and (4.19) are the starting points to derive the quantum BE describing the time evolution of the CP-violating particle density asymmetries.

To proceed, one has to choose a form for the propagators. For a generic fermion ψ (and similarly for scalars) one may adopt the real-time propagator in the form $G^t_\psi(\mathbf{k}, t_x - t_y)$ in terms of the spectral function $\rho_\psi(\mathbf{k}, k_0)$

$$G^t_\psi(\mathbf{k}, t_x - t_y) = \int_{-\infty}^{+\infty} \frac{dk^0}{2\pi} e^{-ik^0(t_x - t_y)} \rho_\psi\left(\mathbf{k}, k^0\right) \times \left\{\left[1 - f_\psi\left(k^0\right)\right]\theta(t_x - t_y) - f_\psi\left(k^0\right)\theta(t_y - t_x)\right\}, \tag{4.20}$$

where $f_\psi(k^0)$ represents the fermion distribution function. Again, particles must be substituted by quasiparticles, dressed propagators are to be adopted, and self-energy corrections to the propagator modify the dispersion relations by introducing a finite width $\Gamma_\psi(k)$. For a fermion with chiral mass m_ψ, one may safely choose

$$\rho_\psi\left(\mathbf{k}, k^0\right) = i(\not{k} + m_\psi)\left[\frac{1}{(k^0 + i\epsilon + i\Gamma_\psi)^2 - \omega^2_\psi(k)} - \frac{1}{(k^0 - i\epsilon - i\Gamma_\psi)^2 - \omega^2_\psi(k)}\right], \tag{4.21}$$

where $\omega^2_\psi(k) = \mathbf{k}^2 + M^2_\psi(T)$ and $M_\psi(T)$ is the effective thermal mass of the fermion in the plasma (not a chiral mass). Performing the integration over k^0 and picking up the poles of the spectral function (which is valid for quasiparticles in equilibrium or very close to equilibrium), one gets

$$\begin{aligned} G^>_\psi(\mathbf{k}, t_x - t_y) &= -\frac{i}{2\omega_\psi}\Big\{(\not{k} + m_\psi)\left[1 - f_\psi(\omega_\psi - i\Gamma_\psi)\right]e^{-i(\omega_\psi - i\Gamma_\psi)(t_x - t_y)} \\ &\quad + \gamma^0(\not{k} - m_\psi)\gamma^0 \bar{f}_\psi(\omega_\psi + i\Gamma_\psi)e^{i(\omega_\psi + i\Gamma_\psi)(t_x - t_y)}\Big\}, \\ G^<_\psi(\mathbf{k}, t_y - t_x) &= \frac{i}{2\omega_\psi}\Big\{(\not{k} + m_\psi)f_\psi(\omega_\psi + i\Gamma_\psi)e^{-i(\omega_\psi - i\Gamma_\psi)(t_x - t_y)} \\ &\quad + \gamma^0(\not{k} - m_\psi)\gamma^0\left[1 - \bar{f}_\psi(\omega_\psi - i\Gamma_\psi)\right]e^{i(\omega_\psi + i\Gamma_\psi)(t_x - t_y)}\Big\}, \end{aligned} \tag{4.22}$$

where $k^0 = \omega_\psi$ and $f_\psi, \bar{f}_\psi$ denote the distribution function of the fermion particles and antiparticles, respectively. The expressions (4.22) are valid for $t_x - t_y > 0$.

The above definitions hold for the lepton doublets (after inserting the chiral LH projector P_L), as well as for the Majorana RH neutrinos, for which one has to assume identical particle and antiparticle distribution functions and insert the inverse of the charge conjugation matrix C in the dispersion relation.

To elucidate further the impact of the CTP approach and to see under which conditions one can obtain the classical BE from the quantum ones, one may consider the dynamics of the lightest RH neutrino N_1. To find its quantum BE we start from (4.18) for the Green's function $G^<_{N_1}$ of the RH neutrino N_1 :

$$\begin{aligned}\left(i\vec{\partial}_x - M_1\right)G^<_{N_1}(x,y) &= -\int d^4z\left[-\Sigma^t_{N_1}(x,z)G^<_{N_1}(z,y) + \Sigma^<_{N_1}(x,z)G^{\bar{t}}_{N_1}(z,y)\right] \\ &= \int d^3z\int_0^t dt_z\left[\Sigma^>_{N_1}(x,z)G^<_{N_1}(z,y) - \Sigma^<_{N_1}(x,z)G^>_{N_1}(z,y)\right].\end{aligned} \tag{4.23}$$

Adopting the corresponding form for the RH neutrino propagator and the center-of-mass coordinates

$$X \equiv \left(t,\vec{X}\right) \equiv \frac{1}{2}(x+y), \qquad (\tau,\vec{r}) \equiv x-y, \tag{4.24}$$

one ends up with the following equation:

$$\begin{aligned}\frac{\partial f_{N_1}(\mathbf{k},t)}{\partial t} &= -2\int_0^t dt_z\int\frac{d^3\mathbf{p}}{(2\pi)^3}\frac{1}{2\omega_\ell(\mathbf{p})}\frac{1}{2\omega_H(\mathbf{k}-\mathbf{p})}\frac{1}{\omega_{N_1}(\mathbf{k})}|\mathcal{M}(N_1\to\ell H)|^2 \\ &\times\left[f_{N_1}(\mathbf{k},t)(1-f_\ell(\mathbf{p},t))(1+f_H(\mathbf{k}-\mathbf{p},t))\right. \\ &\quad\left.-f_\ell(\mathbf{p},t)f_H(\mathbf{k}-\mathbf{p},t)(1-f_{N_1}(\mathbf{k},t))\right] \\ &\times\cos[(\omega_{N_1}(\mathbf{k})-\omega_\ell(\mathbf{p})-\omega_H(\mathbf{k}-\mathbf{p}))(t-t_z)] \\ &\simeq -2\int_0^t dt_z\int\frac{d^3\mathbf{p}}{(2\pi)^3}\frac{1}{2\omega_\ell(\mathbf{p})}\frac{1}{2\omega_H(\mathbf{k}-\mathbf{p})}\frac{1}{\omega_{N_1}(\mathbf{k})}|\mathcal{M}(N_1\to\ell H)|^2 \\ &\times\left(f_{N_1}(\mathbf{k},t)-f_\ell^{\rm eq}(\mathbf{p})f_H^{\rm eq}(\mathbf{k}-\mathbf{p})\right) \\ &\times\cos[(\omega_{N_1}(\mathbf{k})-\omega_\ell(\mathbf{p})-\omega_H(\mathbf{k}-\mathbf{p}))(t-t_z)].\end{aligned} \tag{4.25}$$

This equations holds under the assumption that the relaxation timescale for the distribution functions is longer than the timescale of the nonlocal kernels so that they can be extracted out of the time integral. This allows to think of the distributions as functions of the center-of-mass time only. We have set to zero the damping rates of the particles in (4.22) and retained only those cosines giving rise to energy delta functions that can be satisfied. Under these assumptions, the distribution function may be taken out of the time integral, leading—at large times—to the so-called Markovian description. The kinetic equation (4.25) has an obvious interpretation in terms of gain minus loss processes, but the retarded time integral and the cosine function replace the familiar energy-conserving delta functions. In the second passage, we have also made the usual assumption that all distribution functions are smaller than unity and that those of the Higgs and lepton doublets are in equilibrium and much smaller than unity, $f_\ell f_H \simeq f_\ell^{\rm eq} f_H^{\rm eq}$. Elastic scatterings are typically fast enough to keep kinetic equilibrium. For any distribution function we may write $f = (n/n^{\rm eq}) f^{\rm eq}$, where n denotes the total number density. Therefore, (4.25) can be rewritten as

$$
\begin{aligned}
\frac{\partial n_{N_1}}{\partial t} &= -\langle \Gamma_{N_1}(t) \rangle n_{N_1} + \left\langle \widetilde{\Gamma}_{N_1}(t) \right\rangle n_{N_1}^{\rm eq}, \\
\langle \Gamma_{N_1}(t) \rangle &= \int_0^t dt_z \int \frac{d^3\mathbf{k}}{(2\pi)^3} \frac{f_{N_1}^{\rm eq}}{n_{N_1}^{\rm eq}} \Gamma_{N_1}(t), \\
\Gamma_{N_1}(t) &= 2 \int \frac{d^3\mathbf{p}}{(2\pi)^3} \frac{|\mathcal{M}(N_1 \to \ell H)|^2}{2\omega_\ell 2\omega_H \omega_{N_1}} \cos[(\omega_{N_1} - \omega_\ell - \omega_H)(t - t_z)], \\
\langle \widetilde{\Gamma}_{N_1}(t) \rangle &= \int_0^t dt_z \int \frac{d^3\mathbf{k}}{(2\pi)^3} \frac{f_{N_1}^{\rm eq}}{n_{N_1}^{\rm eq}} \widetilde{\Gamma}_{N_1}(t), \\
\widetilde{\Gamma}_{N_1}(t) &= 2 \int \frac{d^3\mathbf{p}}{(2\pi)^3} \frac{f_\ell^{\rm eq} f_H^{\rm eq}}{f_{N_1}^{\rm eq}} \frac{|\mathcal{M}(N_1 \to \ell H)|^2}{2\omega_\ell 2\omega_H \omega_{N_1}} \cos[(\omega_{N_1} - \omega_\ell - \omega_H)(t - t_z)],
\end{aligned}
\tag{4.26}
$$

where $\langle \Gamma_{N_1}(t) \rangle$ is the time-dependent thermal average of the Lorentz-dilated decay width. Integrating over large times, $t \to \infty$, thereby replacing the cosines by energy-conserving delta functions:

$$
\int_0^\infty dt_z \cos[(\omega_{N_1} - \omega_\ell - \omega_H)(t - t_z)] = \pi \delta(\omega_{N_1} - \omega_\ell - \omega_H), \tag{4.27}
$$

we find that the two averaged rates $\langle \Gamma_{N_1} \rangle$ and $\langle \widetilde{\Gamma}_{N_1} \rangle$ coincide and we recover the usual classical BE for the RH distribution function

$$
\begin{aligned}
\frac{\partial n_{N_1}}{\partial t} &= -\langle \Gamma_{N_1} \rangle \left(n_{N_1} - n_{N_1}^{\rm eq} \right), \\
\langle \Gamma_{N_1} \rangle &= \int \frac{d^3\mathbf{k}}{(2\pi)^3} \frac{f_{N_1}^{\rm eq}}{n_{N_1}^{\rm eq}} \int \frac{d^3\mathbf{p}}{(2\pi)^3} \frac{|\mathcal{M}(N_1 \to \ell H)|^2}{2\omega_\ell 2\omega_H \omega_{N_1}} (2\pi) \delta(\omega_{N_1} - \omega_\ell - \omega_H).
\end{aligned}
\tag{4.28}
$$

Taking the time interval to infinity, namely, implementing Fermi's golden rule, results in neglecting memory effects, which in turn results only in on-shell processes contributing to the rate equation. The main difference between the classical and the quantum BE can be traced to memory effects and to the fact that the time evolution of the distribution function is nonMarkovian. The memory of the past time evolution translates into off-shell processes.

Similarly, one can show that the equation obeyed by the asymmetry reads

$$
\begin{aligned}
\frac{\partial n_{\Delta L_\alpha}(X)}{\partial t} = &- \int d^3z \int_0^t dt_z \, {\rm Tr}\Big[\Sigma_{\ell_\alpha}^{>}(X,z) G_{\ell_\alpha}^{<}(z,X) - G_{\ell_\alpha}^{>}(X,z) \Sigma_{\ell_\alpha}^{<}(z,X) \\
&+ G_{\ell_\alpha}^{<}(X,z) \Sigma_{\ell_\alpha}^{>}(z,X) - \Sigma_{\ell_\alpha}^{<}(X,z) G_{\ell_\alpha}^{>}(z,X) \Big].
\end{aligned}
\tag{4.29}
$$

Proceeding as for the RH neutrino equation one finds (including the moment only for the 1-loop wave contribution to the CP asymmetry ϵ_w^α)

$$
\begin{aligned}
\frac{\partial n_{\Delta L_\alpha}}{\partial t} &= \epsilon_w^\alpha(t)\langle\Gamma_{N_1}\rangle\left(n_{N_1}-n_{N_1}^{\rm eq}\right),\\
\epsilon_w^\alpha(t) &= -\frac{4}{\langle\Gamma_{N_1}\rangle}\sum_{\beta=1}^{3}{\rm Im}\left(Y_{1\alpha}Y_{1\beta}Y_{\beta2}^\dagger Y_{\alpha2}^\dagger\right)\\
&\times\int_0^t dt_z\int_0^{t_z}dt_2\int_0^{t_2}dt_1 e^{-\Gamma_{N_2}(t_z-t_2)}e^{-(\Gamma_{\ell_\beta}+\Gamma_H)(t_2-t_1)}\int\frac{d^3\mathbf{k}}{(2\pi)^3}\frac{f_{N_1}^{\rm eq}}{n_{N_1}^{\rm eq}}\\
&\times\int\frac{d^3\mathbf{p}}{(2\pi)^3}\frac{1-f_{\ell_\beta}^{\rm eq}(\mathbf{p})+f_H^{\rm eq}(\mathbf{k}-\mathbf{p})}{2\omega_{\ell_\beta}(\mathbf{p})2\omega_H(\mathbf{k}-\mathbf{p})\omega_{N_1}(\mathbf{k})}\int\frac{d^3\mathbf{q}}{(2\pi)^3}\frac{1-f_{\ell_\alpha}^{\rm eq}(\mathbf{q})+f_H^{\rm eq}(\mathbf{k}-\mathbf{q})}{2\overline{\omega}_{\ell_\alpha}(\mathbf{q})2\overline{\omega}_H(\mathbf{k}-\mathbf{q})\omega_{N_2}(\mathbf{k})}\\
&\times\sin\left(\omega_{N_1}(t-t_1)+\left(\omega_{\ell_\beta}+\omega_H\right)(t_1-t_2)+\omega_{N_2}(t_2-t_z)+(\overline{\omega}_{\ell_\alpha}+\overline{\omega}_H)(t_z-t)\right)\\
&\times{\rm Tr}(M_1P_L\not{p}M_2\not{q}),
\end{aligned}
\tag{4.30}
$$

where, to avoid double counting, we have not inserted the decay rates in the propagators of the initial and final states and, for simplicity, we have assumed that the damping rates of the lepton doublets and the Higgs field are constant in time. This should be a good approximation as the damping rates are to be computed for momenta of order of the mass of the RH neutrinos. As expected from first principles, we find that the CP asymmetry is a function of time and its value at a given instant depends upon the previous history of the system.

Performing the time integrals and retaining only those pieces which eventually give rise to energy-conserving delta functions in the Markovian limit, we obtain

$$
\begin{aligned}
\epsilon_w^\alpha(t) &= -\frac{4}{\langle\Gamma_{N_1}\rangle}\sum_{\beta=1}^{3}{\rm Im}\left(Y_{1\alpha}Y_{1\beta}Y_{\beta2}^\dagger Y_{\alpha2}^\dagger\right)\\
&\times\int_0^t dt_z\frac{\cos[(\omega_{N_1}-\overline{\omega}_{\ell_\alpha}-\overline{\omega}_H)(t-t_z)]}{\left(\Gamma_{N_2}^2+(\omega_{N_2}-\omega_{N_1})^2\right)\left(\left(\Gamma_{\ell_\beta}+\Gamma_H\right)^2+\left(\omega_{N_1}-\omega_{\ell_\beta}-\omega_H\right)^2\right)}\\
&\times\int\frac{d^3\mathbf{k}}{(2\pi)^3}\frac{f_{N_1}^{\rm eq}}{n_{N_1}^{\rm eq}}\left(\Gamma_{\ell_\beta}+\Gamma_H\right)\\
&\times\left(2(\omega_{N_2}-\omega_{N_1})\sin^2\left[\frac{(\omega_{N_2}-\omega_{N_1})t_z}{2}\right]-\Gamma_{N_2}\sin[(\omega_{N_2}-\omega_{N_1})t_z]\right)\\
&\times\int\frac{d^3\mathbf{p}}{(2\pi)^3}\frac{1-f_{\ell_\beta}^{\rm eq}(\mathbf{p})+f_H^{\rm eq}(\mathbf{k}-\mathbf{p})}{2\omega_{\ell_\beta}(\mathbf{p})2\omega_H(\mathbf{k}-\mathbf{p})\omega_{N_1}(\mathbf{k})}\\
&\times\int\frac{d^3\mathbf{q}}{(2\pi)^3}\frac{1-f_{\ell_\alpha}^{\rm eq}(\mathbf{q})+f_H^{\rm eq}(\mathbf{k}-\mathbf{q})}{2\overline{\omega}_{\ell_\alpha}(\mathbf{q})2\overline{\omega}_H(\mathbf{k}-\mathbf{q})\omega_{N_2}(\mathbf{k})}{\rm Tr}(M_1P_L\not{p}M_2\not{q}).
\end{aligned}
\tag{4.31}
$$

From this expression it is already manifest that the typical timescale for the building up of the coherent CP asymmetry depends crucially on the difference in energy of the two RH neutrinos.

If we now let the upper limit of the time integral take large values, we neglect the memory effects; the CP asymmetry picks contribution only from the on-shell processes. Taking the damping rates of the lepton doublets equal for all the flavours and the RH neutrinos nearly at rest with respect to the thermal bath, the CP asymmetry reads (now summing over all flavour indices)

$$\epsilon_w(t) \simeq -\frac{\mathrm{Im}\,(YY^\dagger)^2_{12}}{(YY^\dagger)_{11}(YY^\dagger)_{22}}\frac{M_1}{M_2}\Gamma_{N_2}\frac{1}{(\Delta M)^2+\Gamma^2_{N_2}} \times\left(2\Delta M\sin^2\left[\frac{\Delta Mt}{2}\right]-\Gamma_{N_2}\sin[\Delta Mt]\right), \tag{4.32}$$

where $\Delta M = (M_2 - M_1)$. The CP asymmetry (4.32) is resonantly enhanced when $\Delta M \simeq \Gamma_{N_2}$, and at the resonance point it is given by

$$\epsilon_w(t) \simeq -\frac{1}{2}\frac{\mathrm{Im}\,(YY^\dagger)^2_{12}}{(YY^\dagger)_{11}(YY^\dagger)_{22}}(1-\sin[\Delta Mt]-\cos[\Delta Mt]). \tag{4.33}$$

The timescale for the building up of the CP asymmetry is $\sim 1/\Delta M$. The CP asymmetry grows starting from a vanishing value, and, for $t \gg (\Delta M)^{-1}$, it averages to the constant standard value. This is true if the timescale for the other processes relevant for leptogenesis is larger than $\sim 1/\Delta M$. In other words, one may define an "average" CP asymmetry

$$\langle\epsilon_w\rangle = \frac{1}{\tau_\mathrm{p}}\int_{t-\tau_\mathrm{p}}^{t} dt'\epsilon^{\mathrm{W}}_{N_1}(t'), \tag{4.34}$$

where τ_p represents the typical timescale of the other processes relevant for leptogenesis, for example, the $\Delta L = 1$ scatterings. If $\tau_\mathrm{p} \gg 1/\Delta M \sim \Gamma^{-1}_{N_2}$, the oscillating functions in (4.33) are averaged to zero and the average CP asymmetry is given by the value used in the literature. However, the expression (4.32) should be used when $\tau_p \lesssim 1/\Delta M \sim \Gamma^{-1}_{N_2}$.

The fact that the CP asymmetry is a function of time is particularly relevant in the case in which the asymmetry is generated by the decays of two heavy states which are nearly degenerate in mass and oscillate into one another with a timescale, given by the inverse of the mass difference and has been studied in [86, 141]. From (4.32) it is manifest that the CP asymmetry itself oscillates with the very same timescale and such a dependence may or may not be neglected depending upon the rates of the other processes in the plasma. If $\Gamma_{N_1} \gtrsim \Gamma_{N_2}$, the time dependence of the CP asymmetry may not be neglected. The expression (4.32) can also be used, once it is divided by a factor 2 (because in the wave diagram also the charged states of Higgs and lepton doublets may propagate) and the limit $M_2 \gg M_1$ is taken, for the CP asymmetry contribution from the vertex diagram:

$$\epsilon_v(t) \simeq -\frac{\mathrm{Im}\,(YY^\dagger)^2_{12}}{16\pi(YY^\dagger)_{11}}\frac{M_1}{M_2}\left(2\sin^2\left[\frac{M_2t}{2}\right]-\frac{\Gamma_{N_2}}{M_2}\sin[M_2t]\right). \tag{4.35}$$

In this case, the timescale for this CP asymmetry is $\sim M_2$ and much larger than any other timescale in the dynamics. Therefore, one can safely average over many oscillations, getting the expression present in the literature.

What has been discussed here provides only one example for which a quantum Boltzmann approach is needed. In general, the lesson is that quantum BEs are relevant when the typical timescale in a quantum physical process, such as the timescale for the unflavour-to-flavour transition or the timescale to build up coherently the CP asymmetry (of the order of $1/\Delta M$), is larger than the timescale for the change of the abundances.

5. Supersymmetric Leptogenesis

5.1. What's New?

Supersymmetric leptogenesis constitutes a theoretically appealing generalization of leptogenesis for the following reason: while the SM equipped with the seesaw provides the simplest way to realize leptogenesis, such a framework is plagued by an unpleasant fine-tuning problem. For a nondegenerate spectrum of heavy Majorana neutrinos, successful leptogenesis requires generically a scale for the singlet neutrino masses that is much larger than the EW scale [61] but at the quantum level the gap between these two scales becomes unstable. Low-energy supersymmetry (SUSY) can naturally stabilize the required hierarchy, and this provides a sound motivation for studying leptogenesis in the framework of the supersymmetrized version of the seesaw mechanism. Supersymmetric leptogenesis, however, introduces a certain conflict between the gravitino bound on the reheat temperature and the thermal production of the heavy singlets neutrinos [142–145]. In this section, we will not be concerned with the gravitino problem, nor with its possible ways out but focus on the new features SUSY brings in for leptogenesis.

The supersymmetric type-I seesaw model is described by the superpotential of the minimal supersymmetric SM (MSSM) with the additional terms:

$$W = \frac{1}{2} M_{pq} N_p^c N_q^c + \lambda_{\alpha p} N_p^c \ell_\alpha H_u, \tag{5.1}$$

where $p, q = 1, 2, \ldots$ label the heavy singlet states in order of increasing mass and $\alpha = e, \mu, \tau$ labels the lepton flavour. In (5.1) ℓ, H_u, and N^c are, respectively, the chiral superfields for the lepton and the up-type Higgs $SU(2)_L$ doublets and for the heavy $SU(2)_L$ singlet neutrinos defined according to usual conventions in terms of their LH Weyl spinor components (e.g., the N^c supermultiplet has scalar component $\widetilde{N}^*$ and fermion component N_L^c). Finally the $SU(2)_L$ index contraction is defined as $\ell_\alpha H_u = \epsilon_{\rho\sigma} \ell_\alpha^\rho H_u^\sigma$ with $\epsilon_{12} = +1$.

Originally, the issue of MSSM leptogenesis was approached in conjunction with SM leptogenesis [49, 51] as well as in dedicated studies [146, 147]. However in this first analysis, several features that are specific of SUSY in the high temperature regime relevant for leptogenesis, in which soft SUSY breaking parameters can be effectively set to zero, had been overlooked. In that case, in spite of the large amount of new reactions, the differences between SM and MSSM leptogenesis can be resumed by means of simple counting of a few numerical factors [109, 130, 148], like, for example, the number of relativistic degrees of freedom in the thermal bath, the number of loop diagrams contributing to the CP asymmetries, and the multiplicities of the final states in the decays of the heavy neutrinos and sneutrinos, and one can estimate [109]

$$\frac{Y_{\Delta B}(\infty)^{\mathrm{MSSM}}}{Y_{\Delta B}(\infty)^{\mathrm{SM}}} \approx \begin{cases} \sqrt{2} & \text{(strong washout)}, \\ 2\sqrt{2} & \text{(weak washout)}. \end{cases} \tag{5.2}$$

Recently it was pointed out in [89] that in fact MSSM leptogenesis is rich in new and nontrivial features, and genuinely different from the simpler realization within the SM. The two important effects are as follows.

(a) If the SUSY breaking scale does not exceed by much 1 TeV, above $T \sim 5 \times 10^7$ GeV the particle and superparticle density asymmetries do not equilibrate [149], and it is mandatory to account in the BE for the differences in the number density asymmetries of the boson and fermion degrees of freedom. This can be given in terms of a nonvanishing gaugino chemical potential.

(b) When soft SUSY breaking parameters are neglected, additional anomalous global symmetries that involve both $SU(2)_L$ and $SU(3)_C$ fermion representations emerge [150]. As a consequence, the EW and QCD sphaleron equilibrium conditions are modified with respect to the SM, and this yields a different pattern of sphaleron-induced lepton-flavour mixing [67, 69, 70]. In addition, a new anomaly-free exactly conserved $\mathcal{R}$-symmetry provides an additional constraint that is not present in the SM, and a careful counting reveals that *four* independent quantities, rather than the *three* of the SM case, are required to give a complete description of the various particle asymmetries in the thermal bath, with the additional quantity corresponding to the number density asymmetry of the heavy scalar neutrinos.

Although the modifications above are interesting from the theoretical point of view, a quantitative comparison with the results obtained when the new effects are ignored shows that the corrections remain below $\mathcal{O}(1)$ [89]. (This modification would be important for certain supersymmetric leptogenesis scenarios which contain new sources of CP violation e.g. soft leptogenesis (see Section 6.2).) Finally, it should be pointed out that in the supersymmetric case, the temperature in which the lepton flavour effects (see Section 3) come into play is enhanced by a factor of $(1 + \tan^2\beta)$ since the charged Yukawa couplings are given by $h_\alpha = m_\alpha/(v_u \cos\beta)$.

The purpose of the following sections is twofold. We describe the chemical equilibrium conditions and conservation laws for MSSM in conjunction to SM. In Section 5.2 we list the constraints that hold independently of assuming a regime in which particle and sparticle chemical potentials equilibrate (superequilibration (SE) regime) or do not equilibrate (nonsuperequilibration (NSE) regime). In Section 5.3 we list the constraints that hold only in the SE regime and in Section 5.4 the ones that hold in NSE regime. The question of NSE is irrelevant in the SM since there are no superparticles. Hence the parts relevant for the SM are Sections 5.2 and 5.3, with the chemical potential of the gaugino set to zero, the chemical potential of the down-type higgsino replaced by the minus of the uptype higgsino (see (5.11) and (5.12)), and all the quantities related to superparticles replaced by the ones for particles.

5.2. General Constraints

We first list in items (1), (2), and (3) the conditions that hold in the whole temperature range $M_W = T \lesssim 10^{14}$ GeV. Conversely, some of the Yukawa coupling conditions given in items (4) and (5) will have to be dropped as the temperature is increased and the corresponding reactions go out of equilibrium. First we will relate the number density asymmetry of a particle $\Delta n \equiv n - \overline{n}$ for which a chemical potential can be defined to its chemical potential. For both bosons (b) and fermions (f) this relation acquires a particularly simple form in the relativistic limit $m_{b,f} \ll T$, and at first order in the chemical potential $\mu_{b,f}/T \ll 1$:

$$\Delta n_b = \frac{g_b}{3} T^2 \mu_b, \qquad \Delta n_f = \frac{g_f}{6} T^2 \mu_f. \tag{5.3}$$

For simplicity of notations, in the following we denote the chemical potentials with the same notation that labels the corresponding field: $\phi \equiv \mu_\phi$.

(1) At scales much higher than M_W, gauge fields have vanishing chemical potential $W = B = g = 0$ [57]. This also implies that all the particles belonging to the same $SU(2)_L$ or $SU(3)_C$ multiplets have the same chemical potential. For example $\phi(I_3 = +(1/2)) = \phi(I_3 = -(1/2))$ for a field ϕ that is a doublet of weak isospin $\vec{I}$, and similarly for color.

(2) Denoting by $\widetilde{W}_R$, $\widetilde{B}_R$, and $\widetilde{g}_R$ the RH winos, binos, and gluinos chemical potentials and by ℓ, Q $(\widetilde{\ell}, \widetilde{Q})$ the chemical potentials of the (s)lepton and (s)quarks LH doublets, the following reactions: $\widetilde{Q} + \widetilde{g}_R \rightarrow Q$, $\widetilde{Q} + \widetilde{W}_R \rightarrow Q$, $\widetilde{\ell} + \widetilde{W}_R \rightarrow \ell$, $\widetilde{\ell} + \widetilde{B}_R \rightarrow \ell$, imply that all gauginos have the same chemical potential: $-\widetilde{g} = Q - \widetilde{Q} = -\widetilde{W} = \ell - \widetilde{\ell} = -\widetilde{B}$, where $\widetilde{W}$, $\widetilde{B}$, and $\widetilde{g}$ denote the chemical potentials of LH gauginos. It follows that the chemical potentials of the SM particles are related to those of their respective superpartners as

$$\begin{aligned} \widetilde{Q}, \widetilde{\ell} &= Q, \ell + \widetilde{g}, \\ H_{u,d} &= \widetilde{H}_{u,d} + \widetilde{g}, \\ \widetilde{u}, \widetilde{d}, \widetilde{e} &= u, d, e - \widetilde{g}. \end{aligned} \tag{5.4}$$

The last relation, in which $u, d, e \equiv u_R, d_R, e_R$ denote the RH $SU(2)_L$ singlets, follows, for example, from $\widetilde{u}^c_L = u^c_L + \widetilde{g}$ for the corresponding LH fields, together with $u^c_L = -u_R$, and from the analogous relation for the $SU(2)_L$ singlet squarks.

(3) Before EW symmetry breaking hypercharge is an exactly conserved quantity, and we can assume a vanishing total hypercharge for the Universe:

$$\sum_i \left(Y_{\Delta Q_i} + 2Y_{\Delta u_i} - Y_{\Delta d_i}\right) - \sum_\alpha \left(Y_{\Delta \ell_\alpha} + Y_{\Delta e_\alpha}\right) + Y_{\Delta \widetilde{H}_u} - Y_{\Delta \widetilde{H}_d} = 0. \tag{5.5}$$

(4) When the reactions mediated by the lepton Yukawa couplings are faster than the Universe expansion rate (see Section 3.1 for the temperature regime when

lepton Yukawa interactions are in thermal equilibrium), the following chemical equilibrium conditions are enforced:

$$\ell_\alpha - e_\alpha + \widetilde{H}_d + \widetilde{g} = 0, \quad (\alpha = e, \mu, \tau). \tag{5.6}$$

If the temperature is not too low lepton flavour equilibration (see Section 3.4) induced by off-diagonal slepton soft masses will not occur. We assume that this is the case, and thus we take the three ℓ_α to be independent quantities.

(5) Reactions mediated by the quark Yukawa couplings enforce the following six chemical equilibrium conditions:

$$Q_i - u_i + \widetilde{H}_u + \widetilde{g} = 0, \quad (u_i = u, c, t), \tag{5.7}$$

$$Q_i - d_i + \widetilde{H}_d + \widetilde{g} = 0, \quad (d_i = d, s, b). \tag{5.8}$$

The upquark Yukawa coupling maintains chemical equilibrium between the LH and RH uptype quarks up to $T \sim 2 \cdot 10^6$ GeV. Note that when the Yukawa reactions of at least two families of quarks are in equilibrium, the mass basis is fixed for all the quarks and squarks. Intergeneration mixing then implies that family-changing charged-current transitions are also in equilibrium: $b_L \rightarrow c_L$ and $t_L \rightarrow s_L$ imply $Q_2 = Q_3$; $s_L \rightarrow u_L$ and $c_L \rightarrow d_L$ imply $Q_1 = Q_2$. Thus, up to temperatures $T \lesssim 10^{11}$ GeV, that are of the order of the equilibration temperature for the charm Yukawa coupling, the three quark doublets have the same chemical potential:

$$Q \equiv Q_3 = Q_2 = Q_1. \tag{5.9}$$

At higher temperatures, when only the third family is in equilibrium, we have instead $Q \equiv Q_3 = Q_2 \neq Q_1$. Above $T \sim 10^{13}$ when (for moderate values of $\tan\beta$) also b-quark (as well as the τ-lepton) $SU(2)_L$ singlets decouple from their Yukawa reactions, all intergeneration mixing becomes negligible and $Q_3 \neq Q_2 \neq Q_1$.

5.3. *Superequilibration Regime*

At relatively low temperatures, additional conditions from reactions in chemical equilibrium hold. Since the constraints below apply only in the SE regime, we number them including this label.

$6_{\rm SE}$: Equilibration of the particle-sparticle chemical potentials $\mu_\phi = \mu_{\widetilde{\phi}}$ [149] is ensured when reactions like $\widetilde{\ell}\widetilde{\ell} \rightarrow \ell\ell$ are faster than the Universe expansion rate. These reactions are induced by gaugino interactions, but since they require a gaugino chirality flip, they turn out to be proportional to its mass $m_{\widetilde{g}}$ and can be neglected in the limit $m_{\widetilde{g}} \rightarrow 0$.

Furthermore, since the μ parameter of the $H_u H_d$ superpotential term is expected to be of the order of the soft gaugino masses, it is reasonable to consider in the same temperature range also the effect of the higgsino mixing term, which implies that the sum of the up- and downhiggsino chemical potentials vanishes. The rates of the corresponding reactions, given approximately by $\Gamma_{\tilde{g}} \sim m_{\tilde{g}}^2/T$ and $\Gamma_\mu \sim \mu^2/T$, are faster than the Universe expansion rate up to temperatures

$$T \lesssim 5 \cdot 10^7 \left(\frac{m_{\tilde{g}}, \mu}{500\,\text{GeV}} \right)^{2/3} \text{GeV}. \tag{5.10}$$

The corresponding chemical equilibrium relations enforce the conditions:

$$\tilde{g} = 0, \tag{5.11}$$

$$\widetilde{H}_u + \widetilde{H}_d = 0. \tag{5.12}$$

7_{SE} : Up to temperatures given by (5.10) the MSSM has the same global anomalies than the SM, that are the EW $SU(2)_L$-$U(1)_{B+L}$ mixed anomaly and the QCD chiral anomaly. They generate the effective operators $O_{\text{EW}} = \Pi_\alpha(QQQ\ell_\alpha)$ [151] and $O_{\text{QCD}} = \Pi_i(QQu^c_{\text{L}i}d^c_{\text{L}i})$ [151–153]. Above the EW phase transition reactions induced by these operators are in thermal equilibrium, and the corresponding conditions read

$$9Q + \sum_\alpha \ell_\alpha = 0, \tag{5.13}$$

$$6Q - \sum_i (u_i + d_i) = 0, \tag{5.14}$$

where we have used the same chemical potential for the three quark doublets (5.9), which is always appropriate in the SE regime below the limit (5.10).

Equations (5.6), (5.7)–(5.9), together with the SE conditions (5.11)-(5.12), the two anomaly conditions (5.13)-(5.14), and the hypercharge neutrality condition equation (5.5), give $11 + 2 + 2 + 1 = 16$ constraints for the 18 chemical potentials. Note however that there is one redundant constraint, that we take to be the QCD sphaleron condition, since by summing up (5.7) and (5.8) and taking into account (5.9), (5.11), and (5.12), we obtain precisely (5.14). Therefore, like in the SM, we have three independent chemical potentials. We can define three linear combinations of the chemical potentials corresponding to the $SU(2)_L$ anomaly free flavour charges $\Delta_\alpha \equiv B/3 - L_\alpha$ that being anomaly-free and perturbatively conserved by the low energy MSSM Lagrangian evolve *slowly* because the corresponding symmetries are violated only by the heavy Majorana neutrino dynamics. Their evolution needs to be computed by means of three independent BEs. In terms of the abundances (2.12) the density of the Δ_α charges normalized to the entropy density can be written as

$$Y_{\Delta_\alpha} = 3\left[\frac{1}{3} \sum_i (2Y_{\Delta Q_i} + Y_{\Delta u_i} + Y_{\Delta d_i}) - (2Y_{\Delta \ell_\alpha} + Y_{\Delta e_\alpha}) - \frac{2}{3} Y_{\Delta \tilde{g}} \right]. \tag{5.15}$$

The expression above is completely general and holds in all temperature regimes, including the NSE regime (see Section 5.4).

The density asymmetries of the doublet leptons and higgsinos, that weight the washout terms in the BE, can now be expressed in terms of the anomaly-free charges by means of the A matrix and C vectors introduced, respectively, in [67, 69] that are defined as:

$$Y_{\Delta \ell_\alpha} = A^{\ell}_{\alpha\beta} Y_{\Delta_\beta}, \qquad Y_{\Delta \widetilde{H}_{u,d}} = C^{\widetilde{H}_{u,d}}_{\alpha} Y_{\Delta_\alpha}. \tag{5.16}$$

Here and in the following we will give results for the A and C matrices for the fermion states. We recall that in the SE regime the density asymmetry of a scalar boson that is in chemical equilibrium with its fermionic partner is given simply by $Y_{\Delta b} = 2Y_{\Delta f}$ with the factor of 2 from statistics.

5.3.1. *First Generation Yukawa Reactions out of Equilibrium (SE Regime)*

As an example let us now consider the temperatures $T \gtrsim 4 \cdot 10^6 (1 + \tan^2\beta)$ GeV, when the d-quark Yukawa coupling can be set to zero (in order to remain within the SE regime we assume $\tan\beta \sim 1$). In this case the equilibrium dynamics is symmetric under the exchange $u \leftrightarrow d$ (both chemical potentials enter only the QCD sphaleron condition equation (5.14) with equal weights) and so must be any physical solution of the set of constraints. Thus, the first condition in (5.8) can be replaced by the condition $d = u$, and again three independent quantities suffice to determine all the particle density asymmetries. The corresponding result is

$$A^{\ell} = \frac{1}{3 \times 2148} \begin{pmatrix} -906 & 120 & 120 \\ 75 & -688 & 28 \\ 75 & 28 & -688 \end{pmatrix}, \qquad C^{\widetilde{H}_u} = -C^{\widetilde{H}_d} = \frac{-1}{2148}(37, 52, 52). \tag{5.17}$$

Note that since in this regime the chemical potentials for the scalar and fermion degrees of freedom of each chiral multiplet equilibrate, the analogous results for $Y_{\Delta \ell_\alpha} + Y_{\Delta \tilde{\ell}_\alpha}$ can be obtained by simply multiplying the A matrix in (5.17) by a factor of 3. This gives the same A matrix obtained in the nonsupersymmetric case in the same regime (see, e.g., (4.12) in [69]). The C matrix (multiplied by the same factor of 3) differs from the nonsupersymmetric result by a factor $1/2$. This is because after substituting $\widetilde{H}_d = -\widetilde{H}_u$ (see (5.12)) all the chemical potential conditions are formally the same than in the SM with $\widetilde{H}_u$ identified with the chemical potential of the scalar Higgs, but since C expresses the result for number densities, in the SM, a factor of 2 from boson statistics appears for the SM Higgs. This agrees with the analysis in [58] and is a general result that holds for SUSY within the SE regime.

5.4. Nonsuperequilibration Regime

At temperatures above the limit given in (5.10) the Universe expansion is fast enough that reactions induced by $m_{\tilde{g}}$ and μ do not occur. Setting to zero in the high temperature effective theory these two parameters has the following consequences

(i) Condition equation (5.11) has to be dropped, and gauginos acquire a nonvanishing chemical potential $\tilde{g} \neq 0$ (corresponding to the difference between the number of LH and RH helicity states). The chemical potentials of the members of the same matter supermultiplets are no more equal (nonsuperequilibration) but related as in (5.4).

(ii) Condition (5.12) also has to be dropped, and the chemical potentials of the up- and downtype Higgs and higgsinos do not necessarily sum up to zero.

(iii) The MSSM gains two new global symmetries: $m_{\tilde{g}} \longrightarrow 0$ yields a global R-symmetry, while $\mu \longrightarrow 0$ corresponds to a global symmetry of the Peccei-Quinn (PQ) type.

5.4.1. Anomalous and Nonanomalous Symmetries

Two linear combinations R_2 and R_3 of R and PQ, having, respectively, only $SU(2)_L$ and $SU(3)_C$ mixed anomalies, have been identified in [150]:

$$R_2 = R - 2PQ, \qquad R_3 = R - 3PQ. \tag{5.18}$$

The authors of [150] have also constructed the effective multifermions operators generated by the mixed anomalies:

$$\tilde{O}_{\mathrm{EW}} = \Pi_\alpha (QQQ\ell_\alpha)\widetilde{H}_u\widetilde{H}_d\widetilde{W}^4, \tag{5.19}$$

$$\tilde{O}_{\mathrm{QCD}} = \Pi_i (QQu^c d^c)_i \tilde{g}^6. \tag{5.20}$$

Given that three global symmetries B, L, and R_2 have mixed $SU(2)_L$ anomalies (but are free of $SU(3)_C$ anomalies) we can construct two anomaly-free combinations, the first one being $B-L$ which is only violated perturbatively by $N^c \ell H_u$ and the second anomaly free combination which is also an exact symmetry of the MSSM+seesaw in the NSE regime [89]:

$$\mathcal{R} = \frac{5}{3}B - L + R_2, \tag{5.21}$$

and is exactly conserved. In the $SU(3)_C$ sector, besides the chiral anomaly we now have also R_3 mixed anomalies. Thus also in this case anomaly-free combinations can be constructed, and in particular we can define one combination for each quark superfield. Assigning to the LH supermultiplets chiral charge $\chi = -1$ these combinations have the form [89]:

$$\chi_{q_L} + \kappa_{q_L} R_3, \tag{5.22}$$

where, for example, $\kappa_{u_L^c} = \kappa_{d_L^c} = 1/3$ and $\kappa_{Q_L} = 2/3$. Note that since R_3 is perturbatively conserved by the complete MSSM+seesaw Lagrangian, when the Yukawa coupling of one quark is set to zero, the corresponding charge (5.22) will be exactly conserved.

5.4.2. Constraints in the Nonsuperequilibration Regime

In the NSE regime, the conditions listed in items 6_{SE} and 7_{SE} of the previous section have to be dropped, but new conditions arise.

6_{NSE}: The conservation law for the $\mathcal{R}$ charge yields the following global neutrality condition:

$$\begin{aligned}\mathcal{R}_{\text{tot}} &= \sum_f \Delta n_f \mathcal{R}_f + \sum_b \Delta n_b \mathcal{R}_b + \Delta n_{\widetilde{N}_1} \mathcal{R}_{\widetilde{N}_1} \\ &= \frac{T^2}{6}\left(\sum_i (2Q_i - 5u_i + 4d_i) + 2\sum_\alpha (\ell_\alpha + e_\alpha) + 5\widetilde{H}_d - \widetilde{H}_u + 31\widetilde{g}\right) - \Delta n_{\widetilde{N}_1} = 0.\end{aligned} \tag{5.23}$$

The last terms in both lines of (5.23) correspond to the contribution to $\mathcal{R}$-neutrality from the lightest sneutrino asymmetry $\Delta n_{\widetilde{N}_1} = n_{\widetilde{N}_1} - n_{\widetilde{N}_1^*}$ with charge $\mathcal{R}_{\widetilde{N}_1} = -\mathcal{R}_{N^c} = -1$. Note that since in general $\widetilde{N}_1$ is not in chemical equilibrium, no chemical potential can be associated to it, and hence this constraint needs to be formulated in terms of its number density asymmetry that has to be evaluated by solving a BE for $Y_{\Delta_{\widetilde{N}}} \equiv Y_{\widetilde{N}_1} - Y_{\widetilde{N}_1^*}$ (see Section 5.5).

7_{NSE}: The operators in (5.19) induce transitions that in the NSE regime are in chemical equilibrium. This enforces the generalized EW and QCD sphaleron equilibrium conditions [150]:

$$\begin{aligned} 3\sum_i Q_i + \sum_\alpha \ell_\alpha + \widetilde{H}_u + \widetilde{H}_d + 4\widetilde{g} &= 0, \\ 2\sum_i Q_i - \sum_i (u_i + d_i) + 6\widetilde{g} &= 0, \end{aligned} \tag{5.24}$$

that replace (5.13) and (5.14).

8_{NSE}: The chiral-R_3 charges in (5.22) are anomaly-free, but clearly they are not conserved by the quark Yukawa interactions. However, when a quark supermultiplet decouples from its Yukawa interactions, an exact conservation law arises. (Note that $h_{u,d} \to 0$ implies u and d decoupling, but Q_1 decoupling is ensured only if also $h_{c,s} \to 0$.) The conservation laws corresponding to these symmetries read

$$\frac{T^2}{6}[3q_R + 6(q_R - \widetilde{g})] + \frac{1}{3}R_{3\text{tot}} = 0, \tag{5.25}$$

$$\frac{T^2}{6}2[3Q_L + 6(Q_L + \widetilde{g})] - \frac{2}{3}R_{3\text{tot}} = 0 \tag{5.26}$$

and hold for $q_R = u_i, d_i$ and $Q_L = Q_i$ in the regimes when the appropriate Yukawa reactions are negligible. Note the factor of 2 for the Q_L chiral charge in front of the first square bracket in (5.26) that is due to $SU(2)_L$ gauge multiplicity. In terms of chemical potentials and $\Delta n_{\widetilde{N}_1}$, the total R_3 charge in (5.25) and (5.26) reads

$$R_{3\text{tot}} = \frac{T^2}{6}\left(82\widetilde{g} - 3\sum_i (2Q_i + 11u_i - 4d_i) + \sum_\alpha (16\ell_\alpha + 13e_\alpha) + 16\widetilde{H}_d - 14\widetilde{H}_u\right) + \Delta n_{\widetilde{N}_1} R_{3\widetilde{N}_1}, \tag{5.27}$$

where $R_{3\widetilde{N}_1} = -1$. As regards the leptons, since they do not couple to the QCD anomaly, by setting $h_e \to 0$, a symmetry under chiral supermultiplet rotations is directly gained for the RH leptons implying $\Delta n_e + \Delta\widetilde{n}_e = 0$ and giving the condition:

$$e - \frac{2}{3}\widetilde{g} = 0. \tag{5.28}$$

No analogous condition arises for the lepton doublets relevant for leptogenesis, since by assumption they remain coupled via Yukawa couplings to the heavy N's.

In the NSE regime there are different flavour mixing matrices for the scalar and fermion components of the leptons and Higgs supermultiplets. To express more concisely all the results, it is convenient to introduce a new C vector to describe the gaugino number density asymmetry per degree of freedom in terms of the relevant charges:

$$Y_{\Delta\widetilde{g}} = C_a^{\widetilde{g}} Y_{\Delta_a}, \quad \text{with} \quad \Delta_a = (\Delta_\alpha, \Delta_{\widetilde{N}}). \tag{5.29}$$

5.4.3. First Generation Yukawa Reactions out of Equilibrium (NSE Regime)

As an example, in the temperature range between 10^8 and 10^{11} GeV, and for moderate values of $\tan\beta$, all the first generation Yukawa couplings can be set to zero. Using for u, d conditions equation (5.25) and for e condition equation (5.28) as implied by $h_{u,d}, h_e \to 0$ we obtain

$$\begin{aligned} A^\ell &= \frac{1}{9\times 162332}\begin{pmatrix} -198117 & 33987 & 33987 & -8253 \\ 26634 & -147571 & 14761 & -8055 \\ 26634 & 14761 & -147571 & -8055 \end{pmatrix}, \\ C^{\widetilde{g}} &= \frac{-11}{162332}(163, 165, 165, -255), \\ C^{\widetilde{H}_u} &= \frac{-1}{162332}(3918, 4713, 4713, 95), \\ C^{\widetilde{H}_d} &= \frac{1}{3\times 162332}(5413, 9712, 9712, -252), \end{aligned} \tag{5.30}$$

where the rows correspond to $(Y_{\Delta_e}, Y_{\Delta_\mu}, Y_{\Delta_\tau}, Y_{\Delta_{\widetilde{N}}})$. For completeness, in (5.30) we have also given the results for $C^{\widetilde{H}_d}$ even if only the uptype Higgs density asymmetry is relevant for the leptogenesis processes. Note that neglecting the contribution of $\Delta n_{\widetilde{N}_1}$ to the global charges $\mathcal{R}_{\text{tot}}$ in (5.23) and $R_{3\text{tot}}$ in (5.27) corresponds precisely to setting to zero the fourth column in all the previous matrices. Then, analogously with the SE and SM cases, within this "3-column approximation" all particle density asymmetries can be expressed just in terms of the three Y_{Δ_α} charge densities.

5.5. Supersymmetric Boltzmann Equations

In order to illustrate how the new effects described above modify the structure of the BE, here we write down a simpler expressions in which only decays and inverse decays are included: (the complete set of BE including decays, inverse decays, and scatterings with top-quark is given in the Appendix of [89].)

$$sHz\frac{dY_{N_1}}{dz} = -\left(\frac{Y_{N_1}}{Y^{\text{eq}}_{N_1}} - 1\right)\gamma_{N_1}, \tag{5.31}$$

$$sHz\frac{dY_{\widetilde{N}_+}}{dz} = -\left(\frac{Y_{\widetilde{N}_+}}{Y^{\text{eq}}_{\widetilde{N}_1}} - 2\right)\gamma_{\widetilde{N}_1}, \tag{5.32}$$

$$sHz\frac{dY_{\Delta_{\widetilde{N}}}}{dz} = -\frac{Y_{\Delta_{\widetilde{N}}}}{Y^{\text{eq}}_{\widetilde{N}_1}}\gamma_{\widetilde{N}_1} - \frac{3}{2}\,\gamma_{\widetilde{N}_1}\sum_a C^{\widetilde{g}}_a\,\frac{Y_{\Delta_a}}{Y^{\text{eq}}_\ell} + \cdots, \tag{5.33}$$

$$\begin{aligned} sHz\frac{dY_{\Delta_\alpha}}{dz} = &-\epsilon_\alpha\left[\left(\frac{Y_{N_1}}{Y^{\text{eq}}_{N_1}} - 1\right)\gamma_{N_1} + \left(\frac{Y_{\widetilde{N}_+}}{Y^{\text{eq}}_{\widetilde{N}_1}} - 2\right)\gamma_{\widetilde{N}_1}\right] \\ &+ P^0_{1\alpha}\left(\gamma_{N_1} + \frac{1}{2}\gamma_{\widetilde{N}_1}\right)\sum_a\left(A^\ell_{\alpha a} + C^{\widetilde{H}_u}_a + C^{\widetilde{g}}_a\right)\frac{Y_{\Delta_a}}{Y^{\text{eq}}_\ell}, \end{aligned} \tag{5.34}$$

where $\gamma_{\widetilde{N}_1}$ is the corresponding thermally averaged decay rate for RH sneutrino $\widetilde{N}_1$. In (5.32) we have introduced the overall sneutrino abundance $Y_{\widetilde{N}_+} = Y_{\widetilde{N}_1} + Y_{\widetilde{N}^*_1}$, while $Y_{\Delta_{\widetilde{N}}} \equiv Y_{\widetilde{N}_1} - Y_{\widetilde{N}^*_1}$ in (5.33) is the sneutrino density asymmetry that was already introduced in Section 5.4. In the washout terms we have normalized the charge densities $Y_{\Delta_a} = (Y_{\Delta_\alpha}, Y_{\Delta_{\widetilde{N}}})$ to the equilibrium density of a fermion with one degree of freedom Y^{eq}_ℓ. In (5.31) and (5.34) we have also neglected for simplicity all finite temperature effects. Taking these effects into account would imply, for example, that the CP asymmetry for $\widetilde{N}$ decays into fermions is different from the one for decays into scalars [49], while we describe both CP asymmetries with ϵ_α. A few remarks regarding (5.33) are in order. In the SE regime $\widetilde{g} = 0$, and thus it would seem that the sneutrino density asymmetry $Y_{\Delta_{\widetilde{N}}}$ vanishes. However, this only happens for decays and inverse decays, and it is no more true when additional terms related to scattering processes, that are represented in the equation by the dots, are also included (see [147] and the Appendix of [89]). Therefore, also in the SE regime $Y_{\widetilde{N}_1}$ and $Y_{\widetilde{N}^*_1}$ in general differ. However, in this case recasting their equations in terms of two equations for $Y_{\widetilde{N}_+}$ and $Y_{\Delta_{\widetilde{N}}}$ is just a convenient parametrization. On the contrary, in the NSE regime this is mandatory, because the sneutrinos carry a globally conserved $\mathcal{R}$-charge and $Y_{\Delta_{\widetilde{N}}}$ is required to formulate properly the corresponding conservation law. As we have seen, this eventually results in $Y_{\Delta_{\widetilde{N}}}$

contributing to the expressions of the lepton flavour density asymmetries in terms of slowly varying quantities.

In [89] a complete numerical analysis was carried out, and it was shown that numerical corrections with respect to the case when NSE effects are neglected remain at the $\mathcal{O}(1)$ level. This is because only spectator processes get affected, while the overall amount of CP violation driving leptogenesis remains the same than in previous treatments.

6. Beyond Type-I Seesaw and Beyond the Seesaw

There exist many variants of leptogenesis models beyond the standard type-I seesaw. In this section, we try to classify them into appropriate groups. Unavoidably there would be some overlap; that is, a hybrid model of leptogenesis which can belong to more than one group, for example, soft leptogenesis (Section 6.2) from resonantly enhanced CP asymmetry could rightly fall under resonant leptogenesis (Section 6.1). However we try our best to categorize them according to the main features of the model, and, when appropriate, they will be quoted in more than one place. Clearly, the number of beyond type-I seesaw leptogenesis models is quite large. We have not attempted in any way to be exhaustive, and we apologize in advance for the unavoidable several omissions.

6.1. Resonant Leptogenesis

A resonant enhancement of the CP asymmetry in N_1 decay occurs when the mass difference between N_1 and N_2 is of the order of the decay widths. Such a scenario has been termed "resonant leptogenesis," and has benefited from many studies in different formalisms (see [154] for a comparison of the different calculations) [50, 52, 75, 85–87, 110, 111, 155–168] (see [169] for a recent review). The resonant effect is related to the self-energy contribution to the CP asymmetry. Consider, for simplicity, the case where only N_2 is quasidegenerate with N_1. Then, the self-energy contribution involving the intermediate N_2 to the total CP asymmetry (we neglect important flavour effects [75]) is given by

$$\epsilon_1(\text{self} - \text{energy}) = -\frac{M_1}{M_2}\frac{\Gamma_{N_2}}{M_2}\frac{M_2^2(M_2^2 - M_1^2)}{(M_2^2 - M_1^2)^2 + M_1^2\Gamma_{N_2}^2}\frac{\text{Im}\left[(\lambda^\dagger\lambda)_{12}^2\right]}{(\lambda^\dagger\lambda)_{11}(\lambda^\dagger\lambda)_{22}}. \tag{6.1}$$

The resonance condition reads

$$|M_2 - M_1| = \frac{\Gamma_{N_2}}{2}. \tag{6.2}$$

In this case

$$|\epsilon_1(\text{resonance})| \simeq \frac{1}{2}\frac{\left|\text{Im}\left[(\lambda^\dagger\lambda)_{12}^2\right]\right|}{(\lambda^\dagger\lambda)_{11}(\lambda^\dagger\lambda)_{22}}. \tag{6.3}$$

Thus, in the resonant case, the asymmetry is suppressed by neither the smallness of the light neutrino masses, nor the smallness of their mass splitting, nor small ratios between the singlet neutrino masses. Actually, the CP asymmetry could be of order one (more accurately, $|\epsilon_1| \leq 1/2$).

With resonant leptogenesis, the BEs are different. The densities of N_1 and N_2 are followed, since both contribute to the asymmetry, and the relevant timescales are different. For instance, the typical timescale to build up coherently the CP asymmetry is unusually long, of order $1/\Delta M$. In particular, it can be larger than the timescale for the change of the abundance of the sterile neutrinos. This situation implies that for resonant leptogenesis quantum effects in the BE can be significant [85, 86, 141, 170] (see Section 4.5).

The fact that the asymmetry could be large, independently of the singlet neutrino masses, opens up the possibility of low scale resonant leptogenesis. Models along these lines have been constructed in [75, 156, 161, 171]. It is a theoretical challenge to construct models where a mass splitting as small as the decay width is naturally achieved. For attempts that utilize approximate flavour symmetries see, for example, [157, 158, 160, 164, 166, 172], while studies of this issue in the framework of minimal flavour violation can be found in [87, 162, 163]. The possibility of observing resonant CP violation due to heavy RH neutrinos at the LHC was studied in [173, 174].

6.2. Soft Leptogenesis

The modifications to standard type-I leptogenesis due to SUSY have been discussed in Section 5. The important parameters are the Yukawa couplings and the singlet neutrino parameters, which appear in the superpotential equation (5.1). SUSY must, however, be broken. In the framework of the MSSM extended to include singlet neutrinos (MSSM+N), there are, in addition to the soft SUSY breaking terms of the MSSM, terms that involve the singlet sneutrinos $\widetilde{N}_i$, in particular bilinear (B) and trilinear (A) scalar couplings. These terms provide additional sources of lepton number violation and of CP violation. Scenarios where these terms play a dominant role in leptogenesis have been termed "soft leptogenesis" [102, 107, 175–196] (see also [54] for a recent review).

Soft leptogenesis can take place even with a single RH neutrino because the presence of the B term implies that $\widetilde{N}$ and $\widetilde{N}^\dagger$ states mix to form two mass eigenstates with mass splitting proportional to B itself. Furthermore when $B \sim \Gamma_{\widetilde{N}}$, the CP asymmetry is resonantly enhanced realizing the resonant leptogenesis scenario (see Section 6.1). In the following we will consider a single generation MSSM+N. The relevant soft SUSY terms involving $\widetilde{N}$, the $SU(2)_L$ gauginos $\widetilde{\lambda}_2^{\pm,0}$, the $U(1)_Y$ gauginos $\widetilde{\lambda}_1$, and the three sleptons $\widetilde{\ell}_\alpha$ in the basis in which charged lepton Yukawa couplings are diagonal are given by

$$\begin{aligned} -\mathcal{L}_{\text{soft}} &= \widetilde{M}^2\widetilde{N}^*\widetilde{N} + \left(AY_\alpha\epsilon_{ab}\widetilde{N}\widetilde{\ell}_\alpha^a H_u^b + \frac{1}{2}BM\widetilde{N}\widetilde{N} + \text{h.c.}\right) \\ &\quad + \frac{1}{2}\left(m_2\overline{\widetilde{\lambda}_2^{\pm,0}}P_L\widetilde{\lambda}_2^{\pm,0} + m_1\overline{\widetilde{\lambda}_1}P_L\widetilde{\lambda}_1 + \text{h.c.}\right), \end{aligned} \tag{6.4}$$

where for simplicity, proportionality of the bilinear and trilinear soft breaking terms to the corresponding SUSY invariant couplings has been assumed: $B_M = BM$ and $A_\alpha = AY_\alpha$. The Lagrangian derived from (5.1) and (6.4) is characterized by only three independent physical phases: $\phi_A \equiv \arg(AB^*)$, $\phi_{g_2} \equiv (1/2)\arg(Bm_2^*)$ and $\phi_{g_Y} \equiv (1/2)\arg(Bm_1^*)$, which can be assigned to A and to the gaugino coupling operators g_2, g_Y, respectively. As mentioned earlier, a crucial role in soft leptogenesis is played by the $\widetilde{N} - \widetilde{N}^{\dagger}$ mixing to form the mass eigenstates

$$\widetilde{N}_{+} = \frac{1}{\sqrt{2}}\left(e^{i\Phi/2}\widetilde{N} + e^{-i\Phi/2}\widetilde{N}^{*}\right),$$
$$\widetilde{N}_{-} = -\frac{i}{\sqrt{2}}\left(e^{i\Phi/2}\widetilde{N} - e^{-i\Phi/2}\widetilde{N}^{*}\right), \tag{6.5}$$

where $\Phi \equiv \arg(BM)$ and the corresponding mass eigenvalues are $M_{\pm}^{2} = M^{2} + \widetilde{M}^{2} \pm |BM|$. Without loss of generality, we can set $\Phi = 0$, which is equivalent to assigning the phases only to A and Y_{α}.

It has been pointed out that the CP asymmetries for the decays of $\widetilde{N}_{\pm}$ into scalars and fermions have opposite sign and cancel each other at the leading order [176, 177, 192], resulting in a strongly suppressed total CP asymmetry $\sim \mathcal{O}(m_{\text{soft}}^{3}/M^{3})$ where m_{soft} is the scale of soft SUSY breaking terms. There are two possibilities that can rescue leptogenesis. Firstly, thermal effects which break SUSY can spoil this cancellation [176, 177, 192]. Secondly, nonsuperequilibration effects (see Section 5.4) which imply that lepton and slepton asymmetries differ can also spoil this cancellation [107]. The CP asymmetries for the decays of $\widetilde{N}_{\pm}$ into scalars and fermions are, respectively, given by

$$\epsilon_{\alpha}^{s}(T) = \overline{\epsilon}_{\alpha}\Delta^{s}(T),$$
$$\epsilon_{\alpha}^{f}(T) = -\overline{\epsilon}_{\alpha}\Delta^{f}(T), \tag{6.6}$$

where $\overline{\epsilon}_{\alpha}$ is the temperature independent term of $\sim\mathcal{O}(m_{\text{soft}}/M)$ which contains contributions from the self-energy correction, vertex correction, and the interference between the two. In the limit $T \longrightarrow 0$, we have $\Delta^{s}(T), \Delta^{f}(T) \longrightarrow 1/2$, and thus the inclusion of thermal effects and/or nonsuperequilibration is mandatory to avoid the cancellation between the asymmetries into scalars and fermions.

We can make a rough estimate of the scale relevant for soft leptogenesis by requiring $|\overline{\epsilon}| \sim m_{\text{soft}}/M \gtrsim 10^{-6}$ which gives $M \lesssim 10^{9}\,\text{GeV}$ for $m_{\text{soft}} \sim 1\,\text{TeV}$. Hence soft leptogenesis always happens in the temperature regime where lepton flavour effects are relevant [190]. In general, the CP asymmetry from self-energy contribution requires $B \ll m_{\text{soft}}$ to be resonantly enhanced. However it was shown in [195] that flavour effects can greatly enhance the efficiency, and eventually $B \sim m_{\text{soft}}$ is allowed. The nice feature of soft leptogenesis is that the tension with the gravitino problem gets generically relaxed and, in the lower temperature window, is completely avoided.

6.3. Dirac Leptogenesis

The extension of the SM with singlet neutrinos allows for two different ways for generating tiny neutrino masses. The first one is the seesaw mechanism which has at least three attractive features

(i) No extra symmetries (and, in particular, no global symmetries) have to be imposed.

(ii) The extreme lightness of neutrino masses is linked to the existence of a high scale of new physics, which is well motivated for various other reasons (e.g., gauge unification).

(iii) Lepton number is violated, which opens the way to leptogenesis.

The second way is to impose lepton number and give to the neutrinos Dirac masses. A priori, one might think that all three attractive features of the seesaw mechanism are lost. Indeed, one must usually impose additional symmetries. But one can still construct natural models where the tiny Yukawa couplings that are necessary for small Dirac masses are related to a small breaking of a symmetry. What is perhaps most surprising is the fact that leptogenesis could proceed successfully even if neutrinos are Dirac particles, and lepton number is not (perturbatively) broken [197, 198]. Such scenarios have been termed "Dirac leptogenesis" [198–207].

An implementation of the idea is the following. A CP-violating decay of a heavy particle can result in a nonzero lepton number for LH particles and an equal and opposite nonzero lepton number for RH particles, so that the total lepton number is zero. For the charged fermions of the SM, the Yukawa interactions are fast enough that they quickly equilibrate the LH and the RH particles, and the lepton number stored in each chirality goes to zero. This is not true, however, for Dirac neutrinos. The size of their Yukawa couplings is $\lambda \lesssim 10^{-11}$, which means that equilibrium between the lepton numbers stored in LH and RH neutrinos will not be reached until the temperature falls well below the electroweak breaking scale. To see this, note that the rate of the Yukawa interactions, given by $\Gamma_\lambda \sim \lambda^2 T$, becomes significant when it equals the expansion rate of the Universe, $H \sim T^2/m_{\rm pl}$. Thus, the temperature of equilibration between LH and RH neutrinos is $T \sim \lambda^2 m_{\rm pl} \sim (\lambda/10^{-11})^2$ MeV, that is, well below the temperature when sphalerons, after having converted part of the LH lepton asymmetry into a net baryon asymmetry, are switched off.

A specific example of a supersymmetric model where Dirac neutrinos arise naturally is presented in [199]. The Majorana masses of the N-superfields are forbidden by $U(1)_L$. The neutrino Yukawa couplings are forbidden by a $U(1)_N$ symmetry where, among all the MSSM+N fields, only the N superfields are charged. The symmetry is spontaneously broken by the vacuum expectation value of a scalar field χ that can naturally be at the weak scale, $\langle\chi\rangle \sim v_u$. This breaking is communicated to the MSSM+N via extra, vector-like lepton doublet fields, $\phi + \overline{\phi}$, that have masses M_ϕ much larger than v_u. Consequently, the neutrino Yukawa couplings are suppressed by the small ratio $\langle\chi\rangle/M_\phi$. The CP violation arises in the decays of the vector-like leptons, whereby $\Gamma(\phi \to NH_u^c) \neq \Gamma(\overline{\phi} \to N^c H_u)$ and $\Gamma(\phi \to L\chi) \neq \Gamma(\overline{\phi} \to L^c\chi^c)$. The resulting asymmetries in N and in L are equal in magnitude and opposite in sign. Finally note that the main phenomenological implication of Dirac leptogenesis is the absence of any signal in neutrinoless double beta decays.

6.4. Triplet Scalar (Type-II) Leptogenesis

One can generate seesaw masses for the light neutrinos by tree-level exchange of $SU(2)_L$-triplet scalars T [44, 208–211]. The relevant new terms in the Lagrangian are

$$\mathcal{L}_T = -M_T^2|T|^2 + \frac{1}{2}\Big([\lambda_L]_{\alpha\beta}\ \ell_\alpha\ell_\beta T + M_T\lambda_\phi\ \phi\phi T^* + \text{h.c.}\Big). \tag{6.7}$$

Here, M_T is a real mass parameter, λ_L is a symmetric 3×3 matrix of dimensionless, complex Yukawa couplings, and λ_ϕ is a dimensionless complex coupling. Since this mechanism necessarily involves lepton number violation and allows for new CP-violating phases, it is interesting to examine it in the light of leptogenesis [81, 180, 185, 212–233]. One obvious problem in this scenario is that, unlike singlet fermions, the triplet scalars have gauge interactions that keep them close to thermal equilibrium at temperatures $T \lesssim 10^{15}$ GeV. It turns out, however, that successful leptogenesis is possible even at a much lower temperature. This subsection is based in large part on [220] where further details and, in particular, an explicit presentation of the relevant BE can be found.

The CP asymmetry that is induced by the triplet scalar decays is defined as follows:

$$\epsilon_T \equiv 2\frac{\Gamma\left(\overline{T} \longrightarrow \ell\ell\right) - \Gamma\left(T \longrightarrow \overline{\ell}\,\overline{\ell}\right)}{\Gamma_T + \Gamma_{\overline{T}}}, \tag{6.8}$$

where the overall factor of 2 comes because the triplet scalar decay produces two (anti)leptons.

To calculate ϵ_T, one should use the Lagrangian in (6.7). While a single triplet is enough to produce three light massive neutrinos, there is a problem in leptogenesis if indeed this is the only source of neutrinos masses. The asymmetry is generated only at higher loops and is unacceptably small. It is still possible to produce the required lepton asymmetry from a single triplet scalar decays if there are additional sources for neutrino masses, such as type I, type III, or type II contributions from additional triplet scalars. Define $m_{\rm II}$ ($m_{\rm I}$) as the part of the light neutrino mass matrix that comes (does not come) from the contributions of the triplet scalar responsible for ϵ_T:

$$m = m_{\rm II} + m_{\rm I}. \tag{6.9}$$

Then, assuming that the particles exchanged to produce $m_{\rm I}$ are all heavier than T, we obtain the CP asymmetry

$$\epsilon_T = \frac{1}{4\pi}\frac{M_T}{v_u^2}\sqrt{B_L B_H}\,\frac{\mathrm{Im}\left[\mathrm{Tr}\left(m_{\rm II}^\dagger m_{\rm I}\right)\right]}{\mathrm{Tr}\left(m_{\rm II}^\dagger m_{\rm II}\right)}, \tag{6.10}$$

where $B_L(B_H)$ is the tree-level branching ratio to leptons (Higgs doublets). If these are the only decay modes, that is, $B_L + B_H = 1$, then $B_L/B_H = \mathrm{Tr}(\lambda_L\lambda_L^\dagger)/(\lambda_H\lambda_H^\dagger)$, and there is an upper bound on the asymmetry

$$|\epsilon_T| \leq \frac{1}{4\pi}\frac{M_T}{v_u^2}\sqrt{B_L B_H \sum_i m_{\nu_i}^2}. \tag{6.11}$$

Note that, unlike the singlet fermion case, $|\epsilon_T|$ increases with larger m_{ν_i}.

As concerns the efficiency factor, it can be close to maximal, $\eta \sim 1$, in spite of the fact that the gauge interactions tend to maintain the triplet abundance very close to thermal equilibrium. There are two necessary conditions that have to be fulfilled by the decay rates $T \to \overline{\ell}\,\overline{\ell}$ and $T \to \phi\phi$ in order that this will happen [220]

(1) One of the two decay rates is faster than the $T\overline{T}$ annihilation rate.

(2) The other decay mode is slower than the expansion rate of the Universe.

The first condition guarantees that gauge scatterings are ineffective: the triplets decay before annihilating. The second condition guarantees that the fast decays do not wash out strongly the lepton asymmetry: lepton number is violated only by the simultaneous presence of $T \to \overline{\ell}\,\overline{\ell}$ and $T \to \phi\phi$.

Combining a calculation of η with the upper bound on the CP asymmetry (6.11), successful leptogenesis implies a lower bound on the triplet mass M_T varying between 10^9 GeV and 10^{12} GeV, depending on the relative weight of $m_{\rm II}$ and $m_{\rm I}$ in the light neutrino mass.

Interestingly, in the supersymmetric framework, "soft leptogenesis" (see Section 6.2) can be successful even with the minimal set of extra fields—a single $T + \overline{T}$—that generates both neutrino masses and the lepton asymmetry [180, 185].

6.5. Triplet Fermion (Type-III) Leptogenesis

One can generate neutrino masses by the tree-level exchange of $SU(2)_L$-triplet fermions T_i^a [234–236] (i denotes a heavy mass eigenstate while a is an $SU(2)_L$ index) with the Lagrangian

$$\mathcal{L}_{T^a} = [\lambda_T]_{\alpha k} \tau^a_{\rho\sigma} \ell^\rho_\alpha \phi^\sigma T^a_k - \frac{1}{2} M_i T^a_i T^a_i + \text{h.c.} \tag{6.12}$$

Here τ^a are the Pauli matrices, M_i are real mass parameters, and λ_T is a 3×3 matrix of complex Yukawa couplings.

This mechanism necessarily involves lepton number violation and allows for new CP-violating phases so we should examine it as a possible source of leptogenesis [228, 237–243]. This subsection is based in large part on [238] where further details and the relevant BE can be found.

As concerns neutrino masses, all the qualitative features are very similar to the singlet fermion case. As concerns leptogenesis there are, however, qualitative and quantitative differences. With regard to the CP asymmetry from the lightest triplet fermion decay, the relative sign between the vertex and self-energy loop contributions is opposite to that of the singlet fermion case. Consequently, in the limit of strong hierarchy in the heavy fermion masses, the asymmetry in triplet decays is three times smaller than in the decays of the singlets. On the other hand, since the triplet has three components, the ratio between the final baryon asymmetry and $\epsilon\eta$ is three times bigger. The decay rate of the heavy fermion is the same in both cases. This, however, means that the thermally averaged decay rate is three times bigger for the triplet, as is the on-shell part of the $\Delta L = 2$ scattering rate.

A significant qualitative difference arises from the fact that the triplet has gauge interactions. The effect on the washout factor η is particularly significant for $\tilde{m} \ll 10^{-3}$ eV, the so-called weak washout regime (note that this name is inappropriate for triplet fermions). The gauge interactions still drive the triplet abundance close to thermal equilibrium. A relic fraction of the triplet fermions survives. The decays of these relic triplets produce a baryon asymmetry, with

$$\eta \approx M_1/10^{13}\,\text{GeV} \quad \left(\text{for } \tilde{m} \ll 10^{-3}\,\text{eV}\right). \tag{6.13}$$

The strong dependence on M_1 results from the fact that the expansion rate of the Universe is slower at lower temperatures. On the other hand, for $\tilde{m} \gg 10^{-3}$ eV, the Yukawa interactions keep the heavy fermion abundance close to thermal equilibrium, so the difference in η between the singlet and triplet case is only $\mathcal{O}(1)$. Ignoring flavour effects, and assuming strong hierarchy between the heavy fermions, [216] obtained the lower bound

$$M_1 \gtrsim 1.5 \times 10^{10}\,\text{GeV}. \tag{6.14}$$

When the triplet fermion scenario is incorporated in a supersymmetric framework, and the soft breaking terms do not play a significant role, the modifications to the previous analysis are by factors of $\mathcal{O}(1)$.

7. Conclusions

During the last few decades, a large set of experiments involving solar, atmospheric, reactor, and accelerator neutrinos have converged to establish that the neutrinos are massive. The seesaw mechanism extends the Standard Model in a way that allows neutrino masses, and it provides a nice explanation of the suppression of the neutrino masses with respect to the electroweak breaking scale. Furthermore, without any addition or modification, it can also account for the observed baryon asymmetry of the Universe. The possibility of giving an explanation of two apparently unrelated experimental facts—neutrino masses and the baryon asymmetry—within a single framework that is a natural extension of the Standard Model, together with the remarkable "coincidence" that the same neutrino mass scale suggested by neutrino oscillation data is also optimal for leptogenesis, makes the idea that baryogenesis occurs through leptogenesis a very attractive one.

Leptogenesis can be quantitatively successful without any fine tuning of the seesaw parameters. Yet, in the nonsupersymmetric seesaw framework, a fine-tuning problem arises due to the large corrections to the mass-squared parameter of the Higgs potential that are proportional to the heavy Majorana neutrino masses. Supersymmetry can cure this problem, avoiding the necessity of fine tuning; however, it brings in the gravitino problem [144] that requires a low reheat temperature after inflation, in conflict with generic leptogenesis models. Thus, constructing a fully satisfactory theoretical framework that implements leptogenesis within the seesaw framework is not a straightforward task.

From the experimental side, the obvious question to ask is if it is possible to test whether the baryon asymmetry has been really produced through leptogenesis. Unfortunately it seems impossible that any direct test can be performed. To establish leptogenesis experimentally, we need to produce the heavy Majorana neutrinos and measure the CP asymmetry in their decays. However, in the most natural seesaw scenarios, these states are simply too heavy to be produced, while if they are light, then their Yukawa couplings must be very tiny, again preventing any chance of direct measurements.

Lacking the possibility of a direct proof, experiments can still provide circumstantial evidence in support of leptogenesis by establishing that (some of) the Sakharov conditions for leptogenesis are realized in nature. Planned neutrinoless double beta decay ($0\nu\beta\beta$) experiments (GERDA [244], MAJORANA [245], and CUORE [246]) aim at a sensitivity to the effective $0\nu\beta\beta$ neutrino mass in the few $\times 10$ meV range. If they succeed in establishing the Majorana nature of the light neutrinos, this will strengthen our confidence that the seesaw mechanism is at the origin of the neutrino masses and, most importantly, will establish that the first Sakharov condition for the dynamical generation of a lepton asymmetry (L violation) is realized. Proposed SuperBeam facilities [247, 248] and second generation off-axis SuperBeam experiments (T2HK [249] and NOνA [250]) can discover CP violation in the leptonic sector. These experiments can only probe the Dirac phase of the neutrino mixing matrix. They cannot probe the Majorana low energy or the high energy phases, but the important point is that they can establish that the second Sakharov condition for the dynamical generation of a lepton asymmetry is satisfied. As regards the third condition, that is, that the heavy neutrino decays occurred out of thermal equilibrium, it might seem the most difficult one to test experimentally. In reality the opposite is true, and in fact we already know that an absolute neutrino mass scale of the order of the solar or atmospheric mass differences is perfectly compatible with sufficiently out of equilibrium heavy neutrinos decays.

Given that we do not know how to prove that leptogenesis is the correct theory, we might ask if there is any chance to falsify it. Indeed, future neutrino experiments could weaken the case for leptogenesis, or even falsify it, mainly by establishing that the seesaw mechanism is not responsible for the observed neutrino masses. By itself, failure in revealing signals of $0\nu\beta\beta$ decays will not disprove leptogenesis. Indeed, with normal neutrino mass hierarchy one expects that the rates of lepton-number-violating processes are below experimental sensitivity. However, if neutrinos masses are quasidegenerate or inversely hierarchical, and future measurements of the oscillation parameters will not fluctuate too much away from the present best fit values, the most sensitive $0\nu\beta\beta$ decay experiments scheduled for the near future should be able to detect a signal [251]. If instead the limit on $|m_{\beta\beta}|$ is pushed below ~10 meV (a quite challenging task), this would suggest that either the mass hierarchy is normal or neutrinos are not Majorana particles. The latter possibility would disprove the seesaw model and standard leptogenesis. Thus, determining the order of the neutrino mass spectrum is extremely important to shed light on the connection between $0\nu\beta\beta$ decay experiments and leptogenesis. In summary, if it is established that the neutrino mass hierarchy is inverted and at the same time no signal of $0\nu\beta\beta$ decays is detected at a level $|m_{ee}| \lesssim 10$ meV, one could conclude that the seesaw is not at the origin of the neutrino masses and that (standard) leptogenesis is not the correct explanation of the baryon asymmetry. As concerns CP violation, a failure in detecting leptonic CP violation will not weaken the case for leptogenesis in a significant way. Instead, it would mean that the Dirac CP phase is small enough to render L-conserving CP-violating effects unobservable.

Finally, the CERN LHC has the capability of providing information that is relevant to leptogenesis, since it can play a fundamental role in establishing that the origin of the neutrino masses is not due to the seesaw mechanism, thus leaving no strong motivation for leptogenesis. This may happen in several different ways. For example (assuming that the related new physics is discovered), the LHC will be able to test if the detailed phenomenology of any of the following models is compatible with an explanation of the observed pattern of neutrino masses and mixing angles: supersymmetric R-parity violating couplings and/or L-violating bilinear terms [252, 253]; leptoquarks [254, 255]; triplet Higgs [256, 257]; new scalar particles of the type predicted in the Zee-Babu [258, 259] types of models [260–262]. It is conceivable that such discoveries can eventually exclude the seesaw mechanism and rule out leptogenesis.

References

[1] G. Steigman, "Observational tests of antimatter cosmologies," *Annual Review of Astronomy and Astrophysics*, vol. 14, pp. 339–372, 1976.

[2] A. G. Cohen, A. De Rújula, and S. L. Glashow, "A matter-antimatter universe?" *Astrophysical Journal Letters*, vol. 495, no. 2, pp. 539–549, 1998.

[3] A. D. Dolgov, "Non-GUT baryogenesis," *Physics Report*, vol. 222, no. 6, pp. 309–386, 1992.

[4] F. Iocco, G. Mangano, G. Miele, O. Pisanti, and P. D. Serpico, "Primordial nucleosynthesis: from precision cosmology to fundamental physics," *Physics Reports*, vol. 472, no. 1–6, pp. 1–76, 2009.

[5] G. Steigman, "Primordial nucleosynthesis in the precision cosmology era," *Annual Review of Nuclear and Particle Science*, vol. 57, pp. 463–491, 2007.

[6] K. Nakamura et al., "Review of particle physics," *Journal of Physics G*, vol. 37, no. 7, Article ID 075021, 2010.

[7] G. Steigman, "Primordial nucleosynthesis: successes and challenges," *International Journal of Modern Physics E*, vol. 15, no. 1, pp. 1–36, 2006.

[8] R. H. Cyburt, B. D. Fields, K. A. Olive, and E. Skillman, "New BBN limits on physics beyond the standard model from 4He," *Astroparticle Physics*, vol. 23, no. 3, pp. 313–323, 2005.

[9] K. A. Olive, G. Steigman, and T. P. Walker, "Primordial nucleosynthesis: theory and observations," *Physics Report*, vol. 333-334, no. 4–6, pp. 389–407, 2000.

[10] W. Hu and S. Dodelson, "Cosmic microwave background anisotropies," *Annual Review of Astronomy and Astrophysics*, vol. 40, pp. 171–216, 2002.

[11] S. Dodelson, *Modern Cosmology*, Academic Press, Amsterdam, Netherlands, 2003.

[12] D. Larson, J. Dunkley, G. Hinshaw et al., "Seven-year wilkinson microwave anisotropy probe (WMAP) observations: power spectra and WMAP-derived parameters," *Astrophysical Journal, Supplement Series*, vol. 192, no. 2, article 16, 2011.

[13] A. D. Sakharov, "Violation of CP invariance, C asymmetry, and baryon asymmetry of the universe," *Zhurnal Eksperimental'noi i Teoreticheskoi Fiziki*, vol. 5, pp. 32–35, 1967.

[14] G. Hooft, "Symmetry breaking through Bell-Jackiw anomalies," *Physical Review Letters*, vol. 37, no. 1, pp. 8–11, 1976.

[15] V. A. Kuzmin, V. A. Rubakov, and M. E. Shaposhnikov, "On anomalous electroweak baryon-number non-conservation in the early universe," *Physics Letters B*, vol. 155, no. 1-2, pp. 36–42, 1985.

[16] M. Kobayashi and T. Maskawa, "CP violation in the renormalizable theory of weak interaction," *Progress of Theoretical Physics*, vol. 49, pp. 652–657, 1973.

[17] C. Jarlskog, "Commutator of the quark mass matrices in the standard electroweak model and a measure of maximal CP nonconservation," *Physical Review Letters*, vol. 55, no. 10, pp. 1039–1042, 1985.

[18] M. B. Gavela, M. Lozano, J. Orloff, and O. Pène, "Standard model CP-violation and baryon asymmetry (I). Zero temperature," *Nuclear Physics B*, vol. 430, no. 2, pp. 345–381, 1994.

[19] M. B. Gavela, P. Hernandez, J. Orloff, O. Pène, and C. Quimbay, "Standard model CP-violation and baryon asymmetry (II). Finite temperature," *Nuclear Physics B*, vol. 430, no. 2, pp. 382–426, 1994.

[20] P. Huet and E. Sather, "Electroweak baryogenesis and standard model CP violation," *Physical Review D*, vol. 51, no. 2, pp. 379–394, 1995.

[21] V. A. Rubakov and M. E. Shaposhnikov, "Electroweak baryon number non-conservation in the early Universe and in high-energy collisions," *Uspekhi Fizicheskikh Nauk*, vol. 166, no. 5, pp. 534–537, 1996.

[22] M. Trodden, "Electroweak baryogenesis," *Reviews of Modern Physics*, vol. 71, no. 5, pp. 1463–1500, 1999.

[23] K. Kajantie, M. Laine, K. Rummukainen, and M. Shaposhnikov, "The electroweak phase transition: a non-perturbative analysis," *Nuclear Physics B*, vol. 466, no. 1-2, pp. 189–258, 1996.

[24] A. Y. Ignatiev, N. V. Krasnikov, V. A. Kuzmin, and A. N. Tavkhelidze, "Universal CP-noninvariant superweak interaction and bayron asymmetry of the universe," *Physics Letters B*, vol. 76, no. 4, pp. 436–438, 1978.

[25] M. Yoshimura, "Unified gauge theories and the baryon number of the universe," *Physical Review Letters*, vol. 41, no. 5, pp. 281–284, 1978.
[26] D. Toussaint, S. B. Treiman, F. Wilczek, and A. Zee, "Matter-antimatter accounting, thermodynamics, and black-hole radiation," *Physical Review D*, vol. 19, no. 4, pp. 1036–1045, 1979.
[27] S. Dimopoulos and L. Susskind, "Baryon number of the universe," *Physical Review D*, vol. 18, no. 12, pp. 4500–4509, 1978.
[28] J. R. Ellis, M. K. Gaillard, and D. V. Nanopoulos, "Baryon number generation in grand unified theories," *Physics Letters B*, vol. 80, no. 4-5, pp. 360–364, 1979.
[29] S. Weinberg, "Cosmological production of baryons," *Physical Review Letters*, vol. 42, no. 13, pp. 850–853, 1979.
[30] M. Yoshimura, "Origin of cosmological baryon asymmetry," *Physics Letters B*, vol. 88, no. 3-4, pp. 294–298, 1979.
[31] S. M. Barr, G. Segrè, and H. A. Weldon, "Magnitude of the cosmological baryon asymmetry," *The Physical Review D*, vol. 20, no. 10, pp. 2494–2498, 1979.
[32] D. V. Nanopoulos and S. Weinberg, "Mechanisms for cosmological baryon production," *Physical Review D*, vol. 20, no. 10, pp. 2484–2493, 1979.
[33] A. Yildiz and P. H. Cox, "Net baryon number and CP nonconservation with unified fields," *Physical Review D*, vol. 21, no. 4, pp. 906–909, 1980.
[34] A. Riotto and M. Trodden, "Recent progress in baryogenesis," *Annual Review of Nuclear and Particle Science*, vol. 49, no. 1, pp. 35–75, 1999.
[35] J. M. Cline, "Baryogenesis," http://arxiv.org/abs/hep-ph/0609145.
[36] I. Affleck and M. Dine, "A new mechanism for baryogenesis," *Nuclear Physics B*, vol. 249, no. 2, pp. 361–380, 1985.
[37] M. Dine, L. Randall, and S. D. Thomas, "Baryogenesis from flat directions of the supersymmetric standard model," *Nuclear Physics B*, vol. 458, no. 1-2, pp. 291–326, 1996.
[38] A. G. Cohen and D. B. Kaplan, "Thermodynamic generation of the baryon asymmetry," *Physics Letters B*, vol. 199, no. 2, pp. 251–258, 1987.
[39] A. G. Cohen and D. B. Kaplan, "Spontaneous baryogenesis," *Nuclear Physics B*, vol. 308, no. 4, pp. 913–928, 1988.
[40] P. Minkowski, "$\mu \rightarrow e\gamma$ at a rate of one out of 109 muon decays?" *Physics Letters B*, vol. 67, no. 4, pp. 421–428, 1977.
[41] T. Yanagida, "Horizontal gauge symmetry and masses of neutrinos," in *Proceedings of the Workshop on the Baryon Number of the Universe and Unified Theories*, Tsukuba, Japan, February 1979.
[42] S. L. Glashow, "Quarks and leptons," in *Cargèse Lectures*, p. 687, Plenum, NY, USA, 1980.
[43] M. Gell-Mann, P. Ramond, and R. Slansky, "Complex spinors and unified theories," in *Supergravity*, P. van Nieuwenhuizen and D. Z. Freedman, Eds., North Holland, 1979.
[44] R. N. Mohapatra and G. Senjanović, "Neutrino masses and mixings in gauge models with spontaneous parity violation," *Physical Review D*, vol. 23, no. 1, pp. 165–180, 1981.
[45] M. Fukugita and T. Yanagida, "Barygenesis without grand unification," *Physics Letters B*, vol. 174, no. 1, pp. 45–47, 1986.
[46] S. Y. Khlebnikov and M. E. Shaposhnikov, "The statistical theory of anomalous fermion number non-conservation," *Nuclear Physics B*, vol. 308, no. 4, pp. 885–912, 1988.
[47] M. Plümacher, "Baryogenesis and lepton number violation," *Zeitschrift fur Physik C-Particles and Fields*, vol. 74, no. 3, pp. 549–559, 1997.
[48] L. Covi, N. Rius, E. Roulet, and F. Vissani, "Finite temperature effects on CP-violating asymmetries," *Physical Review D*, vol. 57, no. 1, pp. 93–99, 1998.
[49] G. F. Giudice, A. Notari, M. Raidal, A. Riotto, and A. Strumia, "Towards a complete theory of thermal leptogenesis in the SM and MSSM," *Nuclear Physics B*, vol. 685, no. 1–3, pp. 89–149, 2004.
[50] M. Flanz, E. A. Paschos, and U. Sarkar, "Baryogenesis from a lepton asymmetric universe," *Physics Letters B*, vol. 345, no. 3, pp. 248–252, 1995.
[51] L. Covi, E. Roulet, and F. Vissani, "CP violating decays in leptogenesis scenarios," *Physics Letters B*, vol. 384, no. 1–4, pp. 169–174, 1996.
[52] A. Pilaftsis, "CP violation and baryogenesis due to heavy Majorana neutrinos," *Physical Review D*, vol. 56, no. 9, pp. 5431–5451, 1997.
[53] E. W. Kolb and S. Wolfram, "Baryon number generation in the early universe," *Nuclear Physics*, vol. 172, pp. 224–284, 1980.
[54] C. S. Fong, M. C. Gonzalez-Garcia, and E. Nardi, "Leptogenesis from soft supersymmetry breaking (Soft Leptogenesis)," *International Journal of Modern Physics A*, vol. 26, pp. 3491–3604, 2011.

[55] W. Buchmüller, P. Di Bari, and M. Plümacher, "Leptogenesis for pedestrians," *Annals of Physics*, vol. 315, no. 2, pp. 305–351, 2005.
[56] C. S. Fong and J. Racker, "On fast CP violating interactions in leptogenesis," *Journal of Cosmology and Astroparticle Physics*, vol. 2010, no. 07, article 001, 2010.
[57] J. A. Harvey and M. S. Turner, "Cosmological baryon and lepton number in the presence of electroweak fermion-number violation," *Physical Review D*, vol. 42, no. 10, pp. 3344–3349, 1990.
[58] T. Inui, T. Ichihara, Y. Mimura, and N. Sakai, "Cosmological baryon asymmetry in supersymmetric standard models and heavy particle effects," *Physics Letters B*, vol. 325, no. 3-4, pp. 392–400, 1994.
[59] D. J. H. Chung, B. Garbrecht, and S. Tulin, "The effect of the sparticle mass spectrum on the conversion of B-L to B," *Journal of Cosmology and Astroparticle Physics*, vol. 0903, p. 008, 2009.
[60] J. A. Casas and A. Ibarra, "Oscillating neutrinos and $\mu \rightarrow e, y$," *Nuclear Physics B*, vol. 618, no. 1-2, pp. 171–204, 2001.
[61] S. Davidson and A. Ibarra, "A lower bound on the right-handed neutrino mass from leptogenesis," *Physics Letters B*, vol. 535, no. 1–4, pp. 25–32, 2002.
[62] W. Buchmuller, P. Di Bari, and M. Plumacher, "Cosmic microwave background, matterantimatter asymmetry and neutrino masses," *Nuclear Physics B*, vol. 643, no. 1–3, pp. 367–390, 2002.
[63] J. R. Ellis and M. Raidal, "Leptogenesis and the violation of lepton number and CP at low energies," *Nuclear Physics B*, vol. 643, no. 1–3, pp. 229–246, 2002.
[64] W. Buchmuller, P. Di Bari, and M. Plumacher, "A bound on neutrino masses from baryogenesis," *Physics Letters B*, vol. 547, no. 3-4, pp. 128–132, 2002.
[65] W. Buchmuller, P. Di Bari, and M. Plumacher, "The neutrino mass window for baryogenesis," *Nuclear Physics B*, vol. 665, pp. 445–468, 2003.
[66] E. Nardi, "Leptogenesis and neutrino masses," *Nuclear Physics B–Proceedings Supplements*, vol. 217, pp. 27–32, 2011.
[67] R. Barbieri, P. Creminelli, A. Strumia, and N. Tetradis, "Baryogenesis through leptogenesis," *Nuclear Physics B*, vol. 575, no. 1-2, pp. 61–77, 2000.
[68] A. Abada, S. Davidson, F. X. Josse-Michaux, M. Losada, and A. Riotto, "Flavour issues in leptogenesis," *Journal of Cosmology and Astroparticle Physics*, vol. 0604, p. 004, 2006.
[69] E. Nardi, Y. Nir, E. Roulet, and J. Racker, "The importance of flavor in leptogenesis," *Journal of High Energy Physics*, vol. 0601, p. 164, 2006.
[70] A. Abada, S. Davidson, A. Ibarra, F.-X. Josse-Michaux, M. Losada, and A. Riotto, "Flavour matters in leptogenesis," *Journal of High Energy Physics*, vol. 0609, p. 010, 2006.
[71] T. Endoh, T. Morozumi, and Z. Xiong, "Primordial lepton family asymmetries in seesaw model," *Progress of Theoretical Physics*, vol. 111, no. 1, pp. 123–149, 2004.
[72] A. Abada, H. Aissaoui, and M. Losada, "A model for leptogenesis at the TeV scale," *Nuclear Physics B*, vol. 728, no. 1–3, pp. 55–66, 2005.
[73] O. Vives, "Flavoured leptogenesis: a successful thermal leptogenesis with N_1 mass below 10^8 GeV," *Journal of Physics: Conference Series*, vol. 171, Article ID 012076, 2009.
[74] T. Fujihara, S. Kaneko, S. Kang, D. Kimura, T. Morozumi, and M. Tanimoto, "Cosmological family asymmetry and CP violation," *Physical Review D*, vol. 72, no. 1, Article ID 016006, 12 pages, 2005.
[75] A. Pilaftsis and T. E. J. Underwood, "Electroweak-scale resonant leptogenesis," *Physical Review D*, vol. 72, no. 11, Article ID 113001, 27 pages, 2005.
[76] G. C. Branco, R. González Felipe, and F. R. Joaquim, "A new bridge between leptonic CP violation and leptogenesis," *Physics Letters B*, vol. 645, no. 5-6, pp. 432–436, 2007.
[77] S. Pascoli, S. T. Petcov, and A. Riotto, "Connecting low energy leptonic CP violation to leptogenesis," *Physical Review D*, vol. 75, no. 8, Article ID 083511, 2007.
[78] S. Pascoli, S. T. Petcov, and A. Riotto, "Leptogenesis and low energy CP-violation in neutrino physics," *Nuclear Physics B*, vol. 774, no. 1–3, pp. 1–52, 2007.
[79] S. Antusch, S. F. King, and A. Riotto, "Flavour-dependent leptogenesis with sequential dominance," *Journal of Cosmology and Astroparticle Physics*, vol. 0611, p. 011, 2006.
[80] S. Antusch and A. M. Teixeira, "Towards constraints on the SUSY seesaw from flavour-dependent leptogenesis," *Journal of Cosmology and Astroparticle Physics*, vol. 0702, p. 024, 2007.
[81] S. Antusch, "Flavour-dependent type II leptogenesis," *Physical Review D*, vol. 76, Article ID 023512, 2007.
[82] S. Blanchet and P. Di Bari, "Flavour effects on leptogenesis predictions," *Journal of Cosmology and Astroparticle Physics*, vol. 0703, p. 018, 2007.
[83] S. Blanchet, P. Di Bari, and G. G. Raffelt, "Quantum Zeno e ect and the impact of avor in leptogenesis," *Journal of Cosmology and Astroparticle Physics*, vol. 0703, p. 012, 2007.

[84] A. De Simone and A. Riotto, "On the impact of flavour oscillations in leptogenesis," *Journal of Cosmology and Astroparticle Physics*, vol. 0702, p. 005, 2007.
[85] A. De Simone and A. Riotto, "Quantum boltzmann equations and leptogenesis," *Journal of Cosmology and Astroparticle Physics*, vol. 0708, p. 002, 2007.
[86] A. De Simone and A. Riotto, "On Resonant Leptogenesis," *Journal of Cosmology and Astroparticle Physics*, vol. 0708, p. 013, 2007.
[87] V. Cirigliano, A. De Simone, G. Isidori, I. Masina, and A. Riotto, "Quantum resonant leptogenesis and minimal lepton flavour violation," *Journal of Cosmology and Astroparticle Physics*, vol. 0801, 004 pages, 2008.
[88] V. Cirigliano, C. Lee, M. J. Ramsey-Musolf, and S. Tulin, "Flavored quantum Boltzmann equations," *Physical Review D*, vol. 81, no. 10, Article ID 103503, 2010.
[89] C. S. Fong, M. C. Gonzalez-Garcia, E. Nardi, and J. Racker, "Supersymmetric leptogenesis," *Journal of Cosmology and Astroparticle Physics*, vol. 1012, p. 013, 2010.
[90] B. A. Campbell, S. Davidson, J. Ellis, and K. A. Olive, "On the baryon, lepton-flavour and right-handed electron asymmetries of the universe," *Physics Letters B*, vol. 297, no. 1-2, pp. 118–124, 1992.
[91] J. M. Cline, K. Kainulainen, and K. A. Olive, "Protecting the primordial baryon asymmetry from erasure by sphalerons," *Physical Review D*, vol. 49, no. 12, pp. 6394–6409, 1994.
[92] S. Weinberg, "Baryon- and lepton-nonconserving processes," *Physical Review Letters*, vol. 43, no. 21, pp. 1566–1570, 1979.
[93] D. A. Sierra, M. Losada, and E. Nardi, "Lepton flavor equilibration and leptogenesis," *Journal of Cosmology and Astroparticle Physics*, vol. 0912, p. 015, 2009.
[94] M. Beneke, B. Garbrecht, C. Fidler, M. Herranen, and P. Schwaller, "Flavoured leptogenesis in the CTP formalism," *Nuclear Physics B*, vol. 843, no. 1, pp. 177–212, 2011.
[95] D. Aristizabal Sierra, L. A. Munoz, and E. Nardi, "Purely Flavored Leptogenesis," *Physical Review D*, vol. 80, Article ID 016007, 2009.
[96] M. C. Gonzalez-Garcia, J. Racker, and N. Rius, "Leptogenesis without violation of B-L," *Journal of High Energy Physics*, vol. 0911, article 079, 2009.
[97] S. Antusch, S. Blanchet, M. Blennow, and E. Fernandez-Martinez, "Non-unitary leptonic mixing and leptogenesis," *Journal of High Energy Physics*, vol. 2010, no. 1, 2010.
[98] J. Racker, M. Pena, and N. Rius, "Leptogenesis with small violation of B-L," *Journal of Cosmology and Astroparticle Physics*, vol. 1207, article 030, 2012.
[99] R. N. Mohapatra and J. W. F. Valle, "Neutrino mass and baryon-number nonconservation in superstring models," *Physical Review D*, vol. 34, no. 5, pp. 1642–1645, 1986.
[100] G. Hooft, "Naturalness, chiral symmetry, and spontaneous chiral symmetry breaking," *NATO Advanced Science Institute B*, vol. 59, article 135, 23 pages, 1980.
[101] A. Basbøll and S. Hannestad, "Decay of heavy Majorana neutrinos studied using the full Boltzmann equation, including its implications for leptogenesis," *Journal of Cosmology and Astroparticle Physics*, vol. 0701, p. 003, 2007.
[102] J. Garayoa, S. Pastor, T. Pinto, N. Rius, and O. Vives, "On the full Boltzmann equations for leptogenesis," *Journal of Cosmology and Astroparticle Physics*, vol. 0909, p. 035, 2009.
[103] F. Hahn-Woernle, M. Plümacher, and Y. Y. Y. Wong, "Full boltzmann equations for leptogenesis including scattering," *Journal of Cosmology and Astroparticle Physics*, vol. 0908, p. 028, 2009.
[104] F. Hahn-Woernle, "Wash-out in N_2-dominated leptogenesis," *Journal of Cosmology and Astroparticle Physics*, vol. 1008, p. 029, 2010.
[105] W. Buchmuller and M. Plumacher, "Spectator processes and baryogenesis," *Physics Letters B*, vol. 511, pp. 74–76, 2001.
[106] E. Nardi, Y. Nir, J. Racker, and E. Roulet, "On Higgs and sphaleron effects during the leptogenesis era," *Journal of High Energy Physics*, no. 1, pp. 1743–1760, 2006.
[107] C. S. Fong, M. C. Gonzalez-Garcia, and E. Nardi, "Early universe effective theories: the soft leptogenesis and R-genesis cases," *Journal of Cosmology and Astroparticle Physics*, vol. 1102, p. 032, 2011.
[108] E. Nardi, J. Racker, and E. Roulet, "CP violation in scatterings, three body processes and the Boltzmann equations for leptogenesis," *Journal of High Energy Physics*, vol. 2007, no. 9, article 090, 2007.
[109] S. Davidson, E. Nardi, and Y. Nir, "Leptogenesis," *Physics Reports*, vol. 466, no. 4-5, pp. 105–177, 2008.
[110] A. Pilaftsis and T. E. J. Underwood, "Resonant leptogenesis," *Nuclear Physics B*, vol. 692, no. 3, pp. 303–345, 2004.

[111] A. Anisimov, A. Broncano, and M. Plümacher, "The C P-asymmetry in resonant leptogenesis," *Nuclear Physics B*, vol. 737, no. 1-2, pp. 176–189, 2006.
[112] C. S. Fong, M. C. Gonzalez-Garcia, and J. Racker, "CP violation from scatterings with gauge bosons in leptogenesis," *Physics Letters B*, vol. 697, no. 5, pp. 463–470, 2011.
[113] K. Kajantie, M. Laine, K. Rummukainen, and M. Shaposhnikov, "Generic rules for high temperature dimensional reduction and their application to the Standard Model," *Nuclear Physics B*, vol. 458, no. 1-2, pp. 90–136, 1996.
[114] J. A. Casas, J. R. Espinosa, A. Ibarra, and I. Navarro, "General RG equations for physical neutrino parameters and their phenomenological implications," *Nuclear Physics B*, vol. 573, no. 3, pp. 652–684, 2000.
[115] S. Antusch, J. Kersten, M. Lindner, and M. Ratz, "Running neutrino masses, mixings and CP phases: analytical results and phenomenological consequences," *Nuclear Physics B*, vol. 674, no. 1-2, pp. 401–433, 2003.
[116] D. Comelli and J. R. Espinosa, "Bosonic thermal masses in supersymmetry," *Physical Review D*, vol. 55, no. 10, pp. 6253–6263, 1997.
[117] P. Elmfors, K. Enqvist, and I. Vilja, "Thermalization of the Higgs field at the electroweak phase transition," *Nuclear Physics B*, vol. 412, no. 1-2, pp. 459–476, 1994.
[118] H. A. Weldon, "Dynamical holes in the quark-gluon plasma," *Physical Review D*, vol. 40, no. 7, pp. 2410–2420, 1989.
[119] H. A. Weldon, "Effective fermion masses of order gT in high-temperature gauge theories with exact chiral invariance," *Physical Review D*, vol. 26, no. 10, pp. 2789–2796, 1982.
[120] V. V. Klimov, "Spectrum of elementary Fermi excitations in quark gluon plasma," *Soviet Journal of Nuclear Physics*, vol. 33, pp. 934–935, 1981 (Russian).
[121] B. Garbrecht, "Leptogenesis: the other cuts," *Nuclear Physics B*, vol. 847, no. 2, pp. 350–366, 2011.
[122] M. Le Bellac, *Thermal Field Theory*, Cambridge University Press, 1996.
[123] N. P. Landsman and C. G. van Weert, "Real- and imaginary-time field theory at finite temperature and density," *Physics Reports*, vol. 145, no. 3-4, pp. 141–249, 1987.
[124] C. P. Kießig, M. Plümacher, and M. H. Thoma, "Decay of a Yukawa fermion at finite temperature and applications to leptogenesis," *Physical Review D*, vol. 82, no. 3, Article ID 036007, 2010.
[125] C. Kiessig and M. Plumacher, "Hard-thermal-loop corrections in leptogenesis I: *CP*-asymmetries," vol. 2012, no. 07, article 014, 2012.
[126] C. Kiessig and M. Plumacher, "Hard-thermal-loop corrections in leptogenesis II: solving the boltzmann equations," *Journal of Cosmology and Astroparticle Physics*, vol. 2012, no. 09, article 012, 2012.
[127] M. Garny, A. Hohenegger, and A. Kartavtsev, "Medium corrections to the CP-violating parameter in leptogenesis," *Physical Review D*, vol. 81, no. 8, Article ID 085028, 2010.
[128] F. Buccella, D. Falcone, C. S. Fong, E. Nardi, and G. Ricciardi, "Squeezing out predictions with leptogenesis from SO(10)," vol. 86, no. 3, Article ID 035012, 15 pages, 2012.
[129] H. B. Nielsen and Y. Takanishi, "Baryogenesis via lepton number violation in anti-GUT model," *Physics Letters B*, vol. 507, no. 1–4, pp. 241–251, 2001.
[130] A. Strumia, "Baryogenesis via eptogenesis," http://arxiv.org/abs/hep-ph/0608347.
[131] G. Engelhard, Y. Grossman, E. Nardi, and Y. Nir, "Importance of the heavier singlet neutrinos in leptogenesis," *Physical Review Letters*, vol. 99, no. 8, Article ID 081802, 2007.
[132] S. Blanchet, P. Di Bari, D. A. Jones, and L. Marzola, "Leptogenesis with heavy neutrino avours: from density matrix to Boltzmann equations," http://arxiv.org/abs/1112.4528.
[133] W. Buchmuller and S. Fredenhagen, "Quantum mechanics of baryogenesis," *Physics Letters B*, vol. 483, pp. 217–224, 2000.
[134] M. Garny, A. Hohenegger, A. Kartavtsev, and M. Lindner, "Systematic approach to leptogenesis in nonequilibrium QFT: vertex contribution to the CP-violating parameter," *Physical Review D*, vol. 80, no. 12, Article ID 125027, 2009.
[135] M. Garny, A. Hohenegger, A. Kartavtsev, and M. Lindner, "Systematic approach to leptogenesis in nonequilibrium QFT: self-energy contribution to the CP-violating parameter," *Physical Review D*, vol. 81, no. 8, Article ID 085027, 2010.
[136] A. Anisimov, W. Buchmüller, M. Drewes, and S. Mendizabal, "Leptogenesis from quantum interference in a thermal bath," *Physical Review Letters*, vol. 104, no. 12, Article ID 121102, 2010.
[137] M. Beneke, B. Garbrecht, M. Herranen, and P. Schwaller, "Finite number density corrections to leptogenesis," *Nuclear Physics B*, vol. 838, no. 1-2, pp. 1–27, 2010.

[138] M. Garny, A. Hohenegger, and A. Kartavtsev, "Medium corrections to the *CP*-violating parameter in leptogenesis," *Physical Review D*, vol. 81, no. 8, Article ID 085028, 2010.
[139] A. Anisimov, W. Buchmüller, M. Drewes, and S. Mendizabal, "Quantum leptogenesis I," *Annals of Physics*, vol. 326, no. 8, pp. 1998–2038, 2011.
[140] K. C. Chou, Z. B. Su, B. L. Hao, and L. Yu, "Equilibrium and nonequilibrium formalisms made unified," *Physics Reports*, vol. 118, no. 1-2, pp. 1–131, 1985.
[141] B. Garbrecht and M. Herranen, "Effective theory of resonant leptogenesis in the closed-time-path approach," *Nuclear Physics B*, vol. 861, no. 1, pp. 17–52, 2012.
[142] H. Pagels and J. R. Primack, "Supersymmetry, cosmology, and new physics at teraelectronvolt energies," *Physical Review Letters*, vol. 48, no. 4, pp. 223–226, 1982.
[143] S. Weinberg, "Cosmological constraints on the scale of supersymmetry breaking," *Physical Review Letters*, vol. 48, no. 19, pp. 1303–1306, 1982.
[144] M. Y. Khlopov and A. D. Linde, "Is it easy to save the gravitino?" *Physics Letters B*, vol. 138, no. 4, pp. 265–268, 1984.
[145] J. Ellis, J. E. Kim, and D. V. Nanopoulos, "Cosmological gravitino regeneration and decay," *Physics Letters B*, vol. 145, no. 3-4, pp. 181–186, 1984.
[146] B. A. Campbell, S. Davidson, and K. A. Olive, "Inflation, neutrino baryogenesis, and (s)neutrino-induced baryogenesis," *Nuclear Physics B*, vol. 399, no. 1, pp. 111–136, 1993.
[147] M. Plümacher, "Baryon asymmetry, neutrino mixing and supersymmetric SO(10) unification," *Nuclear Physics B*, vol. 530, no. 1-2, pp. 207–246, 1998.
[148] P. Di Bari, "Leptogenesis, neutrino mixing data and the absolute neutrino mass scale," http://arxiv.org/abs/hep-ph/0406115.
[149] D. J. H. Chung, B. Garbrecht, M. J. Ramsey-Musolf, and S. Tulin, "Supergauge interactions and electroweak baryogenesis," *Journal of High Energy Physics*, vol. 12, p. 067, 2009.
[150] L. E. Ibáñez and F. Quevedo, "Supersymmetry protects the primordial baryon asymmetry," *Physics Letters B*, vol. 283, no. 3-4, pp. 261–269, 1992.
[151] L. Bento, "Sphaleron relaxation temperatures," *Journal of Cosmology and Astroparticle Physics*, vol. 0311, article 002, 2003.
[152] R. N. Mohapatra and X. Zhang, "QCD sphalerons at high temperature and baryogenesis at the electroweak scale," *Physical Review D*, vol. 45, no. 8, pp. 2699–2705, 1992.
[153] G. D. Moore, "Computing the strong sphaleron rate," *Physics Letters B*, vol. 412, no. 3-4, pp. 359–370, 1997.
[154] R. Rangarajan and H. Mishra, "Leptogenesis with heavy Majorana neutrinos reexamined," *Physical Review D*, vol. 61, no. 4, Article ID 043509, 7 pages, 2000.
[155] L. Covi and E. Roulet, "Baryogenesis from mixed particle decays," *Physics Letters B*, vol. 399, no. 1-2, pp. 113–118, 1997.
[156] T. Hambye, J. March-Russell, and S. M. West, "TeV scale resonant leptogenesis from supersymmetry breaking," *Journal of High Energy Physics*, vol. 0407, p. 070, 2004.
[157] A. Pilaftsis, "Resonant τ leptogenesis with observable lepton number violation," *Physical Review Letters*, vol. 95, no. 8, Article ID 081602, 4 pages, 2005.
[158] C. H. Albright and S. M. Barr, "Resonant leptogenesis in a predictive SO(10) grand unified model," *Physical Review D*, vol. 70, no. 3, Article ID 033013, 2004.
[159] C. H. Albright, "Bounds on the neutrino mixing angles for an SO(10) model with lopsided mass matrices," *Physical Review D*, vol. 72, Article ID 013001, 2005.
[160] Z. Z. Xing and S. Zhou, "Tri-bimaximal neutrino mixing and flavor-dependent resonant leptogenesis," *Physics Letters B*, vol. 653, no. 2–4, pp. 278–287, 2007.
[161] S. M. West, "Neutrino masses and TeV scale resonant leptogenesis from supersymmetry breaking," *Modern Physics Letters A*, vol. 21, no. 21, pp. 1629–1646, 2006.
[162] V. Cirigliano, G. Isidori, and V. Porretti, "CP violation and leptogenesis in models with minimal lepton flavour violation," *Nuclear Physics B*, vol. 763, no. 1-2, pp. 228–246, 2007.
[163] G. C. Branco, A. J. Buras, S. Jager, S. Uhlig, and A. Weiler, "Another look at minimal lepton avour violation, l(i) to l(j) gamma, leptogenesis, and the ratio M(nu)/Lambda(LFV)," *Journal of High Energy Physics*, vol. 0709, p. 004, 2007.
[164] K. S. Babu, A. G. Bachri, and Z. Tavartkiladze, "Predictive model of inverted neutrino mass hierarchy and resonant leptogenesis," *International Journal of Modern Physics A*, vol. 23, no. 11, pp. 1679–1696, 2008.
[165] A. Pilaftsis, "Electroweak resonant leptogenesis in the singlet Majoron model," *Physical Review D*, vol. 78, no. 1, Article ID 013008, 2008.

[166] F. F. Deppisch and A. Pilaftsis, "Lepton flavor violation and $\theta13$ in minimal resonant leptogenesis," *Physical Review D*, vol. 83, no. 7, Article ID 076007, 2011.
[167] S. Iso, N. Okada, and Y. Orikasa, "Resonant leptogenesis in the minimal B-L extended standard model at TeV," *Physical Review D*, vol. 83, no. 9, Article ID 093011, 2011.
[168] P. H. Gu, "Resonant leptogenesis and a verifiable seesaw model from large extra dimensions," *Physical Review D*, vol. 81, no. 7, Article ID 073002, 2010.
[169] A. Pilaftsis, "The little review on leptogenesis," *Journal of Physics*, vol. 171, Article ID 012017, 2009.
[170] M. Garny, A. Kartavtsev, and A. Hohenegger, "Leptogenesis from first principles in the resonant regime," http://arxiv.org/abs/1112.6428.
[171] S. M. West, "Naturally degenerate right handed neutrinos," *Physical Review D*, vol. 71, no. 1, Article ID 013004, 8 pages, 2005.
[172] G. C. Branco, R. González Felipe, F. R. Joaquim, and B. M. Nobre, "Enlarging the window for radiative leptogenesis," *Physics Letters B*, vol. 633, no. 2-3, pp. 336–344, 2006.
[173] S. Bray, J. S. Lee, and A. Pilaftsis, "Resonant CP violation due to heavy neutrinos at the LHC," *Nuclear Physics B*, vol. 786, no. 1-2, pp. 95–118, 2007.
[174] S. Blanchet, Z. Chacko, S. S. Granor, and R. N. Mohapatra, "Probing resonant leptogenesis at the LHC," *Physical Review D*, vol. 82, no. 7, Article ID 076008, 2010.
[175] L. Boubekeur, "Leptogenesis at low scale," http://arxiv.org/abs/hep-ph/0208003.
[176] Y. Grossman, T. Kashti, Y. Nir, and E. Roulet, "Leptogenesis from supersymmetry breaking," *Physical Review Letters*, vol. 91, Article ID 251801, 2003.
[177] G. D'Ambrosio, G. F. Giudice, and M. Raidal, "Soft leptogenesis," *Physics Letters B*, vol. 575, no. 1-2, pp. 75–84, 2003.
[178] E. J. Chun, "Late leptogenesis from radiative soft terms," *Physical Review D*, vol. 69, no. 11, Article ID 117303, 2004.
[179] L. Boubekeur, T. Hambye, and G. Senjanović, "Low-scale leptogenesis and soft supersymmetry breaking," *Physical Review Letters*, vol. 93, no. 11, Article ID 111601, 4 pages, 2004.
[180] G. D'Ambrosio, T. Hambye, A. Hektor, M. Raidal, and A. Rossi, "Leptogenesis in the minimal supersymmetric triplet seesaw model," *Physics Letters B*, vol. 604, no. 3-4, pp. 199–206, 2004.
[181] M. C. Chen and K. T. Mahanthappa, "Lepton flavor violating decays and soft leptogenesis in a supersymmetric SO(10) model," *Physical Review D*, vol. 70, no. 11, Article ID 113013, 2004.
[182] T. Kashti, "Phenomenological consequences of soft leptogenesis," *Physical Review D*, vol. 71, Article ID 013008, 2005.
[183] Y. Grossman, R. Kitano, and H. Murayama, "Natural soft leptogenesis," *Journal of High Energy Physics*, vol. 06, 058 pages, 2005.
[184] J. R. Ellis and S. K. Kang, "Sneutrino leptogenesis at the electroweak scale," http://arxiv.org/abs/hep-ph/0505162.
[185] E. J. Chun and S. Scopel, "Soft leptogenesis in Higgs triplet model," *Physics Letters B*, vol. 636, no. 5, pp. 278–285, 2006.
[186] A. D. Medina and C. E. M. Wagner, "Soft leptogenesis in warped extra dimensions," *Journal of High Energy Physics*, vol. 2006, no. 12, article 037, 2006.
[187] J. Garayoa, M. C. Gonzalez-Garcia, and N. Rius, "Soft leptogenesis in the inverse seesaw model," *Journal of High Energy Physics*, vol. 2007, no. 2, article 021, 2007.
[188] E. J. Chun and L. Velasco-Sevilla, "SO(10) unified models and soft leptogenesis," *Journal of High Energy Physics*, vol. 8, p. 075, 2007.
[189] O. Bahat-Treidel and Z. Surujon, "The Ir(relevance) of initial conditions in soft leptogenesis," *Journal of High Energy Physics*, vol. 11, p. 046, 2008.
[190] C. S. Fong and M. C. Gonzalez-Garcia, "Flavoured soft leptogenesis," *Journal of High Energy Physics*, vol. 0806, p. 076, 2008.
[191] C. S. Fong and M. C. Gonzalez-Garcia, "On quantum effects in soft leptogenesis," *Journal of Cosmology and Astroparticle Physics*, vol. 2008, no. 8, article 008, 2008.
[192] C. S. Fong and M. C. Gonzalez-Garcia, "On gaugino contributions to soft leptogenesis," *Journal of High Energy Physics*, vol. 0903, p. 073, 2009.
[193] K. S. Babu, Y. Meng, and Z. Tavartkiladze, "New ways to leptogenesis with gauged *B-L* symmetry," *Physics Letters B*, vol. 681, no. 1, pp. 37–43, 2009.
[194] Y. Kajiyama, S. Khalil, and M. Raidal, "Electron EDM and soft leptogenesis in supersymmetric *B-L* extension of the Standard Model," *Nuclear Physics B*, vol. 820, no. 1-2, pp. 75–88, 2009.
[195] C. S. Fong, M. C. Gonzalez-Garcia, E. Nardi, and J. Racker, "Flavoured soft leptogenesis and natural values of the B term," *Journal of High Energy Physics*, vol. 1007, p. 001, 2010.

[196] K. Hamaguchi and N. Yokozaki, "Soft leptogenesis and gravitino dark matter in gauge mediation," *Physics Letters B*, vol. 694, no. 4-5, pp. 398–401, 2011.
[197] E. K. Akhmedov, V. A. Rubakov, and A. Y. Smirnov, "Baryogenesis via neutrino oscillations," *Physical Review Letters*, vol. 81, no. 7, pp. 1359–1362, 1998.
[198] K. Dick, M. Lindner, M. Ratz, and D. Wright, "Leptogenesis with dirac neutrinos," *Physical Review Letters*, vol. 84, no. 18, pp. 4039–4042, 2000.
[199] H. Murayama and A. Pierce, "Realistic dirac leptogenesis," *Physical Review Letters*, vol. 89, no. 27, Article ID 271601, p. 4, 2002.
[200] M. Boz and N. K. Pak, "Dirac leptogenesis and anomalous U (1)," *European Physical Journal C*, vol. 37, no. 4, pp. 507–510, 2004.
[201] S. Abel and V. Page, "Affleck-dine (Pseudo)-dirac neutrinogenesis," *Journal of High Energy Physics*, vol. 0605, p. 024, 2006.
[202] D. G. Cerdeño, A. Dedes, and T. E. J. Underwood, "The minimal phantom sector of the standard model: higgs phenomenology and Dirac leptogenesis," *Journal of High Energy Physics*, vol. 0609, p. 067, 2006.
[203] B. Thomas and M. Toharia, "Phenomenology of Dirac neutrinogenesis in split supersymmetry," *Physical Review D*, vol. 73, no. 6, Article ID 063512, 21 pages, 2006.
[204] B. Thomas and M. Toharia, "Lepton flavor violation and supersymmetric Dirac leptogenesis," *Physical Review D*, vol. 75, no. 1, Article ID 013013, 2007.
[205] E. J. Chun and P. Roy, "Dirac leptogenesis in extended nMSSM," *Journal of High Energy Physics*, vol. 0806, p. 089, 2008.
[206] A. Bechinger and G. Seidl, "Resonant dirac leptogenesis on throats," *Physical Review D*, vol. 81, no. 6, Article ID 065015, 2010.
[207] M.-C. Chen, J. Huang, and W. Shepherd, "Dirac leptogenesis with a non-anomalous $U(1)'$ family symmetry," *Journal of High Energy Physics*, vol. 1211, article 059, 2012.
[208] M. Maggand and C. Wetterich, "Neutrino mass problem and gauge hierarchy," *Physics Letters B*, vol. 94, no. 1, pp. 61–64, 1980.
[209] J. Schechter and J. W. F. Valle, "Neutrino masses in $SU(2) \times U(1)$ theories," *Physical Review D*, vol. 22, no. 9, pp. 2227–2235, 1980.
[210] C. Wetterich, "Neutrino masses and the scale of B-L violation," *Nuclear Physics B*, vol. 187, no. 2, pp. 343–375, 1981.
[211] G. Lazarides, Q. Shafi, and C. Wetterich, "Proton lifetime and fermion masses in an SO(10) model," *Nuclear Physics B*, vol. 181, no. 2, pp. 287–300, 1981.
[212] E. Ma and U. Sarkar, "Neutrino masses and leptogenesis with heavy Higgs triplets," *Physical Review Letters*, vol. 80, no. 26, pp. 5716–5719, 1998.
[213] E. J. Chun and S. K. Kang, "Baryogenesis and degenerate neutrinos," *Physical Review D*, vol. 63, no. 9, Article ID 097902, 2001.
[214] T. Hambye, E. Ma, and U. Sarkar, "Supersymmetric triplet Higgs model of neutrino masses and leptogenesis," *Nuclear Physics B*, vol. 602, no. 1-2, pp. 23–38, 2001.
[215] A. S. Joshipura, E. A. Paschos, and W. Rodejohann, "Leptogenesis in left-right symmetric theories," *Nuclear Physics B*, vol. 611, no. 1–3, pp. 227–238, 2001.
[216] T. Hambye and G. Senjanović, "Consequences of triplet seesaw for leptogenesis," *Physics Letters B*, vol. 582, no. 1-2, pp. 73–81, 2004.
[217] W.-l. Guo, "Neutrino mixing and leptogenesis in type II seesaw mechanism," *Physical Review D*, vol. 70, no. 5, Article ID 053009, 7 pages, 2004.
[218] S. Antusch and S. F. King, "Type II leptogenesis and the neutrino mass scale," *Physics Letters B*, vol. 597, no. 2, pp. 199–207, 2004.
[219] S. Antusch and S. F. King, "Leptogenesis in unified theories with type II see-saw," *Journal of High Energy Physics*, vol. 0601, p. 117, 2006.
[220] T. Hambye, M. Raidal, and A. Strumia, "Efficiency and maximal ^{CP}-asymmetry of scalar triplet leptogenesis," *Physics Letters B*, vol. 632, no. 5-6, pp. 667–674, 2006.
[221] E. J. Chun and S. Scopel, "Analysis of leptogenesis in a supersymmetric triplet seesaw model," *Physical Review D*, vol. 75, no. 2, Article ID 023508, 2007.

[222] P.-H. Gu, H. Zhang, and S. Zhou, "A minimal type II seesaw model," *Physical Review D*, vol. 74, no. 7, Article ID 076002, 5 pages, 2006.

[223] N. Sahu and U. Sarkar, "Predictive model for dark matter, dark energy, neutrino masses, and leptogenesis at the TeV scale," *Physical Review D*, vol. 76, no. 4, Article ID 045014, 2007.

[224] J. McDonald, N. Sahu, and U. Sarkar, "Type-II seesaw at collider, lepton asymmetry and singlet scalar dark matter," *Journal of Cosmology and Astroparticle Physics*, vol. 0804, p. 037, 2008.

[225] W. Chao, S. Luo, and Z. Z. Xing, "Neutrino mixing and leptogenesis in type-II seesaw scenarios with left-right symmetry," *Physics Letters B*, vol. 659, no. 1-2, pp. 281–289, 2008.

[226] T. Hällgren, T. Konstandin, and T. Ohlsson, "Triplet leptogenesis in left-right symmetric seesaw models," *Journal of Cosmology and Astroparticle Physics*, vol. 0801, p. 014, 2008.

[227] M. Frigerio, P. Hosteins, S. Lavignac, and A. Romanino, "A new, direct link between the baryon asymmetry and neutrino masses," *Nuclear Physics B*, vol. 806, no. 1-2, pp. 84–102, 2009.

[228] A. Strumia, "Sommerfeld corrections to type-II and type-III leptogenesis," *Nuclear Physics B*, vol. 809, no. 1-2, pp. 308–317, 2009.

[229] P. H. Gu, M. Hirsch, U. Sarkar, and J. W. F. Valle, "Neutrino masses, leptogenesis, and dark matter in a hybrid seesaw model," *Physical Review D*, vol. 79, no. 3, Article ID 033010, 2009.

[230] L. Calibbi, M. Frigerio, S. Lavignac, and A. Romanino, "Flavour violation in supersymmetric SO(10) unification with a type II seesaw mechanism," *Journal of High Energy Physics*, vol. 0912, p. 057, 2009.

[231] C. S. Chen and C. M. Lin, "Type II seesaw Higgs triplet as the inflaton for chaotic inflation and leptogenesis," *Physics Letters B*, vol. 695, no. 1–4, pp. 9–12, 2011.

[232] D. Aristizabal Sierra, F. Bazzocchi, and I. de Medeiros Varzielas, "Leptogenesis in avor models with type I and II seesaws," *Nuclear Physics B*, vol. 858, pp. 196–213, 2012.

[233] C. Arina and N. Sahu, "Asymmetric inelastic inert doublet dark matter from triplet scalar leptogenesis," *Nuclear Physics B*, vol. 854, pp. 666–699, 2012.

[234] R. Foot, H. Lew, X. G. He, and G. C. Joshi, "Seesaw neutrino masses," *Zeitschrift fur Physik C*, vol. 44, p. 441, 1989.

[235] E. Ma, "Pathways to naturally small neutrino masses," *Physical Review Letters*, vol. 81, no. 6, pp. 1171–1174, 1998.

[236] E. Ma and D. P. Roy, "Heavy triplet leptons and new gauge boson," *Nuclear Physics B*, vol. 644, no. 1-2, pp. 290–302, 2002.

[237] B. Brahmachari, E. Ma, and U. Sarkar, "Supersymmetric model of neutrino mass and leptogenesis with string-scale unification," *Physics Letters B*, vol. 520, no. 1-2, pp. 152–158, 2001.

[238] T. Hambye, Y. Lin, A. Notari, M. Papucci, and A. Strumia, "Constraints on neutrino masses from leptogenesis models," *Nuclear Physics B*, vol. 695, no. 1-2, pp. 169–191, 2004.

[239] S. Blanchet and P. Fileviez Pérez, "Baryogenesis via leptogenesis in adjoint SU(5)," *Journal of Cosmology and Astroparticle Physics*, vol. 0808, p. 037, 2008.

[240] S. Blanchet and P. F. PÉrez, "On the role of low-energy CP violation in leptogenesis," *Modern Physics Letters A*, vol. 24, no. 18, pp. 1399–1409, 2009.

[241] S. Patra, A. Sarkar, and U. Sarkar, "Spontaneous left-right symmetry breaking in supersymmetric models with only Higgs doublets," *Physical Review D*, vol. 82, no. 1, Article ID 015010, 2010.

[242] D. A. Sierra, J. F. Kamenik, and M. Nemevšek, "Implications of flavor dynamics for fermion triplet leptogenesis," *Journal of High Energy Physics*, vol. 1010, p. 036, 2010.

[243] K. Kannike and D. V. Zhuridov, "New solution for neutrino masses and leptogenesis in adjoint SU(5)," *Journal of High Energy Physics*, vol. 1107, p. 102, 2011.

[244] I. Abt et al., "GERDA collaboration, proposal to the LNGS," http://www.mpi-hd.mpg.de/ge76/home.html.

[245] C. E. Aalseth, D. Anderson, R. Arthur et al., "The Majorana neutrinoless double-beta decay experiment," *Physics of Atomic Nuclei*, vol. 67, no. 11, pp. 2002–2010, 2004.

[246] R. Ardito et al., "CUORE: a cryogenic underground observatory for rare events," http://arxiv.org/abs/hepex/0501010.

[247] B. Autin, A. Blondel, K . Bongardt et al., "Conceptual design of the SPL, a high-power superconducting H^- linac at CERN," CERN 2000-012, 2000.

[248] M. Mezzetto, "Physics potential of the SPL super beam," *Journal of Physics G*, vol. 29, no. 8, pp. 1781–1784, 2003.

[249] Y. Itow, T. Kajita, K. Kaneyuki et al., "The JHF-Kamioka neutrino project," http://arxiv.org/abs/hep-ex/0106019.

[250] D. S. Ayres et al., "NOvA proposal to build a 30-kiloton o -axis detector to study neutrino oscillations in the Fermilab NuMI beamline," http://arxiv.org/abs/hep-ex/0503053.

[251] A. Strumia and F. Vissani, "Neutrino masses and mixings and...," http://arxiv.org/abs/hepph/0606054.

[252] F. de Campos, O. J. P. Éboli, M. B. Magro et al., "Probing bilinear R-parity violating supergravity at the LHC," *Journal of High Energy Physics*, vol. 0805, p. 048, 2008.

[253] B. C. Allanach, M. A. Bernhardt, H. K. Dreiner, S. Grab, C. H. Kom, and P. Richardson, "R-Parity violating minimal supergravity at the LHC," in *Proceedings of the 15th International Conference on Supersymmetry and the Unification of Fundamental Interactions (SUSY '07)*, pp. 334–337, Karlsruhe, Germany, 2007, http://arxiv.org/abs/0710.2034.

[254] U. Mahanta, "Neutrino masses and mixing angles from leptoquark interactions," *Physical Review D*, vol. 62, no. 7, Article ID 073009, 4 pages, 2000.

[255] D. Aristizabal Sierra, M. Hirsch, and S. G. Kovalenko, "Leptoquarks: neutrino masses and related accelerator signals," *Physical Review D*, vol. 77, no. 5, Article ID 055011, 2008.

[256] J. Garayoa and T. Schwetz, "Neutrino mass hierarchy and Majorana *CP* phases within the Higgs triplet model at the LHC," *Journal of High Energy Physics*, vol. 0803, p. 009, 2008.

[257] M. Kadastik, M. Raidal, and L. Rebane, "Direct determination of neutrino mass parameters at future colliders," *Physical Review D*, vol. 77, no. 11, Article ID 115023, 2008.

[258] A. Zee, "Quantum numbers of Majorana neutrino masses," *Nuclear Physics B*, vol. 264, pp. 99–110, 1986.

[259] K. S. Babu, "Model of "calculable" Majorana neutrino masses," *Physics Letters B*, vol. 203, no. 1-2, pp. 132–136, 1988.

[260] D. A. Sierra and M. Hirsch, "Experimental tests for the Babu-Zee two-loop model of Majorana neutrino masses," *Journal of High Energy Physics*, vol. 2006, no. 12, article 052, 2006.

[261] C. S. Chen, C. Q. Geng, J. N. Ng, and J. M. S. Wu, "Testing radiative neutrino mass generation at the LHC," *Journal of High Energy Physics*, vol. 0708, p. 022, 2007.

[262] M. Nebot, J. F. Oliver, D. Palao, and A. Santamaria, "Prospects for the Zee-Babu model at the CERN LHC and low energy experiments," *Physical Review D*, vol. 77, no. 9, Article ID 093013, 2008.

4

Exclusive Reconstruction of B-Decays with Missing Neutrals

M. Dima

Department of Computational Physics, Horia Hulubei National Institute of Physics and Nuclear Engineering, Atomistilor Street 407, P.O. Box MG-6, 76900 Bucharest, Romania

Correspondence should be addressed to M. Dima, modima@nipne.ro

Academic Editor: Bogdan Mitrica

Often decay channels that are of theoretical interest cannot be reconstructed exclusively due to missing neutrals (such as neutrinos), or due to single-track vertices. This situation appears both in underground astrophysics experiments as well as in conventional accelerator experiments. A method to "recover" such missing particles from their kinematics and reconstruct "exclusively" the modes would benefit both domains in a number of ways. The main idea is to combine 4-momentum conservations in vertices with available geometric information in the event. The paper gives details of such methods on the $B_s^0 \to D_s^- K^+$, $D_s^- \to K^+ K^- \pi^- (\pi^0)$ prototype decay, which also encounters 2-fold ambiguities in its solutions. Such ambiguities can be lifted and the paper shows how, while also addressing the potential the method has in physics analyses and detector studies.

1. Introduction

The kinematics of decay modes with a missing particle, usually an elusive neutrino, are of interest in both underground astrophysics experiments as well as in conventional accelerator experiments. Exclusive reconstruction in today's new generation experiments is important for precision physics, performant background rejections needed in rare decays, and so forth. Some modes that are of considerable theoretical interest, from the experimental point of view, pose inherent difficulties with missing neutrals and single-track vertices. While such is not an impossible situation (since the vertex must be somewhere on the track), the precise location eludes in general the reconstruction of the mode. Nonetheless, in certain cases, information from the rest of the event provides this capability—in general in the form of 4-momentum conservation in subsequent vertices. Work on recovery using dynamic information has been explored before [1, 2], not necessarily in association with vertexing. Code addressing

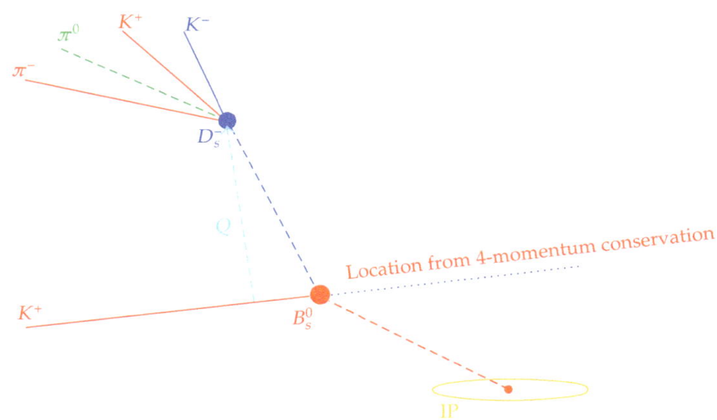

Figure 1: Topology of the mode considered in the absence of magnetic field: $B_s^0 \to D_s^-(K^+K^-\pi^-\pi^0)\ K^+$. The line $\vec{Q}$ is the perpendicular from the D-vertex onto the kaon track.

such methods stems from the days of the CLEO collaboration, KWFIT [3, 4] and of BaBar Collaboration, KinFitter [5].

In underground experiments the geometric equivalent to accelerator B-physics programs are strange-particle decays [6, 7], where kinematic imprecisions associated with the missing neutrino and a traditional lack of some form of vertexing have prevented exclusive reconstruction of the modes. At the new underground astrophysics experiments (such as LAGUNA [8–12]) tracking-vertexing facilities are being discussed in some of the studies. The methods studied could in principle be applied for all "stealth" particles (neutrino, π^0, other neutrals) decaying further downstream.

The objective in such cases is to recover if possible exclusively the event, by combining conservation equations and geometric aspects in the decays. The case made in this paper is for tracking-vertexing assemblies in underground astrophysics experiments and for more precise vertexing in accelerator-based experiments.

Consider the prototype mode $B_s^0 \to D_s^- K^+$, $D_s^- \to K^+K^-\pi^-\ (\pi^0)$, which is of interest for γ_{CKM} measurement, however, with a somewhat fair branching ratio. With π^0 in the final state the branching ratio is roughly equal to that without the π^0 [13], thus putting the two together essentially doubles the statistics. Reconstruction of the π^0 would in principle be possible in the calorimeter, however, with less resolution than using tracking.

Fortunately there is sufficient information in the event for reconstructing the mode without explicit calorimeter information on the π^0—in principle with better resolution also. This information is from 4-momentum conservation in the B- and D-vertices. A particular 2-fold ambiguity from the quadratic equations arises in this (and other) modes. It is important to be able to lift this ambiguity without imposing cuts (which due to detector resolution can eliminate signal). Figure 1 shows the topology of the event (in the absence of magnetic field—such as for the LHCb experiment at the LHC. A small track bending over the vertexing region can also be included, in the form of an $\mathcal{O}_{\text{III}}$ helical correction, giving a solution also for other detectors with full-strength magnetic field in the vertexing region).

Vertexing methods reject reasonably well QCD backgrounds (originating at the interaction point (IP), with some resolution there around); however, other B-modes can constitute significant backgrounds. One such mode is the sister mode with pion in the B-vertex, roughly 4-5 times more abundant. Kinematically and topologically similar modes from B_d^0 also exist—the latter roughly 3 times more abundant than B_s^0. Although there is no exact replica of the mode in B_d^0 version, there are many modes with nonnegligible branching ratios that "losing" one track can mimic this mode. Due to the similarities in topology (displaced vertices) and kinematics (B-masses, center of mass decay energies), the method cannot distinguish between such cases—not with current detector resolution. The only remaining solution is performant Particle ID, making sure that the B-vertex track is a kaon (B_d^0 being less "kaon-productive" than B_s^0) and that there are two kaons present in the D-vertex. All other kinematic parameters offer little protection against B_d^0 modes—due to the high boost factors, mass similarities, and center of mass decay energies. A pleiad of such parameters has been studied: the discriminant of the second order equation, pointback, reconstructed invariant masses, angles between track-combinations or decay-planes, 2-3 parameters combinations of the aforementioned, and so forth. For all the detector resolution was insufficient to tell apart the current mode from its sister mode (best achievable ca. 41% rejection), or from other B_d^0 originating modes (ca. 45% rejection).

2. Dynamic Vertexing

The 4-momentum conservations in the B- and D-vertices are

$$\begin{aligned}(E_K + E_D)^2 &= M_B^2 + \left(p_D^2 + p_K^2 + 2\vec{p}_D\vec{p}_K\right),\\ (E_D - E_v)^2 &= m_\pi^2 + \left(p_D^2 + p_v^2 - 2\vec{p}_D\vec{p}_v\right),\end{aligned} \tag{2.1}$$

where v refers to the "virtual" particle having 4-momentum equal to the sum of the charged tracks in the D-vertex. These can be rewritten as

$$\begin{aligned}\vec{p}_D\vec{p}_K &= \frac{M_K^2 + M_D^2 - M_B^2}{2} + E_K E_D,\\ \vec{p}_D\vec{p}_v &= -\frac{M_D^2 + M_v^2 - m_\pi^2}{2} + E_v E_D.\end{aligned} \tag{2.2}$$

Since $\vec{p}_B = \vec{p}_D + \vec{p}_K$, the three vectors lie in the same plane. Also, the B-vertex is on the kaon track, hence $\vec{p}_D = \lambda\vec{Q} + \mu\vec{p}_K$, the two constants λ and μ being determined from (2.1) as linear functions of E_D with some coefficients:

$$\begin{aligned}\lambda &= \alpha E_D + \beta,\\ \mu &= a E_D + b.\end{aligned} \tag{2.3}$$

However, $p_D^2 = E_D^2 - M_D^2 = (\lambda\vec{Q} + \mu\vec{p}_K)^2$, resulting in a second order equation in E_D:

$$E_D^2 - M_D^2 = E_D^2\left(\alpha\vec{Q} + a\vec{p}_K\right)^2 + E_D \cdot 2\left(\alpha\vec{Q} + a\vec{p}_K\right)\left(\beta\vec{Q} + b\vec{p}_K\right) + \left(\beta\vec{Q} + b\vec{p}_K\right)^2. \tag{2.4}$$

For convenience $\vec{Q}$ has the direction of $\vec{Q}$ in the figure and the absolute value of 1 GeV/c.

The ambiguity raised by the second order equation is solved in favor of the "+" sign. What actually happens is that $(\beta\vec{Q} + b\vec{p}_k)$ is rather small $(\vec{\epsilon})$ and the equation looks like

$$E_D^2 - M_D^2 = \vec{\chi}^2 E_D^2 + \vec{\epsilon}(2\vec{\chi}E_D + \vec{\epsilon}). \tag{2.5}$$

The solution with "−" sign is therefore negative in value, hence unphysical.

Once E_D is known, the coefficients λ and μ are known and so are $\vec{p}_D$ and $\vec{p}_B$.

Having the D-vertex and $-\vec{p}_D$ pointing back at the kaon track, a B-vertex can be now inferred. The problem is formulated as follows: the B-vertex is a point inside the $1\sigma_K$ kaon track error-tube with the property that aiming along the $\vec{p}_D$ direction, the D-vertex is missed by about $1\sigma_D$, while aiming along the $-\vec{p}_B$ direction, the IP is missed by about $1\sigma_{IP}$. The kaon track is parametrised as $\vec{r}_{0K} - \lambda_K\vec{n}_K$ where $\vec{n}_K = \vec{p}_K/p_K$, $\vec{r}_{0K}$, some origin on the track, and λ_K, a running parameter describing the track. Therefore, if $\vec{B}$ is the B-vertex, the above requirements translate into

(1) kaon track tube, σ_K, from $-\vec{r}_{0K} + \lambda_K\vec{n}_K + \vec{B}$,

(2) D-vertex $\pm\ \sigma_D$, from $\vec{B} + \lambda_D\vec{p}_D$,

(3) IP $\pm\ \sigma_{IP}$, from $\vec{B} + \lambda_B\vec{p}_B$.

This implies minimising the above 3 constraints simultaneously:

$$\begin{aligned}&\left\langle \vec{B} + \lambda_K\vec{n}_K - \vec{r}_{0K}\ \middle|\sigma_K^{-2}\middle|\vec{B} + \lambda_K\vec{n}_K - \vec{r}_{0K}\right\rangle \\ &\quad + \left\langle \vec{B} + \lambda_D\vec{p}_D - \vec{v}_D\middle|\sigma_D^{-2}\middle|\vec{B} + \lambda_D\vec{p}_D - \vec{v}_D\right\rangle \\ &\quad + \left\langle \vec{B} + \lambda_B\vec{p}_B - \vec{v}_{IP}\middle|\sigma_{IP}^{-2}\middle|\vec{B} + \lambda_B\vec{p}_B - \vec{v}_{IP}\right\rangle\end{aligned} \tag{2.6}$$

with respect to the three running parameters λ_K (units [m]), λ_D, and λ_B (both in units of [m]/[GeV/c]) and to $\vec{B}$. Annulling the respective derivatives:

$$\begin{aligned}\lambda_K &= \frac{\left\langle \vec{B} - \vec{r}_{0K}|\sigma_K^{-2}|\vec{n}_K\right\rangle}{\langle\vec{n}_K|\sigma_K^{-2}|\vec{n}_K\rangle}, \\ \lambda_D &= \frac{\left\langle \vec{B} - \vec{v}_D|\sigma_D^{-2}|\vec{n}_D\right\rangle}{\langle\vec{n}_D|\sigma_D^{-2}|\vec{n}_D\rangle}, \\ \lambda_B &= \frac{\left\langle \vec{B} - \vec{v}_{IP}|\sigma_{IP}^{-2}|\vec{n}_B\right\rangle}{\langle\vec{n}_B|\sigma_{IP}^{-2}|\vec{n}_B\rangle}.\end{aligned} \tag{2.7}$$

Minimising with respect to the B-vertex $\vec{B}$, the following is obtained:

$$\left|\vec{B}\right\rangle = \left[\sum_{x=K,D,B} \sigma_x^{-2}\left(1 - \frac{|\vec{n}_x\rangle\langle\vec{n}_x|\sigma_x^{-2}}{\langle\vec{n}_x|\sigma_x^{-2}|\vec{n}_x\rangle}\right)\right]^{-1} |\vec{q}\rangle \tag{2.8}$$

with

$$|\vec{q}\rangle = \sum_{x=K,D,B} \sigma_x^{-2} \left(1 - \frac{|\vec{n}_x\rangle\langle\vec{n}_x|\sigma_x^{-2}}{\langle\vec{n}_x|\ \sigma_x^{-2}\ |\vec{n}_x\rangle} \right) |\vec{v}_x\rangle. \tag{2.9}$$

For the kaon track error tube an extended acception of error ellipsoid was used: one with infinite long-axis. In the case of a round cross-section (equal axes) the error-ellipsoid operator reduces to the perpendicular to track projector. Therefore (in $\vec{q}$) only the perpendicular to track error of $\vec{r}_{0K}$ will contribute to the B-vertex error.

Similarly, for spherical error-ellipsoids, the D-vertex and the IP would contribute only with their respective errors perpendicular to $\vec{n}_D$ and $\vec{n}_B$. This is not exactly the case, both ellipsoids being elongated, the IP along z and the D-vertex along its flight direction. This extra elongation simply translates into the addition of a parallel projector along said directions to the respective error-ellipsoid operators. For the D-vertex the elongation being exactly along $\vec{n}_D$ and greater than the spherical part of the operator, the combined effect in $\vec{q}$ is the same as having only the spherical part. For the IP the elongation is along z, which is not $\vec{n}_B$, but almost. So again, to first approximation, only the waist of the IP will play a determinant part in the B-vertex errors. The IP being very accurate waist-wise (for the LHCb-IP: $\sigma_{xy} = 20\,\mu$m versus $\sigma_z = 60\,\mu$m [14]), the main contribution will come from the D-vertex—set by smearing to a typical resolution of this detector [15].

The code was tested using a PYTHIA simulation of the LHC environment for the LHCb experiment (LHC energy $b\bar{b}$ events, "HardQCD:gg2bbbar"), for which the tracks were smeared both in momentum value and direction (like a needle deviating in direction within an error-tube). The momentum value smearing was chosen such as to give 2-body mass resolutions as those reported by this experiment in literature [16, 17] (ca. 12 MeV/c^2), while the error-tube diameter (ca. 15 μm) was chosen such as to give the vertex locator resolutions reported in the literature [17] (ca. 150 μm longitudinally and 25 μm transversally). Once the particle tracks and vertices are obtained in a form simulating detector response, they are fed to the code for reconstruction. Figure 2 shows the reconstructed B-vertex resolution in terms of proper time, determined from $\tau_{\text{proper}} = d_{\text{flight}} M_B / c p_B$. The physically relevant directions are parallel to flight direction and transverse to this. The method uses 3 constraints combined, hence a rudimentary approximation is to allot a gaussian in the resolution fit for each. The top part of the figure shows the residuals with respect to flight-direction (which is almost z-aligned, fitted with 3 Gaussians). The combined resolution is on the order of 32 fs. The bottom part shows residuals transverse to the flight direction (approximately the xy-plane) fitted with 3 half-Gaussians centered at zero, resolution on the order of 5.4 fs. It is evident that the resolution is dominated by the boost direction (flight direction elongated error-ellipsoid). This is to be expected due to the high boosts, where the shallow opening angles among tracks translate pixel detector errors into large longitudinal ones.

The obtained resolution is similar to usual resolutions in B-reconstructions achieved in 4 track decays [18] (40 fs). The obvious advantage here is that although particles are missing, their kinematics helps "recover" them as if they were present and this is an exclusive reconstruction.

The method can also be used for detector studies, such as calibrating Particle ID, or calorimetry, in the lepton (electron/muon) sector. This implies in general using semileptonic modes. Although the traditional leptonic decay of J/ψ is standard, it yields two hard tracks, quite close to each other (which in the electron case makes difficult estimating the individual

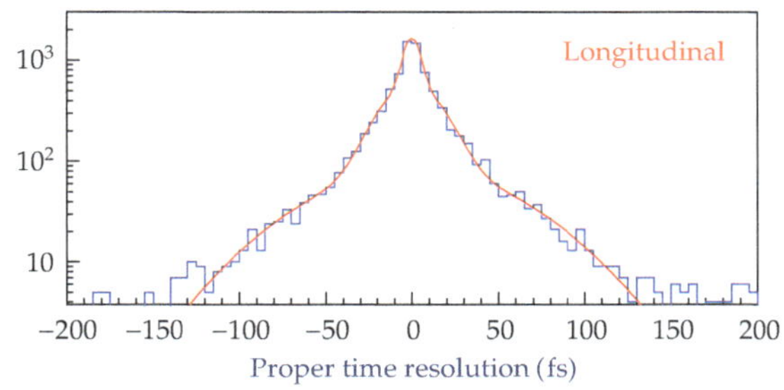

(a)

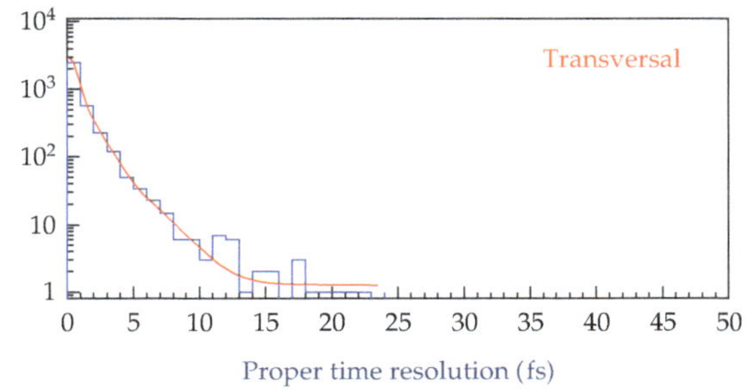

(b)

Figure 2: Residuals of the reconstructed B-vertex with respect to the Monte Carlo position: (a) projected onto the B-flight direction, (b) transverse to the B-flight direction. The dimensions are proper time in fs. Three Gaussians were used in the fits (in the case of the transverse centered at zero), the global resolutions being ca. 32 fs (longitudinally) and 5.4 fs (transversally)—details are in the text.

radiative losses). Also Particle ID performs lesser on tracks very close to one another, the real performance (used in B-physics) being actually better. Therefore, the larger variety given by semileptonic modes would be preferable, if only a pure enough (better than Particle ID miss-ID rates) sample could be selected. Consider, for instance, the mode $B_d^0 \rightarrow D^{*-} (K^-\pi^+l^-\overline{\nu}_l)$ $\pi^+\pi^-\pi^+$. The B-vertex provides 3 "clean" pion tracks useful for Particle-ID studies in the hadron sector. The D-vertex would also be good, if only labels could be attached to the tracks. Owing to the large mass difference between kaon/pion and electron, the equations would prevent a vertexable solution if the masses were wrongly attached. However, in the case $l = \mu$, the pion and lepton tracks are similar in mass. Still, they are quite different in dynamics, one coming from a lepton line, united to the baryon half of the reaction via a W^-, while the other from a continuous quark line. The imbalance in this case would come from the different momentum properties (putting the wrong momentum in $M_{l\nu_l}$, for instance, versus other dynamic quantities in the reaction). In this sense the method can act as a track-labeler in the D-vertex, offering a quite clean sample of B-origin leptons for Particle-ID studies—which is rare. A study was performed for kaon/pion separation power of the method. The mode with no π^0 in the final state was chosen. Of the existing tracks in the D-vertex one was purposely reported "lost" to the program and its mass fed to the code. If the kaon track is reported "lost," but pion mass is given for it to the code, the inner kinematics of the reactions will work out wrong, further influencing the (highly nonlinear) vertexing procedure described aforehand. The vertexing produces now new momenta for all particles implied and an invariant mass can be recomputed for the "lost" particle's mass (different, but close, to the one inputted). This mass error versus pointback is a criterion that gives a ca. 45% kaon-pion separation. As pointed above, for leptons additional handles (due to inner dynamics) exist (i.e., $-M_{l\nu_l}$) and so the rejection power should be better (especially for the electron, where the mass is also much lighter).

3. Conclusions

The method aims to recover modes otherwise nonusable, with missing neutrals lost (neutrino), or less precise (calorimetry). This is very important to both underground based experiments as well as accelerator-based experiments, where background rejection or measurement precision depends on the exclusive reconstruction of the event. The paper lifts the two-fold ambiguity arising in the second order equation involved, which otherwise relied on other criteria (pointback for instance) which cut on signal.

The potential of the method for Particle-ID source studies (lepton sector) is also discussed.

The methods studied could in principle be applied for all "stealth" particles (neutrino, π^0, other neutrals) decaying further downstream. Plans for extending the method could involve these options together with the Particle-ID studies.

Acknowledgment

This work was supported by a Grant of the Romanian National Authority for Scientific Research, CNCS-UEFISCDI, Project no. PN-II-ID-PCE-2011-3-0323.

References

[1] V. I. Goldansky, P. Yu. Nikitin, and I. L. Rozental, *Kinematic Methods in High-Energy Physics*, vol. 2 of *Scientific Reviews Supplement Series, Physics*, Harwood, Chur, Switzerland, 1989.

[2] N. I. Starkov and V. A. Ryabov, "Kinematic methods for analysis of neutrino interactions in calorimetric detectors," *Bulletin of the Lebedev Physics Institute*, vol. 2000N12, pp. 1–6, 2000, Sbornik Kratkie Soobshcheniya Po Fisike, vol. 2000N12, pp.3–8, 2000.

[3] P. Avery, *CSN 98-355 Data Analysis & Kinematic Fitting with KWFIT*, 1998.

[4] P. Avery, *Kinematic Fitting Algorithms and Lessons Learned from KWFIT Padua*, Computing in High Energy and Nuclear Physics, 2000.

[5] BaBar KinFitter, BAD #1061.

[6] B. Mitrica, "Asymmetry of charge ratio for low energetic muons," *AIP Conference Proceedings*, vol. 972, pp. 500–504, 2007.

[7] P. Schreiner and M. Goodman, in *Proceedings of the 30th International Cosmic Ray Conference (ICRC '05)*, vol. 9, p. 97, Pune, India, 2005.

[8] A. Rubbia, "The LAGUNA design study—towards giant liquid based underground detectors for neutrino physics and astrophysics and proton decay searches," *Acta Physica Polonica B*, vol. 41, p. 1727, 2010.

[9] D. Autiero, J. Äystö, A. Badertscher et al., "Large underground, liquid based detectors for astro-particle physics in Europe: scientific case and prospects," *Journal of Cosmology and Astroparticle Physics*, vol. 11, Article ID 011, 2007.

[10] L. Mosca, "Possibilities for the LAGUNA projects at the fréjus site," *AIP Conference Proceedings*, vol. 1304, pp. 303–311, 2010.

[11] LAGUNA FP7 Design Study, Grant Agreement No. 212343.

[12] LAGUNA-LBNO FP7 Design Study, Grant Agreement No. 284518.

[13] C. Amsler, M. Doser, M. Antonelli et al., "Review of particle physics," *Physics Letters B*, vol. 667, pp. 1–6, 2008.

[14] "LHCb Trigger System," *Nuclear Physics B—Proceedings Supplements*, vol. 156, supplement 1, p. 135, 2006.

[15] P. Spradlin, G. Wilkinson, and F. Xinga, "A study of tagged $D^0 \rightarrow hh'$ decays for $D^0 - \overline{D}^0$ mixing measurements," LHCb-2007-049 Report.

[16] S. Borghi, M. Gersabeck, C. Parkes et al., "First spatial alignment of the LHCb VELO and analysis of beam absorber collision data," *Nuclear Instruments and Methods in Physics Research A*, vol. 618, no. 1–3, pp. 108–120, 2010.

[17] A. Augusto Alves, L. M. Andrade Filho, A. F. Barbosa et al., "The LHCb Detector at the LHC," *Journal of Instrumentation*, vol. 3, p. S08005, 2008.

[18] R. Aaij, C. Abellan Beteta, A. Adametz et al., "Measurement of the $\overline{B}_s^0$ Effective Lifetime in the $J/\psi f_0$ final state," *Physical Review Letters*, vol. 109, Article ID 152002, 8 pages, 2012.

5

Status and New Data of the Geochemical Determination of the pp-Neutrino Flux by LOREX

M. K. Pavićević,[1] F. Bosch,[2] G. Amthauer,[1] I. Aničin,[3] B. Boev,[4] W. Brüchle,[5] V. Cvetković,[6] Z. Djurčić,[7] W. F. Henning,[8] R. Jelenković,[6] V. Pejović,[3] and A. Weiss[9]

[1] *Division of Material Sciences and Physics, Department of Mineralogy, University of Salzburg, Hellbrunner Street 34, 5020 Salzburg, Austria*
[2] *GSI Helmholtzzentrum für Schwerionenforschung, Planck Straße 1, 64291 Darmstadt, Germany*
[3] *Institute of Physics, University of Belgrade, Pregrevica 118, 11000 Belgrade, Serbia*
[4] *Faculty of Mining and Geology, Goce Delčev University of Štip, Goce Delčev 89, 92000 Štip, Macedonia*
[5] *GSI Helmholtzzentrum für Schwerionenforschung, Planck Straße. 1, 64291 Darmstadt, Germany*
[6] *Faculty of Mining and Geology, University of Belgrade, Đušina 7/II, 11000 Belgrade, Serbia*
[7] *Physics Division, Argonne National Laboratory (ANL), 9700 South Cass Avenue, Argonne, IL 60439, USA*
[8] *Physics Department E12, Technical University of Munich, James-Franck Street, 85748 Garching, Germany*
[9] *Max-Planck Institute for Astrophysics, Karl-Schwarzschild Street 1, 85741 Garching, Germany*

Correspondence should be addressed to M. K. Pavićević, miodragk.pavicevic@gmail.com

Academic Editor: Bogdan Mitrica

LOREX (LORandite EXperiment) addresses the determination of the solar (pp) neutrino flux during the last four million years by exploiting the reaction $^{205}Tl + \nu_e \rightarrow {}^{205}Pb + e^-$ with an incomparably low-energy threshold of 50 keV for the capture of solar neutrinos. The ratio of $^{205}Pb/^{205}Tl$ atoms in the Tl-bearing mineral lorandite provides, if corrected for the cosmic-ray induced background, the product of the flux of solar neutrinos and their capture probability by ^{205}Tl, averaged over the age of lorandite. To get the mean solar neutrino flux itself, four problems have to be addressed: (1) the geological age of lorandite, (2) the amount of background cosmic-ray-induced ^{205}Pb atoms which strongly depends on the erosion rate of the lorandite-bearing rocks, (3) the capture probability of solar neutrinos by ^{205}Tl and (4) the extraction of lorandite and the appropriate technique to "count" the small number of ^{205}Pb atoms in relation to the number of ^{205}Tl atoms. This paper summarizes the status of items 1 (age) and 3 (neutrino capture probability) and presents in detail the progress achieved most recently concerning the items 2 (background/erosion) and 4 ("counting" of ^{205}Pb atoms in lorandite).

1. Introduction

The determination of the long-time average of the solar neutrino flux Φ_ν with the thallium-bearing mineral lorandite, $TlAsS_2$, from the mine of Allchar, Macedonia, is based on the neutrino capture reaction proposed by Freedman et al., 1976 [1]:

$$^{205}Tl + \nu_e \longrightarrow {}^{205}Pb + e^- \tag{1.1}$$

The average flux $\langle\Phi_\nu\rangle$ over the exposure time a (age of lorandite since its mineralization) follows from the common activation equation:

$$\Phi_\nu = N^{-1}(T - B)(\sigma\varepsilon)^{-1}\,\lambda\,[1 - \exp(-\lambda a)]^{-1}, \tag{1.2}$$

where N is the total number of ^{205}Tl atoms, T the total number of ^{205}Pb atoms, B the number of ^{205}Pb atoms induced by background reactions (mainly ^{205}Tl $(\mu\, p, n)$ ^{205}Pb), σ the solar neutrino capture cross-section of ^{205}Tl, ε the overall detection efficiency, λ the decay constant of ^{205}Pb and a the geological age of lorandite (i.e., the age of mineralization).

The neutrino capture reaction:

$$^{205s}Tl + \nu_e\ (E_{\nu e} \geq 50\,\text{keV}) \longrightarrow^{205} Pb^* + e \tag{1.3}$$

exploits the *by far lowest threshold of* $E_{\nu e} \geq 50$ keV for solar neutrinos.

The last step and the *central problem* of LOREX *is the quantitative determination of* ^{205}Pb atoms in lorandite. This renders finally the mean solar neutrino flux, that is, *the mean luminosity of the sun during the last 4.31 million years,* which is the geological age a of lorandite from Allchar, as given by Neubauer et al. [2]. Before entering the final phase of the experiment, four problems must be reliably addressed; however,

(1) the age a of lorandite has to be known precisely: *age of lorandite,*

(2) the background of ^{205}Pb atoms produced by cosmic radiation (mainly $(\mu\, p, n)$-reactions) and by natural radioactivity must be determined quantitatively. In this context the knowledge of the erosion rate of the overburden rocks during the existence of lorandite is of utmost importance: *background and erosion,*

(3) the ratio $^{205}Pb/^{205}Tl$ provides only the *product* of solar neutrino flux and neutrino capture probability into different nuclear states of ^{205}Pb. However, the capture of neutrinos should populate predominantly the first excited state at $E^* = 2.3$ keV. Hence, to get the neutrino flux itself, one has to determine the capture probability into this low-lying state of ^{205}Pb: *neutrino capture probability into the* 2.3 keV *state of* ^{205}Pb,

(4) how can the expected *ultra-low abundance* of ^{205}Pb be reliably measured: *detection of* ^{205}Pb.

2. Main Directions of Present and Future Investigations

2.1. Age of Lorandite

The age of the Tl mineralization is an essential parameter of the experiment, as it equals the integration time of mineral lorandite as the geochemical pp-solar neutrino detector. As it turned out, another important prerequisite for the realization of LOREX is that the ore deposit has to provide for more than two lorandite-rich ore bodies, located at significantly different depths.

Geological research demonstrated that lorandite in sufficient quantities occurs in the Allchar Sb-As-Tl-Au deposit located in FYR Macedonia at the north-western margins of the Kožuf Mountains, close to the border between FYR Macedonia and Greece. The mine contains the world's largest known concentrations of thallium-bearing minerals and, in particular, the mineral lorandite. The polychronous and polygenetic Sb-As-Tl-Au Allchar deposit was formed by complex physicochemical processes occurring in a heterogeneous geological environment. The deposit originated through polyphase interactions of hydrothermal fluids and the surrounding magmatic, sedimentary, and metamorphic rocks. The genesis of ore mineralization is genetically related to the products of Pliocene magmatic activity. The spatial location of the mineralization was controlled by magmatic, structural, and lithological factors as suggested by Janković and Jelenković [4]. The Allchar deposit comprises several ore bodies of various shapes, textural and structural characteristics, and associations of elements.

Thallium mineralization has been proved in two locations, that is, two ore bodies: Crven Dol and Centralni Deo ore bodies, situated in the northern and central part of the Allchar deposit, respectively, (Figure 1).

Troesch and Frantz determined the geological age of Tl mineralization in Crven Dol (Figure 1(a)) as 4.22 ± 0.07 Ma, using sanidine from adit number 21 [5]. This determination was obtained by isochrone $^{40}Ar/^{39}Ar$ five-degree measurement of argon content with the derivation 2σ of the first phase and 1σ of other four phases.

During the final stages of volcanic activity in the Allchar area occurred hydrothermal alteration of the wall rocks, and this can be considered as the period of initial deposition of ore minerals in this area. The age of K-feldspars (alkali feldspar has a general formula: $(K, Na)AlSi_3O_8$, but varies in crystal structure depending on the temperature it formed at; the formula actually is a blend ranging from pure sodium (albite) to pure potassium end-member orthoclase, microcline, or sanidine) in an altered zone at the Rudina locality is 4.31 ± 0.02 Ma according to Neubauer et al. [2], which documents the beginning of intensive alteration and deposition of thallium mineralization. The close spatial association of mineralization and hydrothermally altered rocks is found by a drill-core B-2 (Figure 1(b)) of the Rio Tinto company in 1999 at the Rudina locality, which is geologically connected with the ore body Centralni Deo. These drill cores revealed the presence of As-, Sb-, and S-mineralization and strongly altered rocks at depth intervals 30–60 m and 160–280 m, respectively, from Rio Tinto Co. [3].

2.2. Background Reactions and Cosmic Ray Contribution

In the case of LOREX more than 30 processes have been identified and analyzed which potentially contribute to the "background" of ^{205}Pb. After careful evaluation only four processes turned out to potentially have nonnegligible contributions:

(1) the $^{205}Tl(\mu p, n)^{205}Pb$ reaction: contribution of *fast* muons,

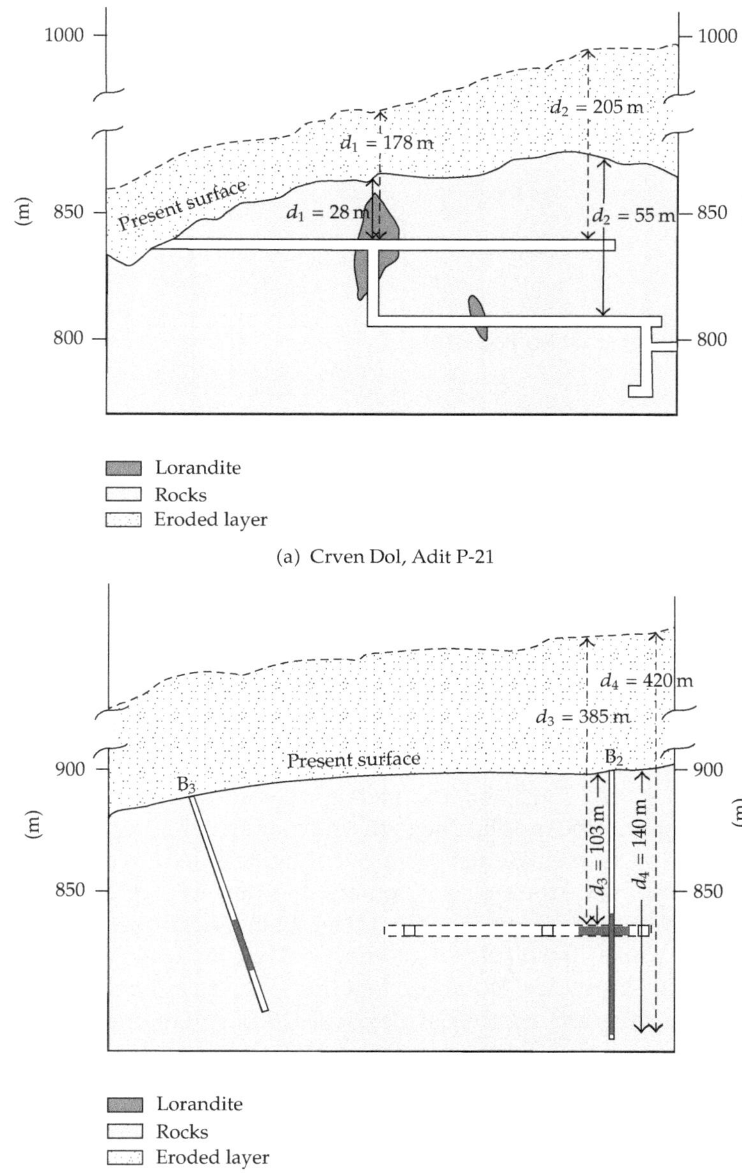

(a) Crven Dol, Adit P-21

(b) Centralni Deo, Adit 823

Figure 1: Geological cross-section of the ore body Crven Dol (a) and Centralni Deo (b) of the Allchar ore deposit together with the presumably eroded layer.

(2) the $^{205}\mathrm{Tl}(\mu p, n)^{205}\mathrm{Pb}$ reaction: contribution of *stopped* muons,

(3) the $^{204}\mathrm{Pb}(n,\gamma)^{205}\mathrm{Pb}$ and $^{206}\mathrm{Pb}(n,2n)^{205}\mathrm{Pb}$ reactions,

(4) $^{205}\mathrm{Pb}$ mobilized from the environment of lorandite.

Figure 2 shows present estimates of different contributions to the production of $^{205}\mathrm{Pb}$ in lorandite on the basis of the erosion rate 70 m/Ma and 130 m/Ma, traces of U and Th

Table 1: Cosmic-ray contributed N_{μ^-} and total number N_t of ^{205}Pb atoms per 10 kg of lorandite from different depths, known to contain significant quantities of lorandite.

Depth of location (m)	Erosion rate (m/Ma)	Paleo-depth d_p of location (mwe)	$N(\mathrm{Pb}^{205})_{\text{fast muons}}$ × 10^4 (at. 10 kg^{-1} lorandite)	$N(\mathrm{Pb}^{205})_{\text{total}}$ × 10^4 (at. 10 kg^{-1} lorandite)	$N(\mathrm{Pb}^{205})_{\nu}$ %	$N(\mathrm{Pb}^{205})_{\text{fast muons}}$ %
$d_1 = 28$	70	455	93.8	115.8	19.0	81.0
$d_2 = 55$	70	525	56.9	78.9	27.9	72.1
$d_3 = 103$	130	980	18.6	40.6	54.2	45.8
$d_{4\,\max} = 140^*$	130	1070	13.0	35.0	62.8	37.2

*Drillings of Rio Tinto Exploration Ltd. [3] (see Figure 1(b)).

in lorandite and surrounding rocks by Pavićević [6], and the method by Heisinger and coworkers [7] as a function of the paleodepth d_p of the deposit.

To determine the erosion rate in Allchar we applied terrestrial in situ cosmogenic nuclides including both radioactive (^{10}Be and ^{26}Al) and stable (^{3}He and ^{21}Ne) nuclides in different minerals and compared the obtained data with the results of geomorphologic analysis. ^{26}Al concentrations of 16 samples have 1σ uncertainties of 26–58% and yield maximum erosion rates from ~70 m/Ma to ~600 m/Ma at different locations. The noble gas analyses included measurements of ^{3}He in one diopside sample and ^{21}Ne in quartz from five different locations and in one sanidine sample, revealing erosion rates between ~20 m/Ma and ~70 m/Ma, with asymmetric 1σ uncertainties between 10 and 200%. The ^{10}Be measurements revealed unrealistically high values and had to be omitted from interpretation. Both the combined ^{21}Ne-^{26}Al data and the geological information indicate that a simple steady-state erosion history is not applicable to the Allchar area. Rather, long periods of burial (~500 ka–3 Ma) by volcaniclastic material and, possibly, glacier ice have interrupted the times of exposure and erosion. Nevertheless, it is remarkable that 85% of all erosion rate values determined from different cosmogenic nuclides (^{3}He, ^{21}Ne, ^{26}Al) in various monitor minerals at various locations, as well as the results of geomorphologic analysis, yield consistent results in the range from ~20 to ~90 m/Ma. We argue that the ^{26}Al concentrations most likely reflect the erosion rate during the last, postburial period (~100 ka). For the two most important Allchar locations with lorandite ore bodies Crven Dol and Centralni Deo we propose the erosion rates of ~70 m/Ma and ~130 m/Ma, respectively, according to Pavićević et al. [8]. The paleodepths for the four confirmed ore-bodies with significant quantity of lorandite are found from their present-day depths and the estimated thickness of eroded layers at corresponding locations (Figure 1 and Table 1).

Table 1 also lists the final estimates of concentrations of ^{205}Pb (atoms per 10 kg lorandite) at different depths; the concentration due to the cosmic-ray fast muons, the concentration due to pp-neutrino captures, and the total concentration. For the shallowest location d_1 the neutrino contribution amounts to ~20%, while for the deepest one d_4 it amounts to ~63%.

As seen in Figure 2, concentrations of ^{205}Pb in samples from different depths reflect—due to the muon contribution which is depth dependent—both the present-day depth of the sample as well as the erosion history of the overburden layers. Curves that correspond to different average erosion rates form a family of exponentials (Aničin et al. [10]) and experimental determination of ^{205}Pb concentrations in samples from a number of different depths allow for the simultaneous determination of both the muon contribution and

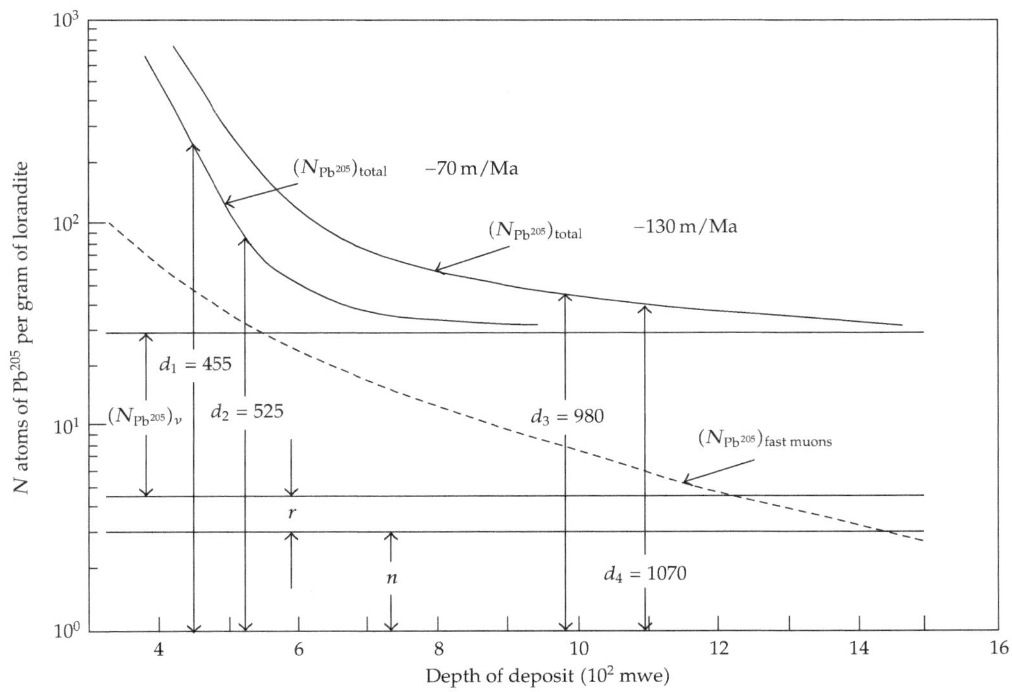

Figure 2: Estimate of the present amount of ^{205}Pb in lorandite due to solar neutrinos and to background reactions in the last $4.31 \cdot 10^6$ years as a function of the paleodepth d_p of lorandite. The values are calculated by Pejović by adopting the value of 0.55 μb for the neutrino capture cross-section for ^{205}Tl from Hartman et al. [9], the contents of U and Th in lorandite and surrounding rocks are from Pavićević [6] and the method of Heisinger et al. [7]. *n*: contributions due to lead mobilized from the rock walls. *r*: contributions from natural radioactivity. $N(Pb^{205})_{fast\ muons}$: contribution due to reactions induced by fast cosmic muons. $N(Pb^{205})_\nu$: number of Pb^{205} due to solar neutrino capture, for a capture rate of 146 SNU (1 SNU = 10^{-36} captures/(target atoms · sec)), obtained after correcting the original 260 SNU value for neutrino flavour-oscillations. $N(Pb^{205})_{total}$: the sum of the neutrino contribution and of all background contributions.

the erosion history of the deposit. The contribution due to the ^{205}Pb mobilized from the surrounding rock at the time of lorandite mineralization, marked by "*n*", and that due to natural radioactivity in lorandite and in the surrounding rocks, marked by "*r*", contribute together by less than 4% to the total concentration (U and Th concentrations are low: marble, dolomite, and quartz latite have contents of U in the range of 0.054–0.45 ppm and Th of 0.061–0.66 ppm, while the mean concentrations in lorandite amount to 0.12 ppm for U and 0.022 ppm for Th, respectively, as given by Pavićević [6].

Estimate of contribution of "*n*," that is, mobilised ^{205}Pb from surrounding rocks at the time of the formation of the Tl-mineralization (age of lorandite) and the natural radioactivity contribution "*r*," as consequence trace concentrations of U and Th amount to less than 4% (see Figure 2). The low contribution of "*n*" is consequence of the very low concentration of U and Th in rocks: marble, dolomite, and Qz-rhyolite, which have the content of U = 0.054–0.45 ppm and Th 0.061–0.66 ppm (Pavićević [6]). The contribution of natural radioactivity "*r*" that is predominantly caused by U and Th in lorandite is low due to low concentrations of these elements in lorandite: U = 0.12 ppm and Th 0.022 ppm

(Pavićević [6]). All background components not specific to lorandite (not originating from thallium) were estimated to amounts of between 3 and 15 SNU (Neumaier et al. [11]) and can experimentally be determined by measuring the ^{205}Pb contents in monitoring minerals (realgar As_2S_2 and orpiment As_2S_3).

From the data for $N(Pb^{205})_\nu$ and $N(Pb^{205})_{\text{fast muons}}$ shown in Table 1 we estimated by a Monte Carlo simulation the mean neutrino flux $\langle \Phi_\nu \rangle$ (1.2) during the geological age a of lorandite as

$$\langle \Phi_\nu \rangle = (3.41 \pm .44) 10^{10}\,\mathrm{m}^{-2}\,\mathrm{sec}^{-1} \tag{2.1}$$

by using the following parameters (see (1.2)):

$$N = 1.25 \cdot 10^{25} \left(2.5 \cdot 10^{23}\right) \text{ atoms of } {}^{205}\mathrm{Tl}. \tag{2.2}$$

This number corresponds to 10 kg of lorandite, collected at a sufficiently deep location, for the background B to be acceptable, namely, at a paleo-depth of $d_p = 980$ m, where the calculated background B of ^{205}Pb atoms amounts to about 46%, (see third row of Table 1)

$$\begin{aligned} \sigma &= 4.3 \cdot 10^{-45} \left(2.15 \cdot 10^{-46}\right) \mathrm{cm}^2, \\ a &= 1.26 \cdot 10^{14} \left(1.26 \cdot 10^{12}\right) \mathrm{s}, \\ \lambda &= 1.48 \cdot 10^{-15} \left(7.4 \cdot 10^{-17}\right) \mathrm{s}^{-1}, \\ \varepsilon &= 1 \cdot 10^{-3} \left(1 \cdot 10^{-4}\right). \end{aligned} \tag{2.3}$$

Whereas the geological age a of lorandite and the decay constant λ of ^{205}Pb are by now satisfactorily well known, the solar neutrino capture cross-section σ of ^{205}Tl is experimentally *unknown*. The value assumed in (2.3) is not more than the best educated guess, based on previous estimates of Bahcall and other theorists (Bahcall [12], Wapstra and Audi [13], Braun and Talmi [14], Freedman and Nolte [15], ana Ogawa and Arita [16]). How the nuclear matrix element for the dominant neutrino capture channel into the first excited state of $^{205}\mathrm{Pb}(E^* = 2.3\,\mathrm{keV})$ can be *experimentally* determined will be outlined in the next Section 2.3. Likewise, the total detection efficiency ε assumed in (2.3) is only a (optimistic) guess. It is obvious that one of the most demanding problems of LOREX is the measurement of the extremely low concentrations of ^{205}Pb in potentially available samples of a generally very scarce material. Methods that might be suited to this difficult task are discussed in Section 2.4.

2.3. Determination of the Neutrino Capture Probability into the 2.3 keV State of ^{205}Pb

From the relative abundance of ^{205}Pb with respect to ^{205}Tl, even corrected for all background effects, one can extract only the time-integrated *product* of solar *neutrino flux* and the *neutrino-capture probability,* leading from the nuclear ground state of ^{205}Tl ($I^{\pi} = 1/2^{+}$) to individual excited nuclear states of ^{205}Pb*, according to

$$^{205}\text{Tl g.s.} \ + \ \nu_e \, (E_{\nu e} \geq 52\,\text{keV}) \longrightarrow {}^{205}\text{Pb}^{*} + \text{e}^{-} \tag{2.4}$$

with an energy threshold at $E_{\nu e} = 52\,\text{keV}$, much lower than the corresponding thresholds of all other radio- or geochemical experiments which are either already performed (e.g., GALLEX) or proposed. The solar pp-neutrinos with a maximum energy of 422 keV can be captured only into the ground state ($I^{\pi} = 5/2^{-}$), the first ($E^{*} = 2.3\,\text{keV}, I^{\pi} = 1/2^{-}$), and the second excited state ($E^{*} = 263\,\text{keV}$, $I^{\pi} = 3/2^{-}$) of ^{205}Pb. It is obvious from the well-known systematic of beta decay that the capture probability into the 2.3 keV state ($\Delta I = 0$, parity change) should be by far larger than the capture into the ground state ($\Delta I = 2$) as well as into the second excited state with $\Delta I = 1$ (Behrens and Jaenecke [17]). The nuclear matrix element for the transition to the 2.3 keV state in ^{205}Pb is *not* known. There is one—and only one—possibility (Kienle [18]) to determine experimentally this unknown nuclear matrix element, namely, by measuring the bound-state beta decay (β_b decay) of bare (or hydrogen-like) ^{205}Tl^{81+} into hydrogen-like (helium-like) ^{205}Pb^{81+}, according to

$$^{205}\text{Tl}^{81+} \longrightarrow {}^{205}\text{Pb}^{81+*} \ (E^{*} = 2.3\,\text{keV}) + \text{e}_{\text{b}}^{-} + \nu_e \ (\text{bar}), \tag{2.5}$$

where a neutron of the ^{205}Tl atomic nucleus transmutes to a proton, and where a monochromatic electron-antineutrino (ν_e (bar)) is being created together with an electron that remains *bound* (e_b^{-}) in an inner shell (mainly K or L-shell) of the daughter ion ^{205}Pb^{81+}. The Q value of this decay amounts to only 52 keV. Since the nuclear parts of neutrino capture and of β_b decay are identical, a measurement of the *bound state beta-decay probability of bare or hydrogen-like* ^{205}Tl *provides the unknown nuclear matrix element* (log ft value) for the transition to the first excited state of ^{205}Pb at $E^{*} = 2.3\,\text{keV}$.

Bound-state beta decay probabilities with comparable small Q values have been measured at the Experimental Storage Ring ESR of GSI in two pioneering experiments, for the examples of bare ^{163}Dy $\rightarrow$ hydrogen-like ^{163}Ho (Jung et al. [20]), and of bare ^{187}Re $\rightarrow$ hydrogen-like ^{187}Os (Bosch et al. [21]), and of bare ^{207}Tl $\rightarrow$ hydrogen-like ^{207}Pb (Ohtsubo et al. [22]). The well-developed technique of detecting β_b decay, as exploited in these experiments, is "Schottky lifetime spectrometry" (Litvinov et al. [23]), where the signals of the stored and electron-cooled parent- and daughter ions, induced on pick-up plates, are steadily recorded, yielding their revolution frequencies which are unambiguously correlated to their corresponding mass-to-charge ratios. From these "Schottky lines," the numbers of both parent and daughter ions are recorded as a function of time, yielding the β_b decay probability and, thus, the nuclear matrix element (log ft value) searched for. However, the revolution frequencies of parent and daughter ions could not be resolved in these experiments due to the very small Q values involved. Therefore, a special technique was applied, namely, the removal of the single bound electron of the daughter ions by means of an internal gas jet acting for a short time.

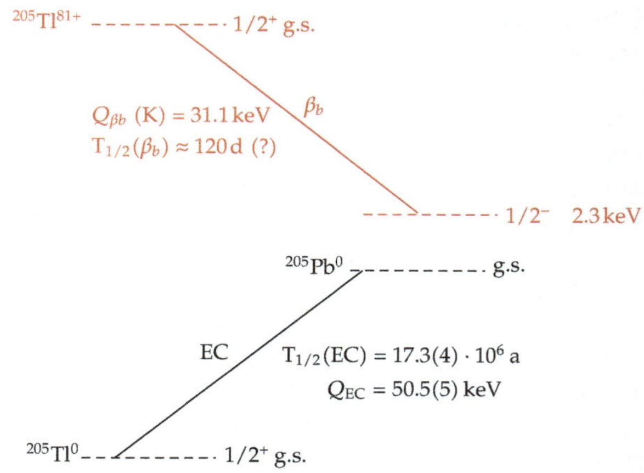

Figure 3: The neutral $^{205}Tl^0$ atom is stable with respect to β^- decay into the continuum since the Q value for this decay is negative, $Q_{\beta-} = -Q_{EC} = -50.5$ keV. However, for bare ($^{205}Tl^{81+}$) or hydrogen-like ^{205}Tl ions bound-state β decay becomes allowed, where the created electron remains bound in the K-shell of the daughter ion $^{205}Pb^{81+}$. Due to the "saved" K-shell binding energy, the Q value for this decay ($Q_{\beta b}$ (K)) gets positive. Both, the capture of solar pp-neutrinos and the bound-state beta decay populate predominantly *the first exited state* ($1/2^-$) of the ^{205}Pb nucleus at 2.3 keV and share the *same* nuclear matrix element for this transition. Therefore, the measurement of the bound-state β^- decay probability of bare or hydrogen-like ^{205}Tl ions provides the unknown capture probability of the solar pp-neutrinos by the ^{205}Tl atoms. Based on the log ft values of comparable β^- decays, the bound-beta half-life $T_{1/2}$ (β_b) of bare ^{205}Tl ions was estimated by Takahashi and Yokoi [19] to about 120 d. Since this extrapolated value has to be regarded as uncertain within (at least) a factor of three, only the measurement of the bound-beta half-life can provide a safe number with a reliable error margin for the capture probability of solar pp-neutrinos by ^{205}Tl.

For the corresponding experiment concerning the β_b decay of bare ^{205}Tl (see Figure 3) with a comparable small Q value, exactly the same detection technique as in the two successful experiments mentioned above has been proposed. However, there were—and still are—serious and too expensive safety restrictions for the use of Tl in an ion source due to its toxicity, which prevented the realisation of this experiment over the last 18 years. Nonetheless, since this proposal remained all the time on the list of high-priority experiments, we were encouraged to search for another way of its realisation. We proposed as the "new technique" getting bare ^{205}Tl ions their *indirect* production by in-flight fragmentation of a primary ^{206}Pb beam in the fragment separator (FRS) together with their subsequent injection into the experimental storage ring (ESR). The decisive point is that *at least* 10^6 *bare* ^{205}Tl *ions* have to be provided and stored, in order to obtain a minimum number of a few 100 β_b decays within storage times of a couple of hours (the β_b half-life of bare ^{205}Tl is estimated to 120 days (Takahashi and Yokoi [19]). By two major achievements of the last years this indispensable number of at least 10^6 stored bare ^{205}Tl ions can now be reached at high confidence (see Figure 4).

(1) By a significant improvement of the output of the lead ion source and the reduction of beam losses, a number of $2 \cdot 10^9$ ^{206}Pb ions could be delivered safely to the FRS already in two experiments. Based on the known cross-section for in-flight fragmentation and the known distribution of atomic charge states, this would yield a number of about 10^5 *bare* ^{205}Tl *ions per injection into the ESR.*

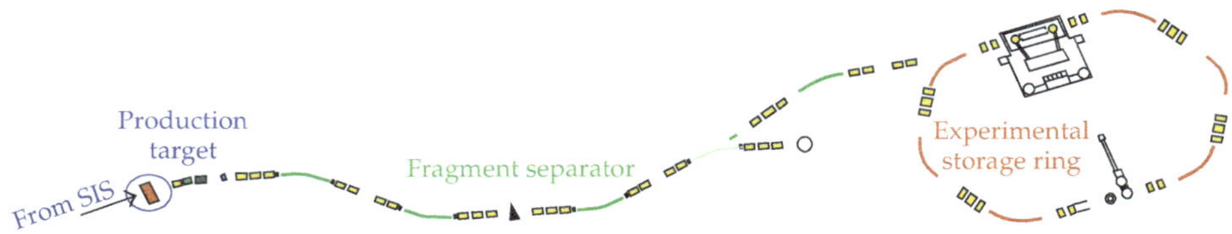

Figure 4: Part of the accelerator facility of GSI, relevant for the proposed measurement of the bound-beta half-life of bare $^{205}Tl^{81+}$ ions. A beam of ^{206}Pb ions, accelerated in the synchrotron SIS to an energy of some 100 A MeV, hits the production target (Be) of the fragment separator FRS, where a plenty of highly-charged ions will be generated by nuclear reactions. With the aid of an appropriate setting of the FRS magnets and by exploiting the Z-dependent stopping power in a degrader placed in the symmetry point of the FRS (black triangle), only bare $^{205}Tl^{81+}$ ions can pass through small slits at the end of the FRS. Those ions will be stored in the experimental storage ring (ESR) and accumulated by repeated injections until a number of at least 10^6 ions will be reached. After the application of electron cooling, the $^{205}Tl^{81+}$ ions are circulating in the ring for a couple of hours, where a few of them will decay to hydrogen-like ^{205}Pb by bound-state β decay.

(2) By the combination of stochastic cooling and the storage of the injected ions at inner orbits several subsequent injections ($10 \cdots 30$) can be applied without any losses of the already stored ions (rf stacking of fragments). This technique has been successfully applied in several experiments within the last years.

Due to these important steps the reach of 10^6 stored (and cooled) bare ^{205}Tl ions is feasible. Therefore, this new procedure for the determination of the β_b decay probability of bare ^{205}Tl—and, therewith, of the solar neutrino capture cross-section by ^{205}Tl—has been approved in June 2010 by the international experimental board of GSI (ESR-proposal 100). Six days of beam time and the highest priority "A" were assigned. A copy of this decision is attached in the supplement of this proposal. The experiment is now in the procedure of scheduling. Its realisation within the next 18 months is reliable, taking into account the "waiting list" for high-priority experiments at the FRS-ESR facility.

2.4. Extraction and Detection of Ultra-Low Amounts of ^{205}Pb in Lorandite

One of the key requirements of the geophysical LOREX project is the ultra-sensitive, quantitative detection of the neutrino-induced trace amounts of ^{205}Pb present in lorandite. For 10 kg lorandite ($TlAsS_2$) one estimates about 3.5–11.6 × 10^5 atoms of ^{205}Pb depending on the depth at Allchar at which the respective lorandite sample is mined. This implies a necessary separation factor of about 10^{20} between ^{205}Pb and ^{205}Tl (70% isotopic abundance). Separation factors of this extreme magnitude are perhaps directly achievable for very short-lived radioisotopes through the measurement of their decay radiation (Low Level Counting, LCC). For ^{205}Pb, with a half-life of about 17 million years and in view of the difficulty to detect electron capture decay, this is impossible.

For the LOREX, therefore, the direct detection and counting of the ^{205}Pb atoms was proposed, either with mass-spectrometric methods or with measurements of characteristic atomic transition radiation. The intrinsic limitation of these methods to small sample sizes is substantially helped with an initial chemical Pb-Tl separation, which has been established to provide a separation factor of 10^{13} (Pavićević and Bruechle [24]). For Allchar lorandite the Pb content has been found to be about 1.5 ppm (Pavićević [6]). Thus, after the initial chemical

Pb-Tl separation, a 20 mg Pb sample with a $^{205}Pb/Pb$-ratio of $\sim 10^{-15}$ will be obtained. The sample will also contain a residual amount of $\sim 10^{12}$ Tl atoms.

The methods that appear suitable, in principle, for the detection of ^{205}Pb at trace amount levels under consideration here are low-energy mass spectrometry (specifically ICP-MS, i.e., inductively coupled plasma mass spectrometry), high-energy mass spectrometry (i.e., AMS, accelerator mass spectrometry), and laser-induced atomic spectroscopy. The latter includes in particular resonance ionization mass spectrometry (RIMS) and the recently developed atom trap trace analysis (ATTA). Quality criteria in each case are the separation factors for Pb/Tl and $^{205}Pb/Pb$ and the overall detection efficiency. The latter is critical for a relevant statistical accuracy of the measured ^{205}Pb concentration. In summary, the unambiguous detection of the ^{205}Pb atoms with statistical uncertainty of less than 10%, requires a method that provides fully resolved separation of ^{205}Pb from ^{205}Tl at the 10^{-7} level, and ^{205}Pb from other lead isotopes at the 10^{-15} level, and detection efficiency for ^{205}Pb atoms of preferably 10^{-2} but at a minimum of 10^{-3}.

High-energy AMS relies on the acceleration of ions to energies where molecules are fully destroyed through breakup in stripper foils, the differential energy loss provides chemical element identification of the remaining ions, and powerful ion-beam detectors could be taken from heavy-ion nuclear physics research. Trace concentrations have been reliably measured for low-mass radioisotopes ($A < 40$) down to values of 10^{-16}. However, for heavier nuclei the nuclear-charge separation is less pronounced at the typical heavy-ion accelerator because of increased energy-loss straggling and decreasing relative nuclear charge resolution. Earlier experiments of $^{205}Pb/^{205}Tl$ separation at about 6 MeV per nucleon beam energy (Ernst et al. [25]; Henning and Schuell [26]) have produced separation factors of 10^2 to 10^3 between ^{205}Pb and ^{205}Tl, by far not sufficient for the 10^7 factor described above. In addition, the overall efficiency in these studies was found to be less than 10^{-6}, and in the most optimistic extrapolations it was difficult to see how this could be brought to values better than 10^{-4}.

A novel variant of AMS was then proposed, based on the recent developments of Schottky mass spectroscopy in a high-energy ion storage ring, following the acceleration of ions to several hundreds of GeV per nucleon energies and full stripping (Radon et al. [27]). The heavy-ion storage-cooler ring ESR at the GSI Helmholtz Zentrum in Darmstadt, Germany, is uniquely suited for providing a clear-cut solution of the "isobar problem" being such a serious obstacle for tandems. After acceleration in a high-energy synchrotron all atomic species can be *fully ionized*, and transferred to the ESR where they will be electron-cooled, affecting a common sharp velocity for all stored ions. In this case, the revolution frequency in the ESR depends only on the mass-to-charge (m/q) ratio of the ions which is—for fully stripped atoms—equal to m/Z. Therefore, isobars of different nuclear charge Z appear at widely separated revolution frequencies and are easily resolved. One also gets rid of any isobaric molecular contamination, which often represents the most serious obstacle for tandem-based AMS.

Moreover, the ESR provides the ultimate detection limit for heavy ($Z > 45$), highly charged ions: even *one single* stored and electron-cooled ion can be unambiguously detected (see Figure 5). The high resolving power and ultimate detection sensitivity of the ESR enabled "Schottky mass spectrometry (SMS)" that delivered new and precise atomic masses for more than 180 nuclides (Litvinov et al. [23]). Related studies on other heavy-mass radioisotopes have demonstrated the needed separation factor, but the overall efficiency cannot be brought above 10^{-4} for a heavy ion around mass 200 due to ion source efficiencies, the need for multiple stripping, and the general transmission losses experienced in the chain of accelerators to reach the required high energies. A variant we shall consider now is to

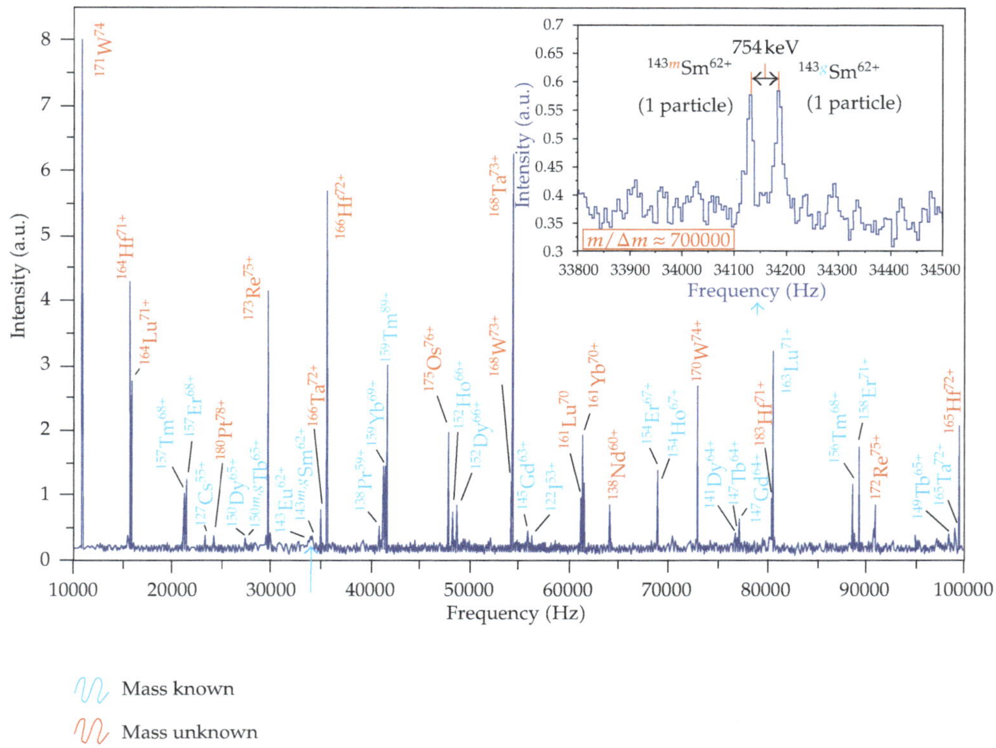

Figure 5: Schottky spectra versus the down-scaled revolution frequencies of stored and cooled nuclei, circulating in the experimental storage ring. Each of the lines shows a different nuclear species. Since for cooled ions a one-to-one correspondence exists between their revolution frequency and their mass-to-charge ratio, those spectra provide a highly precise mass spectroscopy. The inset shows the Schottky signals of *two* Sm^{62+} ions, one of them in the nuclear ground state (at the higher frequency), the other one in a metastable state at an excitation energy of 754 keV. This demonstrates convincingly the capability of the ESR to detect—with ultimate sensitivity—*one single* stored and cooled highly-charged ion together with its accurate mass determination (Figure by courtesy of Litvinov et al. [23]).

inject a lower charge state into the synchrotron accelerator and then to strip at the injection into the storage ring with very high efficiency. This eliminates, in effect, one stripping stage with a resulting broad charge state distribution and thus increases efficiency. Estimated to be around 10^{-4} or better, the exact value needs to be experimentally determined. So there has been considerable progress in this "storage ring variant" of AMS, but the overall detection efficiency that can be finally reached finally is still unknown.

Over the past decades, several methods of trace analysis based on laser spectroscopic techniques were proposed and very successfully developed. The high degree of selectivity obtained by these methods is a result of resonant laser-atom interaction. Atoms of different elements interact resonantly with light at significantly different frequencies owing to differences in their atomic structure. Atoms of different isotopes of the same element exhibit isotope shifts, a small change in their resonance frequencies caused by the variation in the nuclear mass and charge radius. By tuning the laser frequencies precisely onto the resonance frequencies of a particular isotope, one can selectively excite, ionize, or manipulate the atoms of this isotope while having a much smaller effect on the other isotopes and elements.

At the moment the recently developed method of trace analysis with an atom trap appears, with some caveats, a very promising approach. Atom trap trace analysis (ATTA) is a laser-based atom counting method originally developed at Argonne National Laboratory (Chen et al. [28]). Its apparatus consists of lasers and vacuum systems of tabletop size. At its center is a magnetooptical trap, a MOT (Raab et al. [29]). The MOT captures atoms of the desired isotope by using laser beams. When the laser frequency is tuned to within a few natural line widths on the low-frequency side of the resonance of the targeted isotope, only atoms of this particular isotope are trapped. Atoms of other isotopes are either deflected before reaching the trap or allowed to pass through the trap without being captured. The detection method (e.g., a sensitive CCD camera) takes full advantage of the high selectivity of photon burst spectroscopy. A single atom can be trapped and observed for 100 ms or longer, during which 10^6 fluorescence photons can be induced and as many as 10^4 photons can be detected. These advantages allow *single atoms* to be counted with a high signal-to-noise ratio, and, since the selectivity depends exponentially on the number of photons detected in a burst, they also ensure a superb selectivity. Indeed ATTA is immune to contamination from other isotopes, elements, or molecules.

For the present proposal we suggest to focus on the Schottky scan developments at the GSI- Darmstadt ESR heavy-ion storage ring. The crucial question that has to be answered in test and pilot experiments is whether or not an overall detection efficiency (including all steps from the ion source to the storage ring) of 10^{-3} or better (see the reasoning above) for the ^{205}Pb ions can be achieved. In parallel we will carefully observe the progress and the developments in ATTA technology with regard to trace analysis of lead isotopes. The sensitivity improvements are key to sample size. The present proposal to obtain the order of 10 kg lorandite from the Allchahar mine is close to the needed sample size that current trace technologies require. But it is not quite there yet. Any further improvements in the detection techniques, which we believe will occur, have the potential of adding the necessary safety margins. However, this requires serious development and study. Any progress made though might not only benefit the present ^{205}Pb proposal but radioisotope trace analysis in general.

3. Summary

It has been shown that the first two of the four major problems of LOREX, as addressed at the beginning of this work, namely, the age of lorandite and the background-induced amount of ^{205}Pb, have been meanwhile successfully addressed and are near to a final conclusion. In particular, the determination of the erosion rate by cosmogenic nuclides was decisive for getting a first but reliable estimate for paleo-depth and, therefore for the amount of background. Furthermore, it has been demonstrated how the measurement of the bound-beta decay probability of bare ^{205}Tl can provide the unknown neutrino capture probability into the first excited state of ^{205}Pb. The corresponding experiment has been accepted for being performed at the GSI Helmholtzentrum. Finally it has been shown how the small amount of ^{205}Pb atoms to be expected could be determined by the Schottky noise technique at the ion storage ring of GSI which provides single-ion detection sensitivity. Taking into account the present-day state-of-the-art of all the techniques needed to solve the four problems of LOREX, we conclude that it is realistic to expect the first result for the solar pp neutrino flux averaged over the last 4.2 million years in a foreseeable future. This number will have most probably still a large error margin in the order of 50% or even more, at the 68% CL. We expect, however, that this accuracy could be improved with time, and that it might approach a level of $\leq$30% as discussed by Pavićević et al. [30].

Acknowledgment

The authors thank the FWF - Wien for supporting this project by grant P 20594.

References

[1] M. S. Freedman, C. M. Stevens, E. P. Horwitz et al., "Solar neutrinos: proposal for a new test," *Science*, vol. 193, no. 4258, pp. 1117–1119, 1976.

[2] F. Neubauer, M. K. Pavićević, J. Genser, R. Jelenković, B. Boev, and G. Amthauer, "$^{40}Ar/^{39}Ar$ dating of geological events of the Allchar deposit and its host rocks," *Geochimica et Cosmochimica Acta*, vol. 73, no. 13, supplement 1, article A938, 2009.

[3] Rio Tinto, "Internal report," unpublished, 1999.

[4] S. Janković and R. Jelenković, "Thallium mineralization in the Allchar Sb–As–Tl–Au deposit," *Neues Jahrbuch für Mineralogie—Abhandlungen*, vol. 167, pp. 283–297, 1994.

[5] M. Troesch and E. Frantz, "$^{40}Ar/^{39}Ar$ Alter der Tl–As Mine von Crven Dol, Allchar (Macedonia)," *Beiheft zu European Journal of Mineralogy*, vol. 4, no. 1, p. 276, 1992.

[6] M. K. Pavićević, "The "Lorex"- Project, Solar neutrino detection with the mineral lorandite (Progress Report)," *Neues Jahrbuch für Mineralogie—Abhandlungen*, vol. 167, pp. 205–245, 1994.

[7] B. P. Heisinger, D. Lal, A. J. T. Jull et al., "Production of selected cosmogenic radionuclides by muons: 1. Fast muons," *Earth and Planetary Science Letters*, vol. 200, no. 3-4, pp. 345–355, 2002.

[8] M. K. Pavićević and LOREX Collaboration, "Erosion rate study at the Allchar deposit (Macedonia) based on radioactive and stable cosmogenic nuclides," 2012.

[9] F. J. Hartman, B. Heisinger, G. Korschinek et al., "Proposal for an experiment with fast muons at CERN: determination of cross sections of fast muon induced reactions to cosmogenic radionuclides," CERN/SPSLC 95-8, CERN, Geneva, Switzerland, 1995.

[10] I. V. Aničin, V. Pejović, M. K. Pavićević et al., "On the possibility to simultaneously determine the long-term average fluxes of solar pp-NEUTRINOS and cosmic ray muons," *Modern Physics Letters A*, vol. 26, no. 17, pp. 1267–1271, 2011.

[11] S. Neumaier, B. Dockorn, E. Nolte, and H. Morinaga, "Signal to background ratio of the geochemical ^{205}Tl solar neutrino experiment," *Neues Jahrbuch für Mineralogie—Abhandlungen*, vol. 167, pp. 255–263, 1994.

[12] J. N. Bahcall, "Solar neutrino experiments," *Reviews of Modern Physics*, vol. 50, no. 4, pp. 881–903, 1978.

[13] A. H. Wapstra and G. Audi, "The 1983 atomic mass evaluation: (I). Atomic mass table," *Nuclear Physics A*, vol. 432, no. 1, pp. 1–54, 1985.

[14] E. Braun and I. Talmi, "On the β-transition of $^{205}T1$ to ^{205}Pb," in *Weak and Electromagnetic Interactions in Nuclei: Proceedings of the International Symposium*, H. V. Klapdor, Ed., pp. 47–46, Springer Heidelberg, Berlin, Germany, 1986.

[15] M. S. Freedman and E. E. Nolte, "Revue of the priorities of the ^{205}Pb experiment," GSI Report 86-9, GSI, Darmstadt, Germany, 1986.

[16] K. Ogawa and K. Arita, "Shell-model study of the first forbidden $^{205}Tl \rightarrow {}^{205}Pb$ transition and solar neutrino detection," *Nuclear Instruments and Methods in Physics Research A*, vol. 271, no. 2, pp. 280–285, 1988.

[17] H. Behrens and J. Jaenecke, *Numerical Tables for Beta Decay and Electron Capture*, 2001.

[18] P. Kienle, "Studies of radioactive decays of completely ionized nuclei in a heavy ion storage ring," *Nuclear Instruments and Methods in Physics Research A*, vol. 271, no. 2, pp. 277–279, 1988.

[19] K. Takahashi and K. Yokoi, "Beta-decay rates of highly ionized heavy atoms in stellar interiors," *Atomic Data and Nuclear Data Tables*, vol. 36, pp. 375–409, 1987.

[20] M. Jung, F. Bosch, K. Beckert et al., "First observation of bound-state β^- decay," *Physical Review Letters*, vol. 69, no. 15, pp. 2164–2167, 1992.

[21] F. Bosch, T. Faestermann, J. Friese et al., "Observation of bound-state β^- decay of fully ionized ^{187}Re: ^{187}Re-^{187}Os cosmochronometry," *Physical Review Letters*, vol. 77, no. 26, pp. 5190–5193, 1996.

[22] T. Ohtsubo, F. Bosch, H. Geissel et al., "Simultaneous measurement of β^- decay to bound and continuum electron states," *Physical Review Letters*, vol. 95, no. 5, Article ID 052501, 4 pages, 2005.

[23] Y. Litvinov, H. Geissel, T. Radon et al., "Mass measurement of cooled neutron-deficient bismuth projectile fragments with time-resolved Schottky mass spectrometry at the FRS-ESR facility," *Nuclear Physics A*, vol. 756, no. 1-2, pp. 3–38, 2005.

[24] M. K. Pavićević and W. Bruechle, "Bleiextraktion aus Mineralien Lorandit ($TlAsS_2$), Realgar As_2S_2) und Auripiment (As_2S_3)," GSI Internal Report, 2005.

[25] H. Ernst, G. Korschinek, P. Kubik et al., "^{205}Pb: accelerator mass spectrometry of a very heavy radioisotope and the solar neutrino problem," *Nuclear Instruments and Methods in Physics Research B*, vol. 5, no. 2, pp. 426–429, 1984.

[26] W. F. Henning and D. Schuell, "On isobar separation and accelerator mass spectrometry of ^{205}Pb," *Nuclear Instruments and Methods in Physics Research A*, vol. 271, no. 2, pp. 324–327, 1989.

[27] T. Radon, H. Geissel, G. Münzenberg et al., "Schottky mass measurements of stored and cooled neutron-deficient projectile fragments in the element range of $57 \leq Z \leq 84$," *Nuclear Physics A*, vol. 677, no. 1–4, pp. 75–99, 2000.

[28] C. C. Chen, Y. M. Li, K. Bailey, T. P. O'Connor, L. Young, and Z. T. Lu, "Ultrasensitive isotope trace analyses with a magneto-optical trap," *Science*, vol. 286, no. 5442, pp. 1139–1141, 1999.

[29] E. L. Raab, M. Prentiss, A. Cable, S. Chu, and D. E. Pritchard, "Trapping of neutral sodium atoms with radiation pressure," *Physical Review Letters*, vol. 59, no. 23, pp. 2631–2634, 1987.

[30] M. K. Pavićević, F. Bosch, G. Amthauer et al., "New data for the geochemical determination of the solar pp-neutrino flux by means of lorandite mineral," *Nuclear Instruments and Methods in Physics Research A*, vol. 621, no. 1–3, pp. 278–285, 2010.

6

Relativistic and Quantum Revisions of the Multisource Thermal Model in High-Energy Collisions

Fu-Hu Liu, Cai-Xing Tian, Mai-Ying Duan, and Bao-Chun Li

Institute of Theoretical Physics, Shanxi University, Shanxi, Taiyuan 030006, China

Correspondence should be addressed to Fu-Hu Liu, fuhuliu@163.com

Academic Editor: Terry Sloan

The multisource thermal model is revised for relativistic and quantum situations. It is shown that the quantum effect can be neglected due to a very small result. The distributions of particle momenta, momentum components, transverse momenta, kinetic energies, and velocities in both the classical and relativistic situations are presented to give comparisons.

1. Introduction

The multisource thermal model (or previously the multisource ideal gas model) was suggested by us in the past years [1–4]. This model assumes that many emission sources of produced particles and nuclear fragments are formed in high-energy collisions. Each emission source is treated as a thermal equilibrium system of classical ideal gas. Then, the classical Maxwell's ideal gas model and corresponding distributions [5] can be used in the multisource thermal model. Because the classical Maxwell's ideal gas model has no consideration of the relativistic and quantum effects, some revisions are needed when we consider the model's application in high-energy collisions.

In this paper, the relativistic and quantum effects are considered for each emission source. The multisource thermal model [1–4] is then revised for the relativistic and quantum situations. Distributions of particle momenta, momentum components, transverse momenta, kinetic energies, and velocities in both the classical and relativistic situations are given and compared.

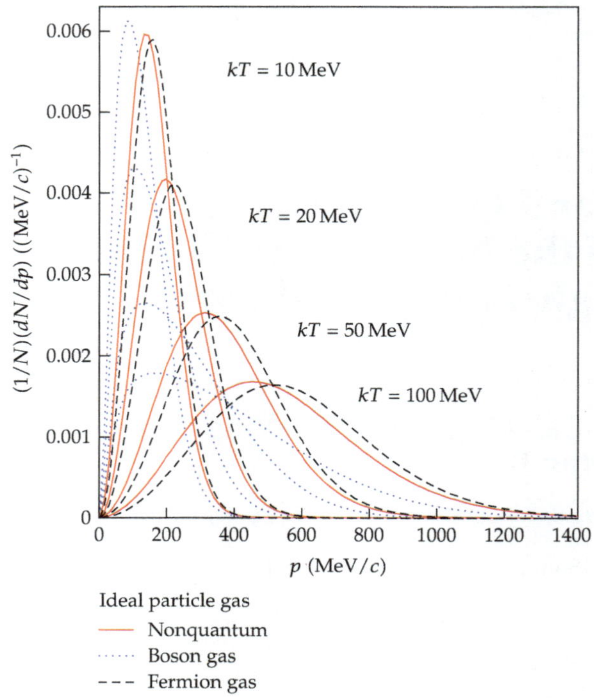

Figure 1: Momentum distributions of relativistic nonquantum particles (solid curves), bosons (dotted curves), and fermions (dashed curves) in ideal gas model calculated by (2.6) and (2.7), respectively, with $m_0 = 938\,\mathrm{MeV}/c^2$ and $\mu = 937\,\mathrm{MeV}$. The results are given for four temperatures indicated in the figure.

2. Formulations and Comparisons

Let k and T denote the Boltzmann constant and temperature of thermal equilibrium system, respectively, and m_0, p, $p_{x,y,z}$, p_T, E_K, v, θ, and φ denote the rest mass, momentum, momentum components, transverse momentum, kinetic energy, velocity, emission angle, and azimuth of a concerned particle, respectively. In our discussions, we define the beam direction to be the oz axis and the reaction plane to be the xoz plane in high-energy collisions.

In the classical Maxwell's ideal gas model [5], we have the distributions of particle momenta, momentum components, transverse momenta, kinetic energies, and velocities to be

$$f_p(p) = \frac{1}{N}\frac{dN}{dp} = \sqrt{\frac{2}{\pi}}\frac{p^2}{(m_0kT)^{3/2}}\exp\left(-\frac{p^2}{2m_0kT}\right), \tag{2.1}$$

$$f_{p_{x,y,z}}(p_{x,y,z}) = \frac{1}{N}\frac{dN}{dp_{x,y,z}} = \frac{1}{\sqrt{2\pi}\sqrt{m_0kT}}\exp\left(-\frac{p_{x,y,z}^2}{2m_0kT}\right), \tag{2.2}$$

$$f_{p_T}(p_T) = \frac{1}{N}\frac{dN}{dp_T} = \frac{p_T}{m_0kT}\exp\left(-\frac{p_T^2}{2m_0kT}\right), \tag{2.3}$$

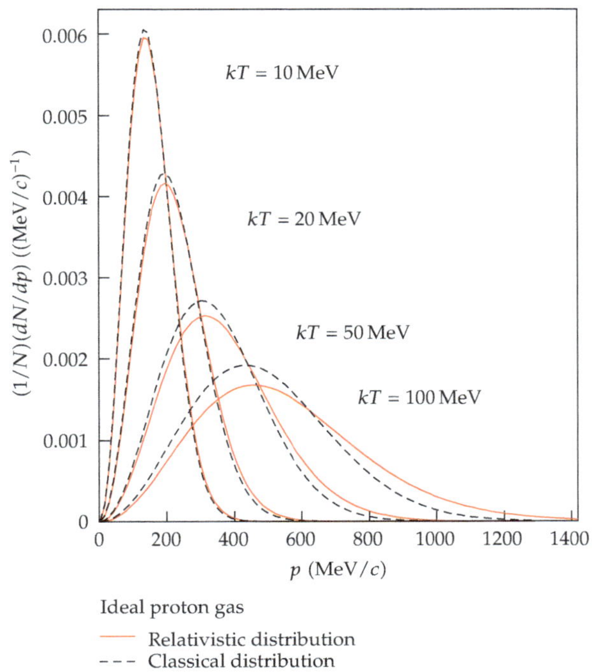

Figure 2: Proton momentum distributions in relativistic (nonquantum) ideal gas model (solid curves) and classical ideal gas model (dashed curves). The different curves are calculated by (2.6) and (2.1), respectively, and are given for four temperatures indicated in the figure. The solid curves in Figure 2 are the same as those in Figure 1.

$$f_{E_K}(E_K) = \frac{1}{N}\frac{dN}{dE_K} = \frac{2}{\sqrt{\pi}}\left(\frac{1}{kT}\right)^{3/2}\sqrt{E_K}\exp\left(-\frac{E_K}{kT}\right), \tag{2.4}$$

$$f_v(v) = \frac{1}{N}\frac{dN}{dv} = \sqrt{\frac{2}{\pi}}\left(\frac{m_0}{kT}\right)^{3/2} v^2 \exp\left(-\frac{m_0 v^2}{2kT}\right), \tag{2.5}$$

respectively.

Considering the relativistic effect [6, 7], we have

$$f_p(p) = \frac{1}{N}\frac{dN}{dp} = Cp^2 \exp\left(-\frac{\sqrt{p^2 + m_0^2}}{kT}\right), \tag{2.6}$$

where $C = (1/m_0^2 kT)(1/K_2(m_0/kT))$ is the normalization constant and $K_2(m_0/kT)$ is the modified Bessel function of order 2. In the above equation, we have taken the natural unit

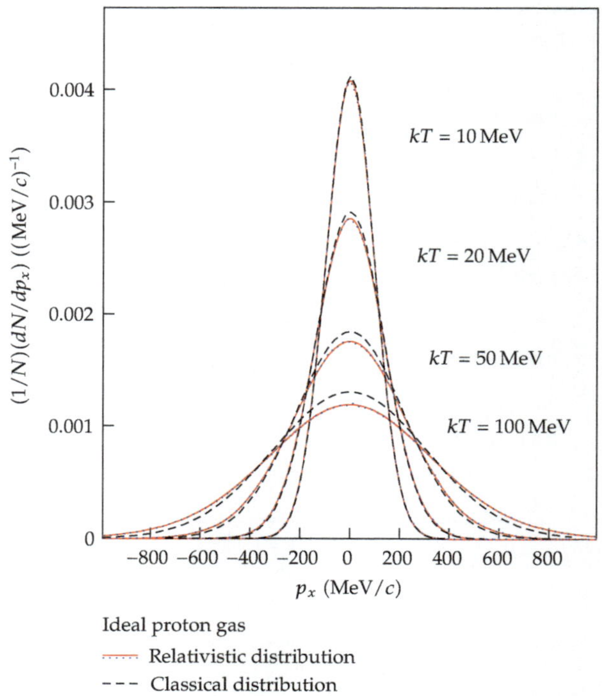

Figure 3: Proton momentum component distributions in relativistic ideal gas model (solid and dotted curves) and classical ideal gas model (dashed curves). The different curves are calculated by (2.8), Monte Carlo method, and (2.2), respectively, and are given for four temperatures indicated in the figure.

system in which the speed c of light in vacuum is 1. Considering the quantum effect [8], we have

$$f_p(p) = \frac{1}{N}\frac{dN}{dp} = C_1 p^2 \left[\exp\left(\frac{\sqrt{p^2 + m_0^2} - \mu}{kT}\right) \pm 1\right]^{-1}, \tag{2.7}$$

where C_1 is the normalization constant, μ is the chemical potential, +1 denotes fermions, and −1 denotes bosons, respectively. Our calculations show that, comparing with (2.6), the quantum effect in (2.7) with μ having large dispersion from m_0 is so small that we do not need to consider it in studying momentum distributions in high-energy collisions. The chemical potential also causes a very small effect in the case of neglecting ±1. Only the combination of quantum effect and $\mu \approx m_0$ causes an obvious dispersion from (2.6). As an example, by taking $m_0 = 938\,\text{MeV}/c^2$ and $\mu = 937\,\text{MeV}$, the calculated momentum distributions for relativistic nonquantum gas (see (2.6)), boson gas (see (2.7) with −1), and fermion gas (see (2.7) with +1) at $kT = 10$, 20, 50, and 100 MeV are shown in Figure 1 by the solid, dotted, and dashed curves, respectively. One can see obvious differences in momentum distributions among the three kinds of ideal gases. Especially at higher temperature, this difference is more obvious.

As the relativistic and quantum revisions of (2.1), (2.7) can be used in the multisource thermal model. To avoid the unbending choice of μ and ±1, in the following we use (2.6) to give comparisons with (2.1)–(2.5). In Figure 2, the particle momentum distributions

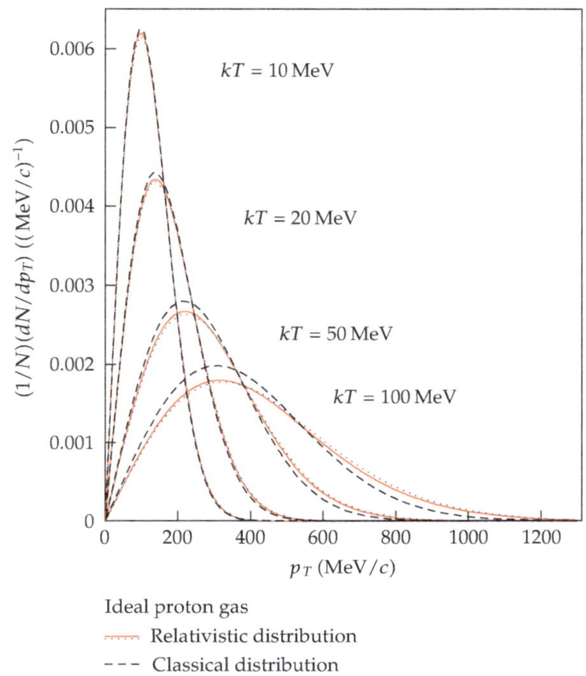

Figure 4: Proton transverse momentum distributions in relativistic ideal gas model (solid and dotted curves) and classical ideal gas model (dashed curves). The different curves are calculated by (2.9), Monte Carlo method, and (2.3), respectively, and are given for four temperatures indicated in the figure.

at $kT = 10$, 20, 50, and 100 MeV calculated by (2.6) and (2.1) are given by the solid and dashed curves, respectively. In the calculation, we take $m_0 = 938\,\text{MeV}/c^2$ which means that an ideal proton gas is considered. One can see that the difference between the classical distribution and relativistic one is not obvious at lower temperatures (10 and 20 MeV), while the difference is obvious at higher temperatures (50 and 100 MeV).

The momentum components can be given by $p_x = p\sin\theta\cos\varphi$, $p_y = p\sin\theta\sin\varphi$, and $p_z = p\cos\theta$, respectively. An isotropic ideal gas model gives the same distribution of three different components, and the emission angular and azimuthal distributions are $f_\theta(\theta) = (1/2)\sin\theta$ and $f_\varphi(\varphi) = 1/2\pi$, respectively. Combining with (2.6), we can use the Monte Carlo method and analytical expression to give the momentum component distributions at relativistic energy. In Figure 3, the dashed curve represents the classical distribution (see (2.2)) of p_x, and the dotted and solid curves represent the relativistic distributions by using the Monte Carlo method and analytical expression

$$f_{p_x}(p_x) = \frac{1}{N}\frac{dN}{dp_x} = C_2\sqrt{p_x^2 + m_0^2}\exp\left(-\frac{\sqrt{p_x^2 + m_0^2}}{kT}\right), \tag{2.8}$$

respectively, where C_2 is the normalization constant. All the distributions are given at four temperatures mentioned in the figure. We see that the difference between the classical and relativistic distributions at lower temperature is not obvious, and the difference at higher

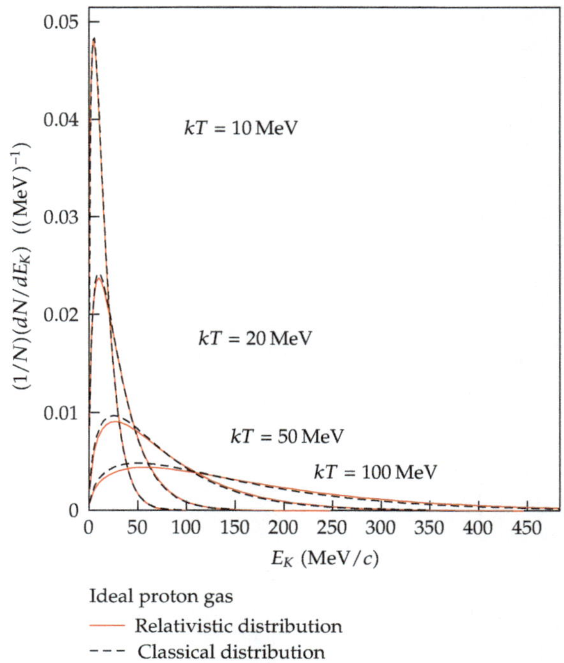

Figure 5: Proton kinetic energy distributions in relativistic ideal gas model (solid curves) and classical ideal gas model (dashed curves). The different curves are calculated by (2.10) and (2.4), respectively, and are given for four temperatures indicated in the figure.

temperature is obvious. The relativistic distributions obtained by the Monte Carlo method and analytical expression are almost the same. This result confirms the two methods all together.

The transverse momentum is given by $p_T = p\sin\theta$. Combining with (2.6), we can use the Monte Carlo method and analytical expression to give the transverse momentum distribution at relativistic energy. In Figure 4, the dashed curve represents the classical distribution (see (2.3)) of p_T, whereas the dotted and solid curves represent the relativistic results by using the Monte Carlo method and Boltzmann distribution

$$f_{p_T}(p_T) = \frac{1}{N}\frac{dN}{dp_T} = C_3 p_T \exp\left(-\frac{\sqrt{p_T^2 + m_0^2}}{kT}\right), \tag{2.9}$$

respectively, where C_3 is the normalization constant. All the distributions are given at four temperatures. We see again that the difference between the classical distribution and relativistic one is small at lower temperature and large at higher temperature. The relativistic results obtained by the Monte Carlo method and Boltzmann distribution are almost the same. This confirms the two methods all together.

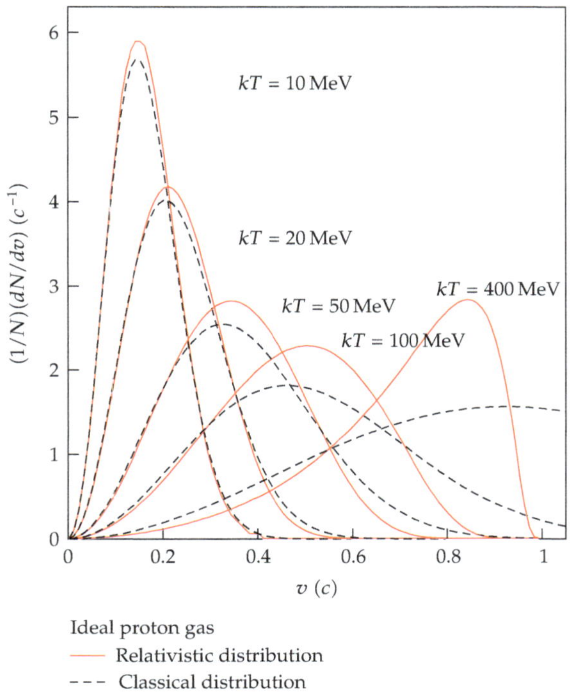

Figure 6: Proton velocity distributions in relativistic ideal gas model (solid curves) and classical ideal gas model (dashed curves). The different curves are calculated by (2.12) and (2.5), respectively, and are given for five temperatures indicated in the figure.

The kinetic energy in relativistic situation can be given by $E_K = \sqrt{p^2 + m_0^2} - m_0$; combining with (2.6), we have the distribution of kinetic energy in relativistic situation to be

$$f_{E_K}(E_K) = \frac{1}{N}\frac{dN}{dE_K} = C(E_K + m_0)\sqrt{(E_K + m_0)^2 - m_0^2}\exp\left(-\frac{E_K + m_0}{kT}\right). \tag{2.10}$$

The total energy $E = \sqrt{p^2 + m_0^2} = E_K + m_0$ is simply given by

$$f_E(E) = \frac{1}{N}\frac{dN}{dE} = CE\sqrt{E^2 - m_0^2}\exp\left(-\frac{E}{kT}\right). \tag{2.11}$$

For ideal gas in classical situation, the energy is only the kinetic energy due to zero potential energy. The comparisons between the kinetic energy distributions in classical situation (dashed curves) and relativistic case (solid curves) are presented in Figure 5. The two curves at lower temperature are almost the same, and those at higher temperature are very similar.

The relationship between momentum and velocity at relativistic energy gives $p = m_0 v/\sqrt{1-v^2}$. Combining with (2.6), we have the distribution of velocity to be

$$f_v(v) = \frac{1}{N}\frac{dN}{dv} = C\frac{m_0^3 v^2}{(1-v^2)^{5/2}} \exp\left(-\frac{m_0}{kT\sqrt{1-v^2}}\right). \tag{2.12}$$

The comparisons between the velocity distributions in classical situation (dashed curves) and relativistic case (solid curves) are presented in Figure 6. There are small differences at lower temperatures and large differences at higher temperatures. One sees $v > c$ in classical situation at around 100 MeV. Especially, the probability of particles with $v > c$ is very large at 400 MeV. Obviously, the relativistic effect has to be considered at these higher temperatures.

3. Conclusion

In conclusion, the multisource thermal model [1–4] is revised in this paper for relativistic and quantum situations. The quantum effect in particle momentum distributions in high-energy collisions is very small when the chemical potential has a large dispersion from the rest mass of the concerned particle. The chemical potential also causes a very small effect when we neglect ±1. Only the combination of quantum effect and nearly the same chemical potential and rest mass causes an obvious dispersion from the classical distribution. In the distributions of particle momenta, momentum components, transverse momenta, kinetic energies, and velocities, the relativistic effect is small at lower temperatures and large at higher temperatures. This is a natural result. Because of the present work, the classical distributions used in the multisource thermal model [1–4] can be replaced by the revised distributions.

Acknowledgments

This work was supported by the National Natural Science Foundation of China under Grant no. 10975095, the National Fundamental Fund of Personnel Training under Grant no. J1103210, the Open Research Subject of the Chinese Academy of Sciences Large-Scale Scientific Facility under Grant no. 2060205, and the Shanxi Scholarship Council of China.

References

[1] F. H. Liu, "Longitudinal and transverse flows of protons in 2–8 A *GeV Au-Au* collisions," *Europhysics Letters*, vol. 63, no. 2, pp. 193–199, 2003.
[2] F. H. Liu, N. N. Abd Allah, D. H. Zhang, and M. Y. Duan, "Angular distributions of target black fragments in nucleus-nucleus collisions at high energy," *International Journal of Modern Physics E*, vol. 12, no. 5, pp. 713–723, 2003.
[3] F. H. Liu, N. N. Abd Allah, and B. K. Singh, "Dependence of black fragment azimuthal and projected angular distributions on polar angle in silicon-emulsion collisions at 4.5A GeV/c," *Physical Review C*, vol. 69, no. 5, Article ID 057601, 4 pages, 2004.
[4] F. H. Liu, J. S. Li, and M. Y. Duan, "Light fragment emission in ^{86}Kr-^{124}Sn collisions at 25 MeV/nucleon," *Physical Review C*, vol. 75, no. 5, Article ID 054613, 5 pages, 2007.
[5] Y. H. Qin and Thermal Physics Higher Education Press, Beijing, China, 2011.
[6] J. L. Synge, *The Relativistic Gas*, North-Holland, Amsterdam, The Netherlands, 1957.

[7] C. D. Dermer, "The production spectrum of a relativistic Maxwell-Boltzmann gas," *The Astrophysical Journal*, vol. 280, no. 1, pp. 328–333, 1984.
[8] P. Z. Ning, L. Li, and D. F. Min, *Foundation of Nuclear Physics: Nucleons and Nuclei*, Higher Education Press, Beijing, China, 2003.

Total Energy of Charged Black Holes in Einstein-Maxwell-Dilaton-Axion Theory

Murat Korunur and Irfan Açıkgöz

Department of Physics, Faculty of Science, Dicle University, 21280 Diyarbakır, Turkey

Correspondence should be addressed to Murat Korunur, muratkorunur@yahoo.com

Academic Editor: Richard J. Szabo

We focus on the energy content (including matter and fields) of the Møller energy-momentum complex in the framework of Einstein-Maxwell-Dilaton-Axion (EMDA) theory using teleparallel gravity. We perform the required calculations for some specific charged black hole models, and we find that total energy distributions associated with asymptotically flat black holes are proportional to the gravitational mass. On the other hand, we see that the energy of the asymptotically nonflat black holes diverge in a limiting case.

1. Introduction

There are many interesting theories aiming to investigate gravitational effects: general relativity and teleparallel gravity. In these theories, calculating the energy-momentum distribution is an old and interesting problem. One can construct a teleparallel equivalent of the general relativity by assuming that curvature and torsion give the equivalent descriptions of the gravitational interactions.

In order to avoid the singularities in the general relativity and to give a general definition of energy momentum, Møller obtained a new expression by using the teleparallel gravity [1, 2]. Pellegrini and Plebanski found the Lagrangian formulation of the teleparallel theory, and this formalism was developed further by Møller [3, 4]. The gauge theory studied in detail by Hayashi [5] for the translational group is formulated by Hayashi and Nakano [6]. After that Hayashi [7] underlined the connection between this theory and the teleparallel theory. Later, Hayashi and Shirafuji [8] tried to unify these two theories. Hehl et al. [9] discussed a generalization of Einstein's gravitational theory with spin and torsion. In the near past Mikhail et al. [10] used the method of the superpotential in the case of the spherical symmetry for the Møller's energy-momentum expression in the teleparallel theory. By using the Møller's super potential method Shirafuji et al. [11, 12] found that energy is equivalent to the gravitational mass. Maluf [13] showed that the energy is equivalent to the gravitational mass by constructing a connection between the gravitational energy related closely to the Sparling two-forms and the teleparallel equivalence of the general relativity. de Andrade et al. [14] obtained a teleparallel equivalence of the Kaluza-Klein theory in five dimensions. Respectively, Pereira et al. [15] and Zhang [16] found the tetrad and torsion fields for the axial symmetric Kerr spacetime and alternative Kerr spacetime by using the teleparallel theory. A new interpretation for torsion in connection with the gravitational interaction is given by Arcos and Pereira [17] as a review. It is pointed out that the teleparallel gravity successfully was applied to describe the rotation spin effect as another application of spin and axial torsion interaction [18]. About the spacetime torsion one can examine the work in [19]. Nashed [20] calculated total energy for some special situations of the most general spherical symmetric and nonsingular black hole solutions in the teleparallel gravity by using the Møller's super potential technique.

The study carried out by Vargas [21] for the Friedman-Robertson-Walker (FRW) universe possessing a cartesian geometry has been a starting point for the other studies done after that. Later, the energy densities for the Einstein, Bergman-Thomson, and Landau-Lifshitz energy momentum expressions written in cartesian coordinates are obtained in the teleparallel theory. Therefore the teleparallel equivalence of general relativistic case is obtained by Nashed [22] for some solutions of the models of the universe having charged and spherical symmetric solutions. Furthermore, the energy-momentum and angular momentum are calculated for two different fields by using energy-momentum tensor [23]. In the framework of Riemann's geometry it is found that the energy density in Kerr-Nut space-time model is proportional to the gravitational mass by using Møller's energy-momentum complex [24]. During the last five years it is stressed that the energy-momentum densities calculated both in general relativity theory and teleparallel theory are equivalent. In this point of view it is concluded that the teleparallel theory is an alternative one for the general relativity.

This paper is organized as follows. In the next section we introduce Møller energy-momentum complex in the teleparallel theory. In the Section 4, the total energy is found for the EMDA theory. Last section is devoted to discussion. Throughout this paper, Latin indices $(i, j, k, \ldots)$ denote the vector numbers and Greek indices $(\mu, \nu, \sigma, \ldots)$ represent the vector components. All indices run from 0 to 3, and we use units in which $G = 1$ and $c = 1$.

2. The Møller Energy-Momentum Complex

The metric tensor can be written in the tetrad form:

$$g^{\mu\nu} = \eta^{ij} h_i{}^{\mu} h_j{}^{\nu}, \tag{2.1}$$

where η^{ij} is the Minkowski metric defined by $\text{Diag}\{-1, +1, +1, +1\}$. The torsion tensor in Møller's theory is

$$T^{\mu}{}_{\nu\lambda} = \Gamma^{\mu}{}_{\lambda\nu} - \Gamma^{\mu}{}_{\nu\lambda}, \tag{2.2}$$

and here $\Gamma^{\mu}{}_{\lambda\nu}$ is Weitzenböck connection [25] given by

$$\Gamma^{\mu}{}_{\nu\lambda} = h_a{}^{\mu} \partial_{\lambda} h^a{}_{\nu}. \tag{2.3}$$

A general expression for an energy-momentum complex was found by Møller by using the method of infinitesimal transformations in the superpotential $\Im_\beta{}^{\mu\nu}$ form [10]:

$$M_\beta{}^\mu \equiv \sqrt{-g}(T_\beta{}^\mu + t_\beta{}^\mu) = \Im_{\beta;\nu}{}^{\mu\nu}. \tag{2.4}$$

Here, $T_\beta{}^\mu$, $t_\beta{}^\mu$ energy-momentum tensor and energy-momentum pseudotensor arise from matter and gravitational field, respectively. $\Im_\beta{}^{\mu\nu}$ is given by

$$\Im_\beta{}^{\mu\nu} = \frac{\sqrt{-g}}{2\kappa} P_{\chi\rho\sigma}{}^{\tau\mu\nu}\left[\Phi^\rho g^{\sigma\chi} g_{\beta\tau} - \lambda g_{\tau\beta}\xi^{\chi\rho\sigma} - (1-2\lambda) g_{\tau\beta}\xi^{\sigma\rho\chi}\right]. \tag{2.5}$$

Here λ is a free dimensionless parameter. Φ^ρ is defined by

$$\Phi_\rho = \xi^\sigma{}_{\rho\sigma}, \tag{2.6}$$

and $\xi_{\alpha\beta\mu} = h_{i\alpha} e^i{}_{\beta;\mu}$ is the con-torsion tensor. $P_{\chi\rho\sigma}{}^{\tau\mu\nu}$ is the tensor of the form

$$\begin{aligned} P_{\chi\rho\sigma}^{\tau\mu\nu} &= \delta_\chi{}^\tau(\delta_\rho{}^\mu\delta_\sigma{}^\nu - \delta_\sigma{}^\mu\delta_\rho{}^\nu) + \delta_\rho{}^\tau(\delta_\sigma{}^\mu\delta_\chi{}^\nu - \delta_\chi{}^\mu\delta_\sigma{}^\nu) \\ &\quad - \delta_\sigma{}^\tau(\delta_\chi{}^\mu\delta_\rho{}^\nu - \delta_\rho{}^\mu\delta_\chi{}^\nu). \end{aligned} \tag{2.7}$$

The total energy is given by the surface integral below

$$E = \lim_{r\to\infty} \int_{r=\text{const.}} \Im_0{}^{0\alpha} \eta_\alpha dS, \tag{2.8}$$

where dS is the surface element and η_α is the unit 3-vector normal to the surface.

3. Energy Contents of Charged Black Holes in EMDA Theory

In Einstein's frame, Einstein-Hilbert-Maxwell action coupled with a string in four dimensions is given [26]:

$$S = \int d^4x \frac{\sqrt{-g}}{2\kappa}\left[R - \frac{1}{2}\partial_\mu\varphi\partial^\mu\varphi - \frac{1}{2}\omega(\varphi)\partial_\sigma\zeta\partial^\sigma\zeta - \alpha(\varphi,\zeta)F_{\gamma\eta}F^{\gamma\eta} - \beta(\varphi,\zeta)F_{\gamma\eta}\tilde{F}^{\gamma\eta}\right], \tag{3.1}$$

where κ is the four-dimensional gravitational coupling constant, R is the curvature scalar, and $F_{\mu\nu}$ is the field strength of the Maxwell field. Here φ, ζ are scalar and pseudoscalar fields, respectively. Additionally, α and β functions describe how the Maxwell field is coupled with φ and ζ. Also, the Maxwell field strength is defined as $\tilde{F}^{\gamma\eta} = (1/2)\epsilon^{\gamma\eta e\tau}F_{e\tau}$ (ϵ is the fourth Levi-Civita tensor).

Taking $\omega(\varphi) = e^{2a\varphi}$, $\alpha(\varphi) = e^{-a\varphi}$, and $\beta(\zeta) = b\zeta$ for the four dimensional Einstein-Maxwell theory the generalized action coupled with massless scalar dilaton φ and pseudoscalar axion ζ is given as

$$S = \int d^4x\sqrt{-g}\left[\frac{1}{2\kappa}\left(R - \frac{1}{2}\partial_\mu\varphi\partial^\mu\varphi - \frac{1}{2}e^{2a\varphi}\partial_\sigma\zeta\partial^\sigma\zeta\right) - e^{-a\varphi}F_{\gamma\eta}F^{\gamma\eta} - b\zeta F_{\gamma\eta}\widetilde{F}^{\gamma\eta}\right], \tag{3.2}$$

where a and b are two constant-free parameters.

The general metric for this theory is written as

$$ds^2 = -G^2(r)dt^2 + \frac{1}{G^2(r)}dr^2 + F^2(r)\left(d\theta^2 + \sin^2\theta d\phi\right). \tag{3.3}$$

For this line element, the metric tensor and its inverse are obtained as

$$g_{\mu\nu} = \operatorname{diag}\left\{-G^2(r), \frac{1}{G^2(r)}, F^2(r), F^2(r)\sin^2\theta\right\},$$
$$g^{\mu\nu} = \operatorname{diag}\left\{-\frac{1}{G^2(r)}, G^2(r), \frac{1}{F^2(r)}, \frac{1}{F^2(r)\sin^2\theta}\right\}. \tag{3.4}$$

Having spherical symmetry for the general form of the tetrad and using the coordinate transformation, the tetrad components can be written in a matrix form as

$$h_a{}^\mu = \begin{pmatrix} \frac{1}{G(r)} & 0 & 0 & 0 \\ 0 & G(r)\sin\theta\cos\phi & \frac{\cos\theta\cos\phi}{F(r)} & -\frac{\sin\phi}{F(r)\sin\theta} \\ 0 & G(r)\sin\theta\sin\phi & \frac{\cos\theta\sin\phi}{F(r)} & \frac{\cos\phi}{F(r)\sin\theta} \\ 0 & G(r)\cos\theta & -\frac{\sin\theta}{F(r)} & 0 \end{pmatrix}. \tag{3.5}$$

3.1. *Asymptotically Flat Black Holes*

For the asymptotically flat black holes, the line element (3.3) becomes [26]

$$ds^2 = -\frac{(r - r_+)(r - r_-)}{(r - r_0)^{2-2n}(r + r_0)^{2n}}dt^2 + \frac{(r - r_0)^{2-2n}(r + r_0)^{2n}}{(r - r_+)(r - r_-)}dr^2 + \frac{(r + r_+)^{2n}}{(r - r_0)^{2n}}d\Omega^2, \tag{3.6}$$

where $r_\pm$, r_0 are constants, $0 \le n \le 1$, and $d\Omega^2 = d\theta^2 + \sin^2\theta d\phi$.

Case $a = b = 1$

Taking $a = b = 1$ in the action (3.2) the metric becomes [26]

$$ds^2 = -\left(1 - \frac{2m}{r}\right)dt^2 + \left(1 - \frac{2m}{r}\right)^{-1} dr^2 + r(r - 2r_0)d\Omega^2. \quad (3.7)$$

Taking $G^2(r) = (1 - (2m/r))$ and $F^2(r) = r(r - 2r_0)$ in (3.5) and then using it in (2.5) one can obtain the Freud superpotential as follows:

$$\Im_0{}^{01} = \frac{2\sin\theta}{\kappa}\left[2m + r_0 - \frac{2r_0}{r} - r + (r - 2m)^{1/2}(r - 2r_0)^{1/2}\right]. \quad (3.8)$$

Using this result in the energy integral (2.8) the energy density is found as

$$E(r) = 2m + \frac{e^{-\varphi_0}Q^2}{4m}\left(1 - \frac{2m}{r}\right) - r + (r - 2m)^{1/2}\left(r - \frac{e^{-\varphi_0}Q^2}{2m}\right)^{1/2}, \quad (3.9)$$

where φ_0 is a scalar field when $r \rightarrow \infty$. Finally the total energy when $r \rightarrow \infty$ for the asymptotically flat black holes ($a = b = 1$) is

$$E = m. \quad (3.10)$$

Here $r_0 = Q^2 e^{-\varphi_0}/(2m)$, Q_e is electrical charge, Q_m is magnetic charge, and the Q is total charge ($Q^2 = Q_e^2 + Q_m^2$).

Case $a = 1, b \ll 1$

In this case the line element is [26]

$$ds^2 = -\frac{(r - r_+)(r - r_-)}{(r^2 - r_0^2)}dt^2 + \frac{(r^2 - r_0^2)}{(r - r_+)(r - r_-)}dr^2 + \left(r^2 - r_0^2\right)d\Omega^2. \quad (3.11)$$

If one can use $G^2(r) = (r - r_+)(r - r_-)/(r^2 - r_0^2)$ and $F^2(r) = (r^2 - r_0^2)$ in (3.5), is (2.5) obtained as follows:

$$\Im_0{}^{01} = \frac{2\sin\theta}{\kappa}\left[(r - r_+)^{(1/2)}(r - r_-)^{1/2} - \frac{r(r - r_+)(r - r_-)}{r^2 - r_0^2}\right]. \quad (3.12)$$

The energy density obtained by (2.8) is

$$E(r) = \left[r^2 - 2mr + \left(Q_e^2 + Q_m^2\right)e^{-\phi_0} - r_0^2\right]^{1/2} - r\left[1 - \frac{2mr - (Q_e^2 + Q_m^2)e^{-\phi_0}}{r^2 - r_0^2}\right], \quad (3.13)$$

where $r_0 = (Q_e^2 - Q_m^2)e^{-\phi_0}/2m$ and $r_\pm = m \pm [m^2 + r_0^2 - (Q_e^2 + Q_m^2)e^{-\phi_0}]^{1/2}$. One can easily see that, for $Q_m = 0$, $Q_e = Q$ and under the coordinate transformation $r \pm r_0 \to r$, the case transforms in to a Garfinkle-Horowitz-Strominger (GHS) dilaton black hole [27]. Therefore the total energy for GHS dilaton black hole is [28]

$$E = m - \frac{Q^2 e^{-\varphi_0}}{r}. \tag{3.14}$$

The total energy obtained by taking $Q_e = Q_m = 0$ (Schwarzschild solution) for electrically or magnetically charged black holes is Schwarzschild solution

$$E = m. \tag{3.15}$$

Case $a = b \gg 1$

In this case the line element is written by [26]

$$ds^2 \approx -\left(1 - \frac{2m}{r - 2r_0}\right)dt^2 + \left(1 - \frac{2m}{r - 2r_0}\right)^{-1} dr^2 + (r - 2r_0)^2 d\Omega^2. \tag{3.16}$$

Using $G^2(r) = (1 - (2m/(r - 2r_0)))$ and $F(r) = (r - 2r_0)$ in the (3.5), (2.5) is obtained as

$$\Im_0{}^{01} = \frac{2\sin\theta}{\kappa}\left\{2m + (r - 2r_0)\left[\left(1 - \frac{2m}{r - 2r_0}\right)^{1/2} - 1\right]\right\}. \tag{3.17}$$

The energy density is obtained as

$$E(r) = 2m + \left(r - \frac{a^2 e^{-a\phi} Q^2}{2m_0}\right)\left[\left(1 - \frac{4mm_0}{2rm_0 - a^2 e^{-a\phi} Q^2}\right)^{1/2} - 1\right], \tag{3.18}$$

and the total energy for $a = b \gg 1$ is found as

$$E = m, \tag{3.19}$$

where $r_0 \approx a^2 e^{-a\phi} Q^2/4m_0$ and $m_0 \approx m + r_0$.

3.2. *Asymptotically Nonflat Black Holes*

When we consider asymptotically non-flat black holes, the line element (3.3) becomes [26]

$$\begin{aligned} ds^2 = &-\left(\frac{r}{2r_0}\right)^{2n}\left[1 - \frac{2m}{(1-n)r}\right]dt^2 + \left(\frac{2r_0}{r}\right)^{2n}\left[1 - \frac{2m}{(1-n)r}\right]^{-1} dr^2 \\ &+ r^2\left(\frac{2r_0}{r}\right)^{2n} d\Omega^2, \end{aligned} \tag{3.20}$$

where $r_+ = 2m/(1+n)$, $r_- = 0$, and $n = 1/(1+a^2)$.

Case $a = b = 1$

Now, the metric has the form [26]

$$ds^2 = -\left(\frac{r-4m}{2r_0}\right)dt^2 + \left(\frac{r-4m}{2r_0}\right)^{-1}dr^2 + 2rr_0 d\Omega^2. \tag{3.21}$$

Considering $G^2(r) = ((r-4m)/2r_0)$ and $F^2(r) = 2rr_0$ in (3.5) and then using it in (2.5) one can obtain the necessary component of the Freud superpotentials as follows:

$$\Im_0{}^{01} = \frac{2\sin\theta}{\kappa}\left[2m - \frac{r}{2} + r^{1/2}(r-4m)^{1/2}\right]. \tag{3.22}$$

Substituting this result into the (2.8) the energy distribution is found as

$$E(r) = 2m - \frac{r}{2} + r^{1/2}(r-4m)^{1/2}. \tag{3.23}$$

When we take the limit $r \to \infty$, we see that the energy distribution diverges.

Case $a = b \ll 1$

At this point the metric becomes [26]

$$ds^2 = -\left(\frac{r}{2r_0}\right)^2\left[1 - \frac{2m}{a^2 r}\right]dt^2 + \left(\frac{2r_0}{r}\right)^2\left[1 - \frac{2m}{a^2 r}\right]^{-1}dr^2 + 4r_0^2 d\Omega^2. \tag{3.24}$$

If we use $G^2(r) = (r/2r_0)^2[1 - (2m/a^2 r)]$ and $F(r) = 2r_0$ in (3.5) and (2.5), the calculated component of the Freud superpotentials is

$$\Im_0{}^{01} = \frac{2r\sin\theta}{\kappa}\left(1 - \frac{2m}{a^2 r}\right)^{1/2}. \tag{3.25}$$

From (2.8) the energy is found

$$E(r) = r\left(1 - \frac{2m}{a^2 \mathrm{r}}\right)^{1/2}. \tag{3.26}$$

At large distances, the total energy diverges.

Case $a = b \gg 1$

Here the metric becomes [26]

$$ds^2 = -\left(\frac{r}{2r_0}\right)^{2/a^2}\left[1-\frac{2a^2m}{(a^2-1)r}\right]dt^2 + \left(\frac{2r_0}{r}\right)^{2/a^2}\left[1-\frac{2a^2m}{(a^2-1)r}\right]^{-1}dr^2 + r^2\left(\frac{2r_0}{r}\right)^{2/a^2}d\Omega^2. \tag{3.27}$$

If we have $G(r) = (r/2r_0)^{1/a^2}[1-(2a^2m/(a^2-1)r)]^{1/2}$ and $F(r) = r(2r_0/r)^{1/a^2}$ in (3.5), then (2.5) is obtained as follows:

$$\Im_0{}^{01} = \frac{2\sin\theta}{\kappa}r^{(a^2+1)/a^2}\left(1-\frac{2a^m}{(a^2-1)r}\right)^{1/2}\left[1-\left(1-\frac{2a^2m}{(a^2-1)r}\right)^{1/2}r^{1/a^2}\right]. \tag{3.28}$$

Using Freud superpotential in the (2.8) the obtained energy distribution is

$$E(r) = r^{(a^2+1)/a^2}\left(1-\frac{2a^m}{(a^2-1)r}\right)^{1/2}\left[1-\left(1-\frac{2a^2m}{(a^2-1)r}\right)^{1/2}r^{1/a^2}\right]. \tag{3.29}$$

When $a \to \infty$ and $r \to \infty$, the total energy becomes

$$E = m. \tag{3.30}$$

Case $|a| \neq |b|$

For this final case the metric becomes [26]

$$ds^2 = -\frac{(r-r_+)(r-r_-)}{2rr_0}dt^2 + \frac{2rr_0}{(r-r_+)(r-r_-)}dr^2 + 2rr_0 d\Omega^2. \tag{3.31}$$

If we consider $G^2(r) = (r-r_+)(r-r_-)/2rr_0$ and $F^2(r) = 2rr_0$ the component of the Freud superpotentials is calculated as

$$\Im_0{}^{01} = \frac{\sin\theta}{\kappa}\left\{\left[r_+ + r_- + 2(r-r_+)^{1/2}(r-r_-)^2\right] - \frac{r_+r_-}{r} - r\right\}. \tag{3.32}$$

Now the energy distribution is

$$E(r) = 2m - \frac{q_e^2q_m^2}{2rr_0^2} - \frac{r}{2} + \left(r^2 - 4mr + \frac{q_e^2q_m^2}{r_0^2}\right)^{1/2}. \tag{3.33}$$

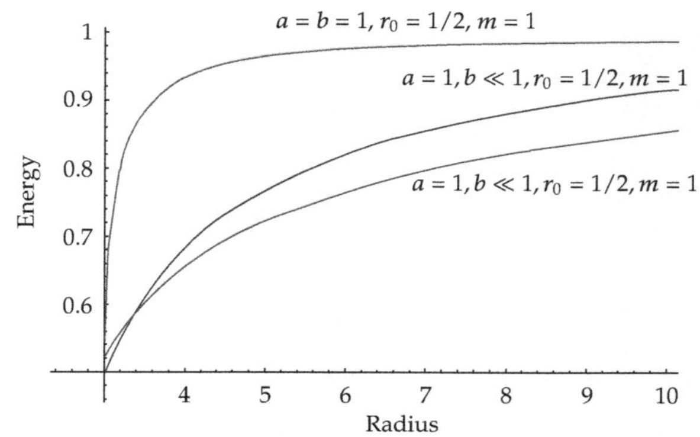

Figure 1: The plot for the Möller energy versus radius, in the case of the asymptotically flat black holes.

It can be easily checked that, at large distances, the total energy diverges. Here, q_e and q_m are electromagnetic charges which are related to components of the electromagnetic field strength given by $F_{tr} = (1/2q_e)dt \wedge dr$ and $F_{\theta\varphi} = q_m \sin\theta d\theta \wedge d\varphi$.

4. Summary and Discussion

According to Lessner [29] perspective, the Møller energy-momentum complex can be evaluated in any coordinate system. Hence, this framework is the most powerful one in calculating the energy and momentum distributions associated with spacetime.

In the present work, in order to compute the energy distribution associated with some specific black holes in the Einstein-Maxwell-Dilaton Axion theory, we focus on the Møller energy-momentum distribution in the teleparallel gravity. For the asymptotically flat charged black holes, for all the cases ($a = b = 1$, $a = b \gg 1$, and $a = 1, b \ll 1$), it is found that the energy distribution depends on the mass m and the charge Q. The corresponding teleparallel Møller total energy is obtained proportional to gravitational mass (Schwarzschild mass) when

$$\lim_{r \to \infty} E(r) = m. \tag{4.1}$$

It is given the plot for the Möller energy versus radius, in the case of the asymptotically flat black holes in Figure 1. According to graph it is seen that the energy of the case $a = b = 1$ approaches to the gravitational mass faster than the others.

Considering asymptotically non-flat charged black holes, for both the cases $a = b = 1$ and $a = b \ll 1$ the energies diverge. Only for the case $a = b \gg 1$ the total energy is proportional to m. The energy distribution for the case $|a| \neq |b|$ diverges as well. For small values of a and b it is seen that the energy distributions tend toward infinity. If the values of a and b are large enough the corresponding teleparallel energy will be equal to the gravitational mass.

These results agree well with the previous results [30–35] obtained by using the general relativity version of the Møller energy-momentum complex. The energy is confined to the region of nonvanishing energy-momentum tensor of matter and all nongravitational fields [36]. The results are quite important in the theory of teleparallel gravity, since this theory provides more satisfactory solution of the energy-momentum problem than general relativity [10].

Acknowledgment

This work is supported by the project DÜBAP-07-02-017.

References

[1] C. Møller, "On the localization of the energy of a physical system in the general theory of relativity," *Annals of Physics*, vol. 4, no. 4, pp. 347–371, 1958.
[2] C. Møller, "Further remarks on the localization of the energy in the general theory of relativity," *Annals of Physics*, vol. 12, no. 1, pp. 118–133, 1961.
[3] C. Pellegrini and J. Plebanski, "Tetrad fields and gravitional fields," *Matematisk-Fysiske Skrifter udg. af det Kongelige Danske Videnskabernes Selskab*, vol. 2, no. 4, pp. 1–38, 1963.
[4] C. Møller, "On the crisis in the theory of gravitation and a possible solution," *Det Kongelige Danske Videnskabernes Selskab Matematisk-Fysiske Meddelelser*, vol. 89, no. 89, pp. 1–32, 1978.
[5] K. Hayashi, "Gravitational interactions of the proton and the electron: the possible existence of a massless scalar particle," *Nuovo Cimento A*, vol. 16, no. 4, pp. 639–673, 1973.
[6] K. Hayashi and K. Nakano, "Extended translation invariance and associated gauge fields," *Progress of Theoretical Physics*, vol. 38, no. 2, pp. 491–507, 1967.
[7] K. Hayashi, "The gauge theory of the translation group and underlying geometry," *Physics Letters B*, vol. 69, no. 4, pp. 441–444, 1977.
[8] K. Hayashi and T. Shirafuji, "New general relativity," *Physical Review D*, vol. 19, no. 12, pp. 3524–3553, 1979.
[9] F. W. Hehl, P. Von Der Heyde, G. D. Kerlick, and J. M. Nester, "General relativity with spin and torsion: foundations and prospects," *Reviews of Modern Physics*, vol. 48, no. 3, pp. 393–416, 1976.
[10] F. I. Mikhail, M. I. Wanas, A. Hindawi, and E. I. Lashin, "Energy-momentum complex in Møller's tetrad theory of gravitation," *International Journal of Theoretical Physics*, vol. 32, no. 9, pp. 1627–1642, 1993.
[11] T. Shirafuji, G. G. L. Nashed, and K. Hayashi, "Energy of general spherically symmetric solution in the tetrad theory of gravitation," *Progress of Theoretical Physics*, vol. 95, no. 3, pp. 665–678, 1996.
[12] T. Shirafuji and G. G. L. Nashed, "Energy and momentum in the tetrad theory of gravitation," *Progress of Theoretical Physics*, vol. 98, no. 6, pp. 1355–1370, 1997.
[13] J. W. Maluf, "Sparling two-forms, the conformal factor and the gravitational energy density of the teleparallel equivalent of general relativity," *General Relativity and Gravitation*, vol. 30, no. 3, pp. 413–423, 1998.
[14] V. C. de Andrade, L. C. T. Guillen, and J. G. Pereira, "Teleparallel equivalent of the Kaluza-Klein theory," *Physical Review D*, vol. 61, no. 8, Article ID 084031, 4 pages, 2000.
[15] J. G. Pereira, T. Vargas, and C. M. Zhang, "Axial-vector torsion and the teleparallel Kerr spacetime," *Classical and Quantum Gravity*, vol. 18, no. 5, p. 833, 2001.
[16] C. M. Zhang, "Torsion axial-vector in an alternative Kerr tetrad field," *Communications in Theoretical Physics*, vol. 44, no. 2, p. 279, 2005.
[17] H. I. Arcos and J. G. Pereira, "Torsion gravity: a reappraisal," *International Journal of Modern Physics D*, vol. 13, no. 10, pp. 2193–2240, 2004.
[18] C. M. Zhang, "Description of dirac spin–rotation effect with relativistic factor," *International Journal of Modern Physics D*, vol. 16, no. 4, pp. 737–744, 2007.
[19] R. Aldrovandi and J. G. Pereira, *An Introduction to Geometrical Physics*, World Scientific, 1995.
[20] G. G. L. Nashed, "General spherically symmetric nonsingular black hole solutions in a teleparallel theory of gravitation," *Physical Review D*, vol. 66, no. 6, Article ID 064015, 8 pages, 2002.
[21] T. Vargas, "The energy of the universe in teleparallel gravity," *General Relativity and Gravitation*, vol. 36, no. 6, pp. 1255–1264, 2004.
[22] G. G. L. Nashed, "Charged axially symmetric solution and energy in teleparallel theory equivalent to general relativity," *The European Physical Journal C*, vol. 49, no. 3, pp. 851–857, 2007.
[23] G. G. L. Nashed, "Charged dilaton, energy, momentum and angular-momentum in teleparallel theory equivalent to general relativity," *The European Physical Journal C*, vol. 54, no. 2, pp. 291–302, 2008.
[24] G. G. L. Nashed, "Møller's Energy of Kerr-NUT Metric," *Chinese Physics Letters*, vol. 25, no. 4, pp. 1202–1204, 2008.
[25] R. Weitzenbock, *Invariantten Theorie*, P. Noordhoff, Groningen, The Netherlands, 1923.
[26] S. Sur, S. Das, and S. SenGupta, "Charged black holes in generalized dilaton-axion gravity," *Journal of High Energy Physics*, vol. 510, p. 64, 2005.

[27] D. Garfinkle, G. T. Horowitz, and A. Strominger, "Charged black holes in string theory," *Physical Review D*, vol. 43, no. 10, pp. 3140–3143, 1991.
[28] O. Aydogdu, M. Salti, M. Korunur, and I. Acikgoz, "Energy associated with the Gibbons-Maeda dilaton spacetime," *Foundations of Physics Letters*, vol. 19, no. 7, pp. 709–721, 2006.
[29] G. Lessner, "Møller's energy-momentum complex—once again," *General Relativity and Gravitation*, vol. 28, no. 5, pp. 527–535, 1996.
[30] I. Radinschi, F. Rahaman, and A. Ghosh, "On the energy of charged black holes in generalized dilaton-axion gravity," *International Journal of Theoretical Physics*, vol. 49, no. 5, pp. 943–956, 2010.
[31] A. Chamorro and K. S. Virbhadra, "Energy associated with charged dilaton black holes," *International Journal of Modern Physics D*, vol. 5, no. 3, pp. 251–256, 1996.
[32] S. S. Xulu, "Møller energy for the kerr–newman metric," *Modern Physics Letters A*, vol. 15, no. 24, pp. 1511–1517, 2000.
[33] S. S. Xulu, "Møller energy of the nonstatic spherically symmetric metrics," *Astrophysics and Space Science*, vol. 283, no. 1, pp. 23–32, 2003.
[34] E. C. Vagenas, "Energy distribution in the dyadosphere of a reissner–nordstrom black hole in Møller's prescription," *Modern Physics Letters A*, vol. 21, no. 25, pp. 1947–1956, 2006.
[35] M. Salti and O. Aydogdu, "Møller's energy in the dyadosphere of a charged black hole," *Pramana*, vol. 67, no. 2, pp. 239–247, 2006.
[36] F. I. Cooperstock and S. A. Richardson, "Energy localization and the Kerr-Newman metric," in *Proceedings of the 4th Canadiand Conference on General 8 Relativity and Relativistic Astrophysics*, World Scientific, Singapore, 1991.

On Possible Reduction of Equilibrium Radius of a Neutron Star Influenced by Superstrong Magnetic Field

Vladimir V. Skobelev

Moscow State Industrial University, Faculty of Applied Mathematics and the Technical Physics, 115280, Russia

Correspondence should be addressed to Vladimir V. Skobelev, v.skobelev@inbox.ru

Academic Editor: C. Q. Geng

Necessary condition for β-decay suppression of a neutron in degenerate magnetized electron gas is formulated. Based on this, it is shown that, in superstrong magnetic field, equilibrium radius of a neutron star is approximately several times smaller than without the field influence. Therefore, we can make a prediction that in short-period pulsars, such fields can be observed. In fact, possible existence of new class of stellar objects is noted, the objects with superstrong magnetic field and supersmall radius about 1 km which we named *minimagnetars*. They can be detected by gravitational red shift of their radiation.

1. Introduction

It is well known that the main contribution in total pressure in neutron star is made by degenerate neutron gas [1], and equilibrium radius about 10 km is formed only under this condition available. On the other hand, existence of superstrong magnetic field is possible there, B up to 10^{17} G [2, 3]. Therefore, the interest for exploration of such field influence onto equilibrium radius by interaction of neutron star matter, in particular, with degenerate electron gas, is arising. The latter might be connected with star dynamics in implicit way, by posing influence on neutron β-decay $n \rightarrow p + e^- + \overline{\nu}_e$ and on reverse reaction $e^- + p \rightarrow n + \nu_e$ (the so called *direct URCA processes*, that dominate in case of rather high concentration of electrons and protons [4]), that is, on the process of *neutronization* and equilibrium radius (in this research we do not cover the role of modified URCA processes [5] with one extra nucleon in such reactions).

We underline the fact that in the nucleus of the neutron star direct process dominates over modified process, being more specific; in the nucleus, there is superstrong magnetic field

[3]—with higher probability. As for the influence of magnetic field on the neutron star radius, we are familiar only with one dedicated paper—our own paper [6, 7], where we consider interaction of the magnetic field with anomalous magnetic moment of the neutron, leading to increasing of neutron pressure and, consequently, leading to radius increase. But the effect is too insignificant for observations to be performed. In other papers, this problem under the aspect considered has not been discussed.

We start from the fact that the wave function of the electron in the homogeneous magnetic field (oriented to the third axis) on the ground Landau level is the following [8] (see [9] also):

$$\Psi = \frac{(\gamma/\pi)^{1/4}}{\sqrt{2EL_2L_3}} \exp\left[i(p_3 z - p_2 y) - \frac{1}{2}\left(x\sqrt{\gamma} - \frac{p_2}{\sqrt{\gamma}}\right)^2\right] u, \tag{1.1}$$

where u is the two-component spinor in the subspace $(0,3), \gamma = |eB|$, p_3 is the momentum along the field, p_2 is the quasimomentum that determines the position of the center of the wave package on the x-axis:

$$X = \frac{p_2}{\gamma}. \tag{1.2}$$

As a result, the integration over p_2 gives

$$\int_{-\infty}^{\infty} dp_2 \longrightarrow \int_0^{L_1} \gamma dX = \gamma L_1. \tag{1.3}$$

Further, we will formulate the condition according to which the neutron decay is impossible in the degenerate magnetized electron gas.

Here, it is necessary to note that the total number of electrons can be represented in the form of a sum over all quantum states [10] (spin projection on ground Landau level is fixed):

$$N = \sum_{\text{states}} f_e \longrightarrow \frac{L_2}{2\pi}\frac{L_3}{2\pi}\int_{-\infty}^{\infty} dp_2 \int_{-\infty}^{\infty} f_e dp_3, \tag{1.4}$$

where

$$f_e = \left[\exp\left(\frac{E-\mu}{T}\right) + 1\right]^{-1} \tag{1.5}$$

is the distribution function and

$$E = \sqrt{m^2 + p_3^2} \tag{1.6}$$

the relativistic electron energy on the ground Landau level.

By using the ratio (1.3), we receive the following for concentration:

$$n \equiv \frac{N}{V} = \frac{\gamma}{2\pi^2}\int_0^{\infty} f_e dp_3. \tag{1.7}$$

For completely degenerate electron gas ($T = 0, f_e \to 1, E < E_F \equiv \mu|_{T=0}$), we find

$$n = \frac{\gamma p_F}{2\pi^2}, \tag{1.8a}$$

where

$$p_F = \frac{2\pi^2 n}{\gamma} \tag{1.8b}$$

is the Fermi momentum.

It follows from the general formula for the electron energy in the magnetic field

$$E = \sqrt{m^2 + p_3^2 + 2\gamma\tilde{n}}, \quad \tilde{n} = 0, 1, 2, \ldots, \tag{1.9}$$

that the states $\tilde{n} = 1, 2, \ldots$ do not energize if $B > B_{\rm cr}$, at the same time, the "critical" field can be found from the following equation

$$p_F^2 = 2\gamma \tag{1.10}$$

or

$$4\pi^2 n = (2\gamma)^{3/2}. \tag{1.10a}$$

On the basis of this equation, we can find the following:

$$B_{\rm cr} = \frac{B_0}{2}\left(4\pi^2 n \lambda_C^3\right)^{2/3}, \qquad B_0 = \frac{m^2}{e} = 4.41 * 10^{13}\,\mathrm{G} \tag{1.11}$$

(λ_C is the Compton electron wavelength), and for the standard electron concentration $n \sim 10^{35}\,\mathrm{cm}^{-3}$, which is 10^{-3} [11] from the neutron concentration in the center of star under nuclear density, we obtain $B_{\rm cr} \approx 8 * 10^{16}\,\mathrm{G}$. That is why under out assumptions and under $B \approx 10^{17}\,\mathrm{G}$, we can consider that the majority of electrons is concentrated on the ground Landau level.

Furthermore, we can add to the above-mentioned "field" condition with "temperature" condition

$$\sqrt{m^2 + 2\gamma} - m \gg T, \tag{1.12}$$

which is valid for "old" neutron stars with temperature $T \sim 1\,\mathrm{MeV}$ [12].

The neutron decay is impossible when $n > n_{\text{min}}$. At the same time, the minimum concentration can be found on the basis of the following equation:

$$\Delta = \sqrt{m^2 + p_F^2}, \tag{1.13}$$

and the formula (1.8b). Here, $\Delta = m_n - m_p \approx 1.3\,\text{MeV}$ stands for the energy output in the neutron β-decay. With the use of (1.8b) and (1.13) equations and maximum field value $B \sim 10^{17}\,\text{G}$, we have

$$n_{\text{min}}{}^{(17)} \approx 4.7 * 10^{33}\,\text{cm}^{-3}. \tag{1.14}$$

On the other hand, in the absence of the magnetic field, the following assessment can be received from (1.13) and standard expression $p_F = (3\pi^2 n)^{1/3}$ [10]:

$$n_{\text{min}}{}^{(0)} \approx 7.6 * 10^{30}\,\text{cm}^{-3}. \tag{1.15}$$

Thus, in the process of gravitational compression of the star, its substance is "neutronized" in the absence of the magnetic field with larger value of the star radius (R_0) as compared to the condition when the magnetic field is present (R_{17}). In particular, we have the following result for star radii at the beginning of "neutronization" (under other equal conditions):

$$\frac{R_0}{R_{17}} \approx \left(\frac{n_{\text{min}}{}^{(17)}}{n_{\text{min}}{}^{(0)}}\right)^{1/3} \cdot \approx 8.5. \tag{1.16}$$

We can also note that the interaction of the anomalous magnetic moment of the neutron with the magnetic field is characterized by the dimensionless parameter $y \pm x$ (see the appendix to the paper [13]) which is equal to 1 in the absence of the magnetic field and has the same order up to the field value 10^{17} G. That is why we can abandon the effect which the magnetic field exercises over the pressure of neutrons and, in this sense, over the value of the equilibrium radius $R^{(\text{eq})}$ taken as the final result of the gravitational compression. If we also assume that the equilibrium radius is proportional to the neutronization radius and the superstrong magnetic field is generated prior to the neutronization, this and (1.16) take us to the final result:

$$\frac{R_0{}^{(\text{eq})}}{R_{17}{}^{(\text{eq})}} \sim 10. \tag{1.17}$$

This assessment is more likely to be overstated since the gravitational compression continues after the neutronization as well, but still it seems that the significant difference in the sizes is still present and we can observe it empirically (see below). In particular pulsars with short period and, therefore, according to the angular momentum rule ($R^2\omega = \text{const}$), with small radius can have superstrong magnetic field. On the other hand, pulsars with large period must have large radius without superstrong magnetic field.

Furthermore, on the basis of equations (1.8b) and (1.13) and the condition of suppression of energized states of the final electron ($\Delta < \sqrt{m^2 + 2\gamma}$, see (1.9)), we can generalize (1.14) and (1.16) in the following way:

$$n_{\min}{}^{(B)} \approx 0.47 * 10^{30} B_{13}\,\mathrm{cm}^{-3}, \qquad B_{13} = \frac{B}{10^{13}\,\mathrm{G}}, \tag{1.18}$$

$$\frac{R_0}{R_B} \approx 0.4 B_{13}{}^{1/3}, \qquad B > 1.2 * 10^{14}\,\mathrm{G}. \tag{1.19}$$

In the particular case $B \approx 10^{17}$ G, we obtain the maximum results (1.14) and (1.16) again.

It will be also observed that our approximation of completely degenerate electron gas features an additional limitation (in addition to (1.12)) for the temperature value on the ground Landau level:

$$K_F \equiv E_F - m \gg T. \tag{1.20}$$

Since the minimum value K_F is equal to $K_F{}^{(\min)} = \Delta - m \approx 0.8\,\mathrm{MeV}$, the condition (1.20) amounts to the limitation

$$\left(\frac{T}{m}\right) \ll 1.57, \tag{1.21}$$

which is broadly fulfilled for values $T \le m$ but not for $T \sim 1\,\mathrm{MeV}$. However, even in the last case, formulas (1.16) and (1.17) are approximately correct because $n_{\min}{}^{(17)}, n_{\min}{}^{(0)}$ are increased and $R_{17}{}^{(\mathrm{eq})}, R_0{}^{(\mathrm{eq})}$ are decreased at $T \neq 0$ to the same extent, so our main conclusion remains unchanged.

As regards suppression of neutron decay with the influence of nonrelativistic proton gas, this phenomenon occurs (if it does at all) at $n_{\min}{}^{(0,17)} \sim 10^{35}\,\mathrm{cm}^{-3}$. This results from formulas for calculation of the minimum concentration in the absence of the magnetic field

$$\frac{\left(3\pi^2 n_{\min}{}^{(0)}\right)^{2/3}}{2M} = \Delta - m \tag{1.22a}$$

and in the presence thereof (for protons at the ground Landau level)

$$\frac{1}{2M}\left(\frac{2\pi^2 n_{\min}{}^{(B)}}{\gamma}\right)^2 = \Delta - m, \tag{1.22b}$$

where M is proton mass.

Therefore, the electron component of a neutron star is critical under the aspect considered since the relevant suppression occurs "earlier" with larger radius of a neutron star and our conclusions remain unchanged. Thus, there is a double suppression of neutron decay at $n_e \approx n_p \sim 10^{35}\,\text{cm}^{-3}$ both by electron and by proton gas, so we may conclude based on our computations that neutron stars with superstrong magnetic field may shrink to such concentration in the process of a gravitational collapse.

The stellar objects considered herein may be called minimagnetars whose radius (about 1 km instead of 10 km in the absence of the magnetic field, e.g., Crab nebula pulsar), according to equations (1.16), (1.17), and (1.19), is close to the gravitational one that is why these objects may be identified by gravitational red shift in their radiation.

References

[1] S. L. Shapiro and S. A. Teukolsky, *Black Holes, White Dwarfs and Neutron Stars*, Wiley-Interscience, New York, NY, USA, 1982.

[2] R. C. Duncan and C. Thompson, "Formation of very strongly magnetized neutron stars: implications for gamma-ray bursts," *Astrophysical Journal Letters*, vol. 392, no. 1, part 2, pp. L9–L13, 1992.

[3] S. Dall'Osso, S. N. Shore, and L. Stella, "Early evolution of newly born magnetars with a strong toroidal field," *Monthly Notices of the Royal Astronomical Society*, vol. 398, no. 4, pp. 1869–1885, 2009.

[4] J. M. Lattimer, C. J. Pethick, M. Prakash, and P. Haensel, "Direct URCA process in neutron stars," *Physical Review Letters*, vol. 66, no. 21, pp. 2701–2704, 1991.

[5] H. Y. Chiu and E. E. Salpeter, "Surface X-Ray emission from neutron stars," *Physical Review Letters*, vol. 12, no. 15, pp. 413–415, 1964.

[6] V. V. Skobelev, "Thermodynamic functionsof a nonrelativistic degenerate neutron gas in a magnetic field," *Journal of Experimental and Theoretical Physics*, vol. 111, no. 6, pp. 962–966, 2010.

[7] V. V. Skobelev, "Temperature corrections to the thermodynamic functions of a degenerate neutron gas in a magnetic field," *Journal of Experimental and Theoretical Physics*, vol. 113, no. 2, pp. 227–231, 2011.

[8] V. V. Skobelev, "Photogeneration of neutrinos and axions under the stimulating effect of a strong magnetic field," *Journal of Experimental and Theoretical Physics*, vol. 93, no. 4, pp. 685–694, 2001.

[9] L. D. Landau and E. M. Lifshits, *Course of Theoretical Physics*, vol. 3 of *Quantum Mechanics: Non-Relativistic Theory*, Butterworth-Heinemann, Oxford, UK, 1977.

[10] L. D. Landau and E. M. Lifshits, *Statistical Physics*, vol. 5, part 1 of *Course of Theoretical Physics*, Butterworth-Heinemann, Oxford, UK, 3rd edition, 2000.

[11] S. Weinberg, *Gravitation and Cosmology*, John Wiley & Sons, New York, NY, USA, 1972.

[12] D. G. Yakovlev, O. Y. Gnedin, A. D. Kaminker, and A. Y. Potekhin, "Theory of cooling neutron stars versus observations," in *Proceedings of the AIP Conference*, vol. 983, pp. 379–387, 2008.

[13] V. V. Skobelev, "Emission and absorption of neutrinos by nonrelativistic neutrons in a magnetic field," *Journal of Experimental and Theoretical Physics*, vol. 111, no. 1, pp. 64–69, 2010.

A 125 GeV Higgs and Its Diphoton Signal in Different SUSY Models: A Mini Review

Zhaoxia Heng

Department of Physics, Henan Normal University, Xinxiang 453007, China

Correspondence should be addressed to Zhaoxia Heng, zxheng@htu.cn

Academic Editor: Stefano Moretti

In this paper we briefly review our recent studies on a 125 GeV Higgs and its diphoton signal rate in different low-energy supersymmetric models, namely, the minimal supersymmetric standard model (MSSM), the next-to-minimal supersymmetric standard model (NMSSM), the nearly minimal supersymmetric standard model (nMSSM), and the constrained MSSM. Our conclusion is as follows (i) in the allowed parameter space the SM-like Higgs boson can easily be 125 GeV in the MSSM, NMSSM, and nMSSM, while it is hard to realize in the constrained MSSM; (ii) the diphoton Higgs signal rate in the nMSSM and constrained MSSM is suppressed relative to the prediction of the SM, while the signal rate can be enhanced in the MSSM and NMSSM; (iii) the NMSSM may allow for a lighter top squark than the MSSM, which can thus ameliorate the fine-tuning problem.

1. Introduction

Considering the important role of the Higgs boson in particle physics, hunting for it has been one of the major tasks of the running Large Hadron Collider (LHC). Recently, both the ATLAS and CMS collaborations have reported some evidence for a light Higgs boson near 125 GeV [1–12] with a diphoton signal rate slightly above the SM prediction [13].

As is well known, in new physics beyond the SM model several Higgs bosons are predicted, among which the SM-like one may be near 125 GeV [14–43]. Recently, in our studies [42, 43] we examined the mass of the SM-like Higgs boson in several supersymmetric (SUSY) models including the minimal supersymmetric standard model (MSSM) [44–47], the next-to-minimal supersymmetric standard model (NMSSM) [48–61], and the constrained MSSM [62–68]. At tree level, these SUSY models are hard to predict a Higgs boson near 125 GeV, and sizable radiative corrections, which mainly come from the top and top-squark loops, are necessary to enhance the Higgs boson mass [69–72]. Due to the different properties of these SUSY models, the loop contributions to the Higgs boson mass are different for giving a 125 GeV Higgs boson. Therefore, different models have different lower bounds on the top-squark mass which is associated with the fine-tuning problem [73–75]. On the other hand, since the di-photon Higgs signal is the most promising discovery channel for a light Higgs boson at the LHC [76–80], in our recent study [81] we performed a comparative study for the di-photon Higgs signal in different SUSY models, namely, the MSSM, NMSSM, and the nearly minimal supersymmetric standard model (nMSSM) [82–90]. In this note we briefly review these studies on a 125 GeV Higgs boson and its di-photon signal rate in different SUSY models.

This note is organized as follows. In the next section we briefly describe the Higgs sector and the di-photon Higgs signal in these SUSY models. Then we present the numerical results and discussions in Section 3. Finally, the conclusions are given in Section 4.

2. The Higgs Sector and Di-Photon Signal Rate in SUSY Models

2.1. The Higgs Sector in SUSY Models

Different from the SM, the Higgs sector in the supersymmetric models is usually extended by adding Higgs doublets and/or singlets. The most economical realization is the MSSM, which consists of two Higgs doublet H_u and H_d. In order to solve the μ-problem and the little hierarchy problem in the MSSM, the singlet extension of MSSM, such as the NMSSM [48–51] and nMSSM [82–90], has been intensively studied [91]. The differences between these models come from their superpotentials and the corresponding soft-breaking terms, which are given by

$$\begin{aligned}
W_{\text{MSSM}} &= W_F + \mu \widehat{H}_u \cdot \widehat{H}_d, \\
W_{\text{NMSSM}} &= W_F + \lambda \widehat{H}_u \cdot \widehat{H}_d \widehat{S} + \frac{1}{3}\kappa \widehat{S}^3, \\
W_{\text{nMSSM}} &= W_F + \lambda \widehat{H}_u \cdot \widehat{H}_d \widehat{S} + \xi_F M_n^2 \widehat{S}, \\
V_{\text{soft}}^{\text{MSSM}} &= \widetilde{m}_u^2 |H_u|^2 + \widetilde{m}_d^2 |H_d|^2 + (B\mu H_u \cdot H_d + h.c.), \\
V_{\text{soft}}^{\text{NMSSM}} &= \widetilde{m}_u^2 |H_u|^2 + \widetilde{m}_d^2 |H_d|^2 + \widetilde{m}_S^2 |S|^2 + \left(A_\lambda \lambda S H_u \cdot H_d + \frac{A_\kappa}{3}\kappa S^3 + h.c.\right), \\
V_{\text{soft}}^{\text{nMSSM}} &= \widetilde{m}_u^2 |H_u|^2 + \widetilde{m}_d^2 |H_d|^2 + \widetilde{m}_S^2 |S|^2 + \left(A_\lambda \lambda S H_u \cdot H_d + \xi_S M_n^3 S + h.c.\right),
\end{aligned} \tag{2.1}$$

where W_F is the MSSM superpotential without the μ term, λ, κ and ξ_F are the dimensionless parameters, and $\widetilde{m}_u$, $\widetilde{m}_d$, $\widetilde{m}_S$, B, A_λ, A_κ, and $\xi_S M_n^3$ are soft-breaking parameters. Note that in the NMSSM and nMSSM the μ-term is replaced by the $\mu_{\text{eff}} = \lambda s$ when the singlet Higgs field $\widehat{S}$ develops a VEV s. The differences between the NMSSM and nMSSM reflect the last term in the superpotential, where the cubic singlet term $\kappa \widehat{S}^3$ in the NMSSM is replaced by a tadpole term $\xi_F M_n^2 \widehat{S}$ in the nMSSM. This replacement in the superpotential makes the nMSSM have no discrete symmetry and thus free of the domain wall problem that the NMSSM suffers from. Actually, due to that the tadpole term $\xi_F M_n^2$ does not induce any interaction, the nMSSM is

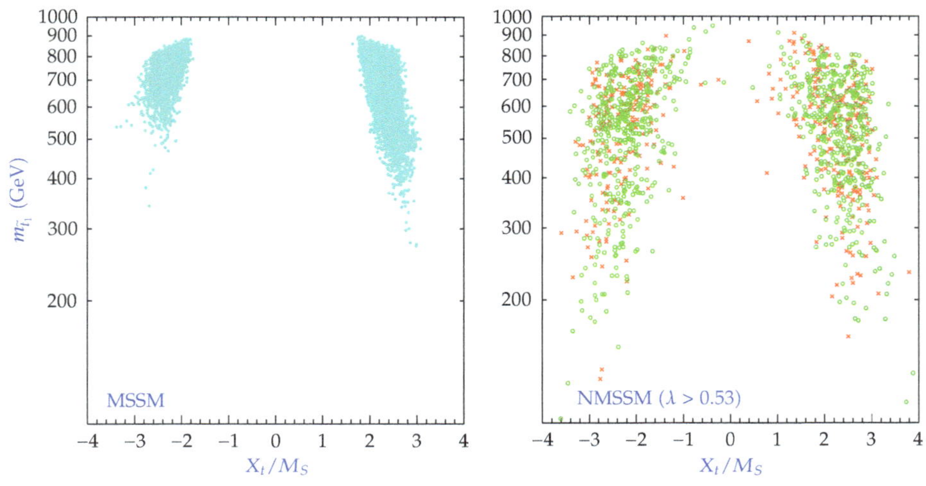

Figure 1: The scatter plots of the samples in the MSSM and NMSSM (with $\lambda > 0.53$) satisfying various constraints listed in the text (including $123\,\text{GeV} \le m_h \le 127\,\text{GeV}$), showing the correlation between the mass of the lighter top squark and X_t/M_S with $M_S \equiv \sqrt{m_{\tilde{t}_1} m_{\tilde{t}_2}}$ and $X_t \equiv A_t - \mu \cot\beta$. In the right panel the circles (green) denote the pull-down case (the lightest Higgs boson being the SM-like Higgs), and the times (red) denote the push-up case (the next-to-lightest Higgs boson being the SM-like Higgs).

identical to the NMSSM with $\kappa = 0$, except for the minimization conditions of the Higgs potential and the tree-level Higgs mass matrices.

With the superpotentials and the soft-breaking terms giving above, one can get the Higgs potentials of these SUSY models and then can derive the Higgs mass matrics and eigenstates. At the minimum of the potential, the Higgs fields H_u, H_d, and S are expanded as

$$H_u = \begin{pmatrix} H_u^+ \\ v_u + \dfrac{\phi_u + i\varphi_u}{\sqrt{2}} \end{pmatrix}, \qquad H_d = \begin{pmatrix} v_d + \dfrac{\phi_d + i\varphi_d}{\sqrt{2}} \\ H_d^- \end{pmatrix}, \qquad S = s + \frac{1}{\sqrt{2}}(\sigma + i\xi), \tag{2.2}$$

with $v = \sqrt{v_u^2 + v_d^2} = 174\,\text{GeV}$. By unitary rotation the mass eigenstates can be given by

$$\begin{pmatrix} h_1 \\ h_2 \\ h_3 \end{pmatrix} = S \begin{pmatrix} \phi_u \\ \phi_d \\ \sigma \end{pmatrix}, \qquad \begin{pmatrix} a \\ A \\ G^0 \end{pmatrix} = P \begin{pmatrix} \varphi_u \\ \varphi_d \\ \xi \end{pmatrix}, \qquad \begin{pmatrix} H^+ \\ G^+ \end{pmatrix} = U \begin{pmatrix} H_u^+ \\ H_d^+ \end{pmatrix}, \tag{2.3}$$

where h_1, h_2, and h_3 are physical CP-even Higgs bosons ($m_{h_1} < m_{h_2} < m_{h_3}$), a, A are CP-odd Higgs bosons, H^+ is the charged Higgs boson, and G^0 G^+ are Goldstone bosons eaten by Z and W^+. Due to the absence of the singlet field S, the MSSM only has two CP-even Higgs bosons and one CP-odd Higgs bosons, as well as one pair of charged Higgs bosons.

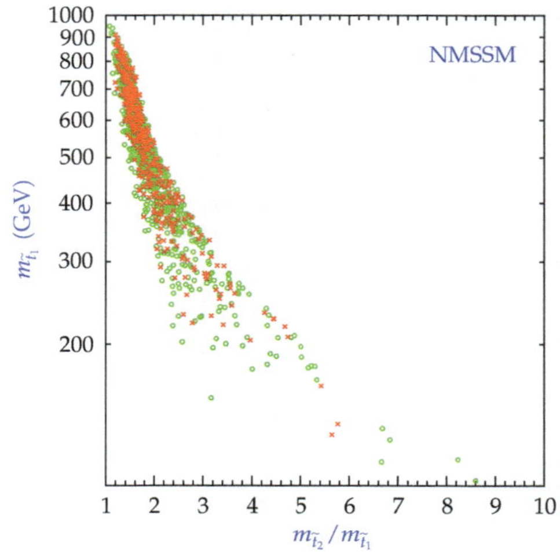

Figure 2: Same as Figure 1, but only for the NMSSM, showing the correlation between $m_{\tilde{t}_1}$ and the ratio $m_{\tilde{t}_2}/m_{\tilde{t}_1}$.

At the tree level, the Higgs masses in the MSSM are conventionally parameterized in terms of the mass of the CP-odd Higgs boson (m_A) and $\tan\beta \equiv v_u/v_d$ and the loop corrections typically come from top and stop loops due to their large Yukawa coupling. For small splitting between the stop masses, an approximate formula of the lightest Higgs boson mass is given by [92]

$$m_h^2 \simeq M_Z^2 \cos^2 2\beta + \frac{3m_t^4}{4\pi^2 v^2} \ln \frac{M_S^2}{m_t^2} + \frac{3m_t^4}{4\pi^2 v^2} \frac{X_t^2}{M_S^2} \left(1 - \frac{X_t^2}{12 M_S^2} \right), \tag{2.4}$$

where $M_S = \sqrt{m_{\tilde{t}_1} m_{\tilde{t}_2}}$ and $X_t \equiv A_t - \mu \cot\beta$. The formula manifests that larger M_S or $\tan\beta$ is necessary to push up the Higgs boson mass. And the Higgs boson mass can reach a maximum when $X_t/M_S = \sqrt{6}$ for given M_S (i.e., the so-called $m_h^{\max}$ scenario). Note that the lightest Higgs boson is the SM-like Higgs boson h (with the largest coupling to vector bosons) in most of the MSSM parameter space.

Different from the case in the MSSM, the Higgs sector in the NMSSM depends on the following six parameters:

$$\lambda, \quad \kappa, \quad M_A^2 = \frac{2\mu(A_\lambda + \kappa s)}{\sin 2\beta}, \quad A_\kappa, \quad \tan\beta = \frac{v_u}{v_d}, \quad \mu = \lambda s, \tag{2.5}$$

and in the nMSSM the input parameters in the Higgs sector are

$$\lambda, \quad \tan\beta, \quad \mu \quad A_\lambda, \quad \tilde{m}_S, \quad M_A^2 = \frac{2(\mu A_\lambda + \lambda \xi_F M_n^2)}{\sin 2\beta}. \tag{2.6}$$

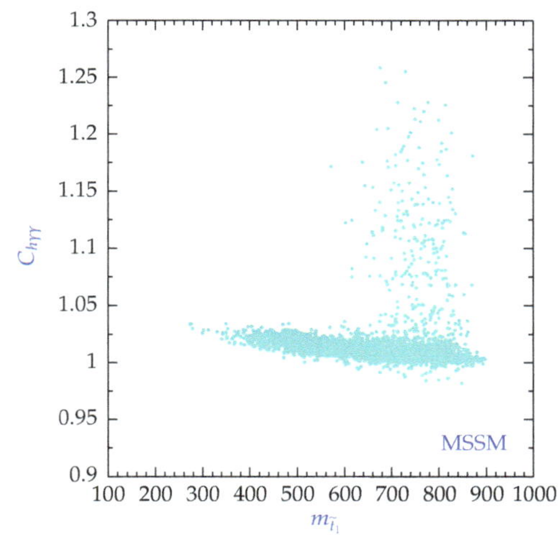

(a)

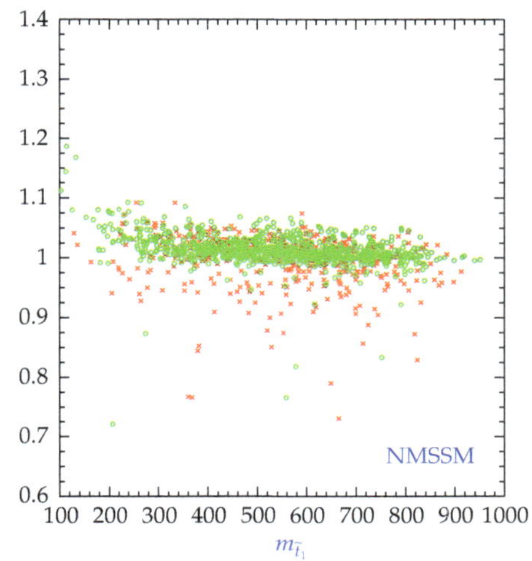

(b)

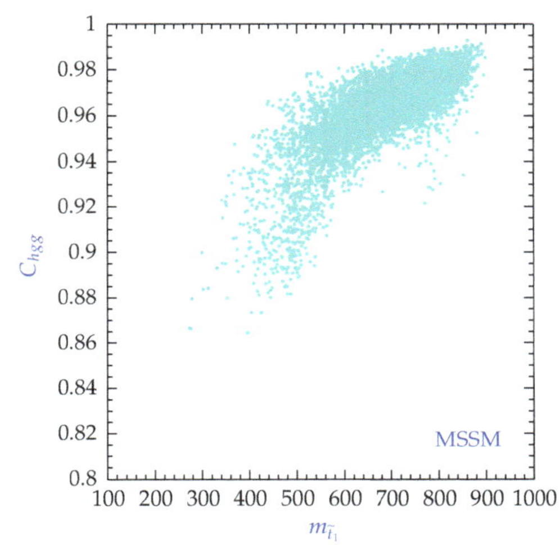

(c)

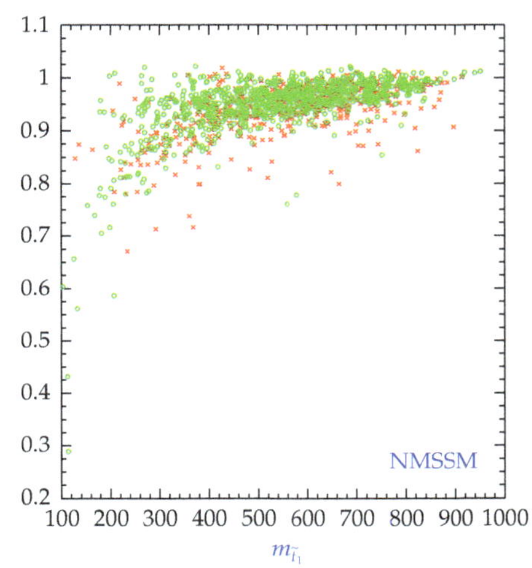

(d)

Figure 3: Same as Figure 1, projected in the planes of $m_{\tilde{t}_1}$ versus the reduced couplings $C_{h\gamma\gamma}$ and C_{hgg}, respectively.

Because of the coupling $\lambda \widehat{H}_u \cdot \widehat{H}_d \widehat{S}$ in the superpotential, the tree-level Higgs boson mass has an additional contribution in the NMSSM and nMSSM

$$\Delta m_h^2 = \lambda^2 v^2 \sin^2 2\beta. \tag{2.7}$$

In order to push up the tree-level Higgs boson mass, λ has to be as large as possible, and $\tan\beta$ has to be small. The requirement of the absence of a landau singularity below the GUT scale implies that $\lambda \lesssim 0.7$ at the weak scale, and the upper bound on λ at the weak scale depends strongly on $\tan\beta$ and grows with increasing $\tan\beta$ [93]. However, this can still lead to a larger

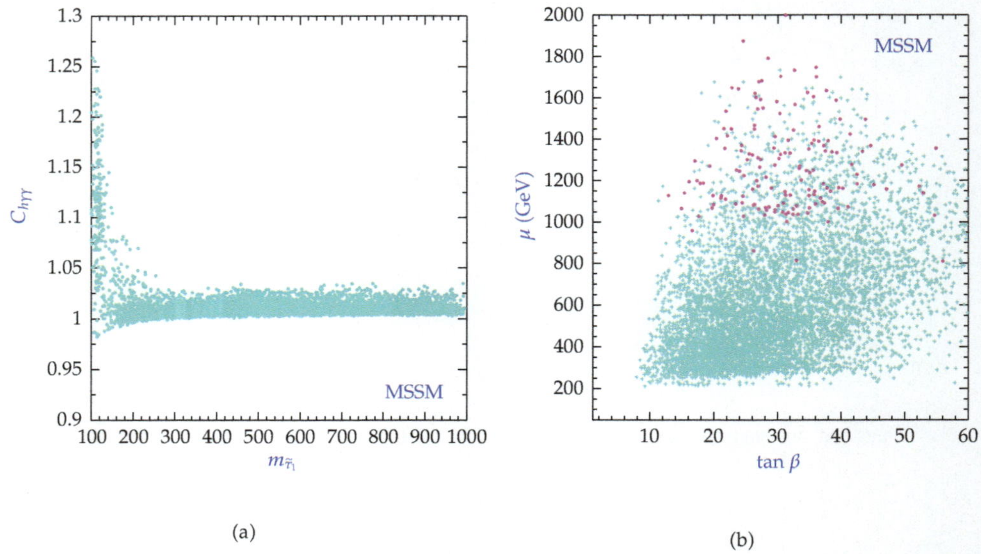

Figure 4: Same as Figure 1, but only for the MSSM, showing the correlation between $m_{\tilde{\tau}_1}$ and the reduced coupling $C_{h\gamma\gamma}$, μ, and $\tan\beta$, respectively. The purple points correspond to $R_{\gamma\gamma} > 1$.

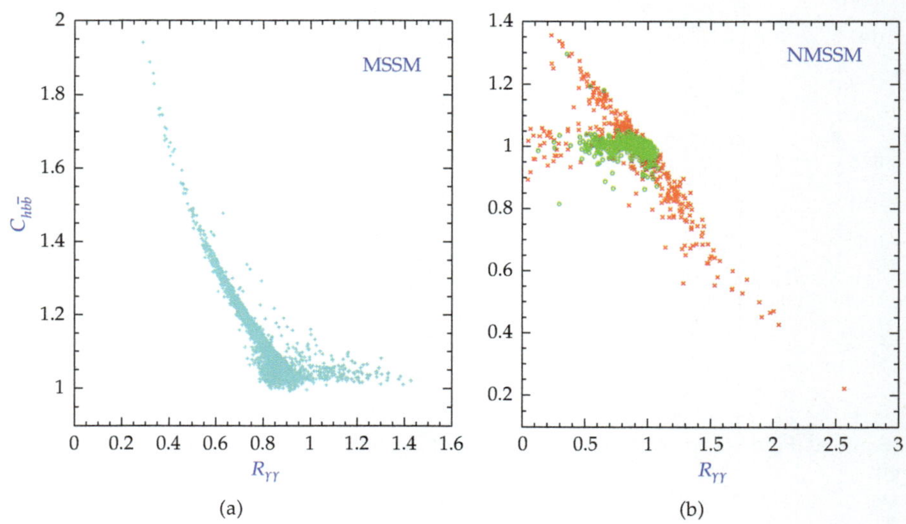

Figure 5: Same as Figure 1, but showing the dependence of the di-photon signal rate $R_{\gamma\gamma}$ on the effective $hb\bar{b}$ coupling $C_{hb\bar{b}} \equiv C^{\text{SUSY}}_{hb\bar{b}}/C^{\text{SM}}_{hb\bar{b}}$ (taken for [42]).

tree-level Higgs boson mass than in the MSSM. Therefore, the radiative corrections to m_h^2 may be reduced in the NMSSM and nMSSM, which may induce light top squark and ameliorate the fine-tuning problem [94, 95].

In the NMSSM and nMSSM, due to the mixing between the doublet Higgs fields and the singlet Higgs field, the SM-like Higgs boson h may either be the lightest CP-even Higgs

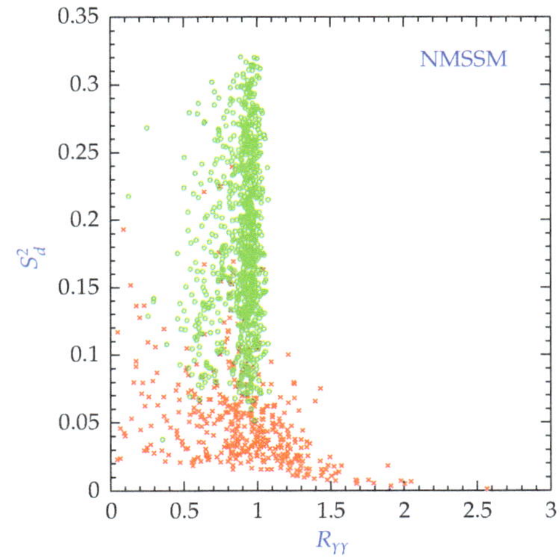

Figure 6: Same as Figure 2, showing the signal rate $R_{\gamma\gamma}$ versus S_d^2 with $S_d = C_{hb\bar{b}} \cos\beta$.

boson or the next-to-lightest CP-even Higgs boson, which corresponds to the so-called pull-down case or the push-up case [42], respectively. Although the mass of the SM-like Higgs boson in the nMSSM is quite similar to that in the NMSSM, the Higgs signal is quite different. This is because of the peculiarity of the neutralino sector in the nMSSM, where the lightest neutralino $\tilde{\chi}_1^0$ as the lightest supersymmetric particle (LSP) acts as the dark matter candidate, and its mass takes the form [96]

$$m_{\tilde{\chi}_1^0} \simeq \frac{2\mu\lambda^2 v^2}{\mu^2 + \lambda^2 v^2} \frac{\tan\beta}{\tan^2\beta + 1}. \tag{2.8}$$

This expression implies that $\tilde{\chi}_1^0$ must be lighter than about 60 GeV for $\lambda < 0.7$ (perturbativity bound) and $\mu > 100\,\text{GeV}$ (from lower bound on chargino mass). And $\tilde{\chi}_1^0$ must annihilate by exchanging a resonant light CP-odd Higgs boson to get the correct relic density. For such a light neutralino, the SM-like Higgs boson around 125 GeV tends to decay predominantly into light neutralinos or other light Higgs bosons [89, 90].

2.2. The Diphoton Higgs Signal

Considering the di-photon signal is of prime importance to searching for Higgs boson near 125 GeV, it is necessary to estimate its signal rate, and we define the normalized production rate as

$$\begin{aligned} R_{\gamma\gamma} &\equiv \frac{\sigma_{\mathrm{SUSY}}(pp \longrightarrow h \longrightarrow \gamma\gamma)}{\sigma_{\mathrm{SM}}(pp \longrightarrow h \longrightarrow \gamma\gamma)} \\ &\simeq \frac{[\Gamma(h \longrightarrow gg)\mathrm{Br}(h \longrightarrow \gamma\gamma)]}{[\Gamma(h_{\mathrm{SM}} \longrightarrow gg)\mathrm{Br}(h_{\mathrm{SM}} \longrightarrow \gamma\gamma)]} \\ &= \frac{[\Gamma(h \longrightarrow gg)\Gamma(h \longrightarrow \gamma\gamma)]}{[\Gamma(h_{\mathrm{SM}} \longrightarrow gg)\Gamma(h_{\mathrm{SM}} \longrightarrow \gamma\gamma)]} \times \frac{\Gamma_{\mathrm{tot}}(h_{SM})}{\Gamma_{\mathrm{tot}}(h)} \\ &= C_{hgg}^2 C_{h\gamma\gamma}^2 \times \frac{\Gamma_{\mathrm{tot}}(h_{\mathrm{SM}})}{\Gamma_{\mathrm{tot}}(h)}, \end{aligned} \tag{2.9}$$

where C_{hgg} and $C_{h\gamma\gamma}$ are the couplings of Higgs to gluons and photons in SUSY with respect to their SM values, respectively. In SUSY, the hgg coupling arises mainly from the loops mediated by the third generation quarks and squarks, while the $h\gamma\gamma$ coupling has additional contributions from loops mediated by W-boson, charged Higgs boson, charginos, and the third generation leptons and sleptons. Their decay widths are given by [47]

$$\begin{aligned} \Gamma(h \longrightarrow gg) &= \frac{G_F \alpha_s^2 m_h^3}{36\sqrt{2}\pi^3} \left| \frac{3}{4} \sum_q g_{hqq} A_{1/2}^h(\tau_q) + \frac{3}{4} \mathcal{A}^{gg} \right|^2 \\ \Gamma(h \longrightarrow \gamma\gamma) &= \frac{G_F \alpha^2 m_h^3}{128\sqrt{2}\pi} \left| \sum_f N_c Q_f^2 g_{hff} A_{1/2}^h(\tau_f) + g_{hWW} A_1^h(\tau_W) + \mathcal{A}^{\gamma\gamma} \right|^2, \end{aligned} \tag{2.10}$$

with $\tau_i = m_h^2/(4m_i^2)$, and

$$\begin{aligned} \mathcal{A}^{gg} &= \sum_i \frac{g_{h\tilde{q}_i\tilde{q}_i}}{m_{\tilde{q}_i}^2} A_0^h(\tau_{\tilde{q}_i}), \\ \mathcal{A}^{\gamma\gamma} &= g_{hH^+H^-} \frac{m_W^2}{m_{H^\pm}^2} A_0^h(\tau_{H^\pm}) + \sum_f \frac{g_{h\tilde{f}\tilde{f}}}{m_{\tilde{f}}^2} A_0^h\left(\tau_{\tilde{f}}\right) + \sum_i g_{h\chi_i^+\chi_i^-} \frac{m_W}{m_{\chi_i}} A_{1/2}^h(\tau_{\chi_i}), \end{aligned} \tag{2.11}$$

where $m_{\tilde{f}}$ and m_{χ_i} are the mass of sfermion and chargino, respectively. In the limit $\tau_i \ll 1$, the asymptotic behaviors of A_i^h are given by

$$A_0^h \longrightarrow -\frac{1}{3}, \qquad A_{1/2}^h \longrightarrow -\frac{4}{3}, \qquad A_1^h \longrightarrow +7. \tag{2.12}$$

One can easily learn that the W-boson contribution to $h\gamma\gamma$ is by far dominant, however, for light stau or squarks with large mixing, the $h\gamma\gamma$ coupling can be enhanced, while light squarks with large mixing can suppress the hgg coupling. Therefore, light stau with large mixing may enhance the di-photon signal rate [92], while light squarks with large mixing have little effect on the di-photon signal rate.

3. Numerical Results and Discussions

In our work the Higgs boson mass is calculated by the package NMSSMTools [97, 98], which include the dominant one-loop and leading logarithmic two-loop corrections. Considering the Higgs hints at the LHC, we focus on the Higgs boson mass between 123 GeV and 127 GeV, furthermore, we consider the following constraints:

(1) the constraints from LHC experiment for the nonstandard Higgs boson,

(2) the constraints from LEP and Tevatron on the masses of the Higgs boson and sparticles, as well as on the neutralino pair productions,

(3) the indirect constraints from B-physics (such as the latest experimental result of $B_s \to \mu^+\mu^-$) and from the electroweak precision observables such as M_W, $\sin^2\theta^{\ell}_{\rm eff}$, and ρ_ℓ, or their combinations $\epsilon_i (i = 1,2,3)$ [99, 100],

(4) the constraints from the muon $g-2$: $a_\mu^{\rm exp} - a_\mu^{SM} = (25.5 \pm 8.2) \times 10^{-10}$ [101]. We require SUSY to explain the discrepancy at 2σ level,

(5) the dark matter constraints from WMAP relic density ($0.1053 < \Omega h^2 < 0.1193$) [102] and the direct detection exclusion limits on the scattering cross-section from XENON100 experiment (at 90% C.L.) [103].

Note that most of the above constraints have been encoded in the package NMSSMTools.

Natural supersymmetry is usually characterized by a small superpotential parameter μ, and the third generation squarks with mass $\lesssim$0.5–1.5 TeV [104]. Therefore, we only consider the case with

$$100\,\text{GeV} \le (M_{Q_3}, M_{U_3}) \le 1\,\text{TeV}, \qquad |A_t| \le 3\,\text{TeV}. \tag{3.1}$$

For the case with $\lambda < 0.2$ in the NMSSM, the property of the NMSSM is similar to the case in the MSSM [42]. In order to distinguish the features between MSSM and NMSSM, we only consider the case with $\lambda > m_Z/v \simeq 0.53$ in the NMSSM. We scan over the parameter space of the MSSM and NMSSM under the above experimental constraints and study the property of the Higgs boson for the samples surviving the constraints.

In Figure 1 we display the surviving samples in the MSSM and NMSSM (with $\lambda >$ 0.53), showing the correlation between the lighter top-squark mass and the ratio X_t/M_S with $M_S \equiv \sqrt{m_{\tilde{t}_1} m_{\tilde{t}_2}}$. From the figure we see that for a moderate light $\tilde{t}_1$, large X_t is necessary to satisfy $m_h \sim 125$ GeV, and, for large $m_{\tilde{t}_1}$, the ratio X_t/M_S decreases. In the MSSM, $|X_t/M_S| >$ 1.6 for $m_{\tilde{t}_1} < 1$ TeV; that is, nomixing scenario ($X_t = 0$) cannot survive, and the top-squark mass is usually larger than 300 GeV. This implies that a large top-squark mass or a near-maximal stop mixing is necessary to satisfy the Higgs mass near 125 GeV. However, the case is very different in the NMSSM, $X_t \approx 0$ may also survive, and the lighter top-squark mass can be as light as about 100 GeV, which may alleviate the fine-tuning problem and make the NMSSM seems more natural. In the case of light $m_{\tilde{t}_1}$, $|X_t/M_S|$ is usually larger than $\sqrt{6}$, which corresponds to a large splitting between $m_{\tilde{t}_1}$ and $m_{\tilde{t}_2}$, as the Figure 2 shown.

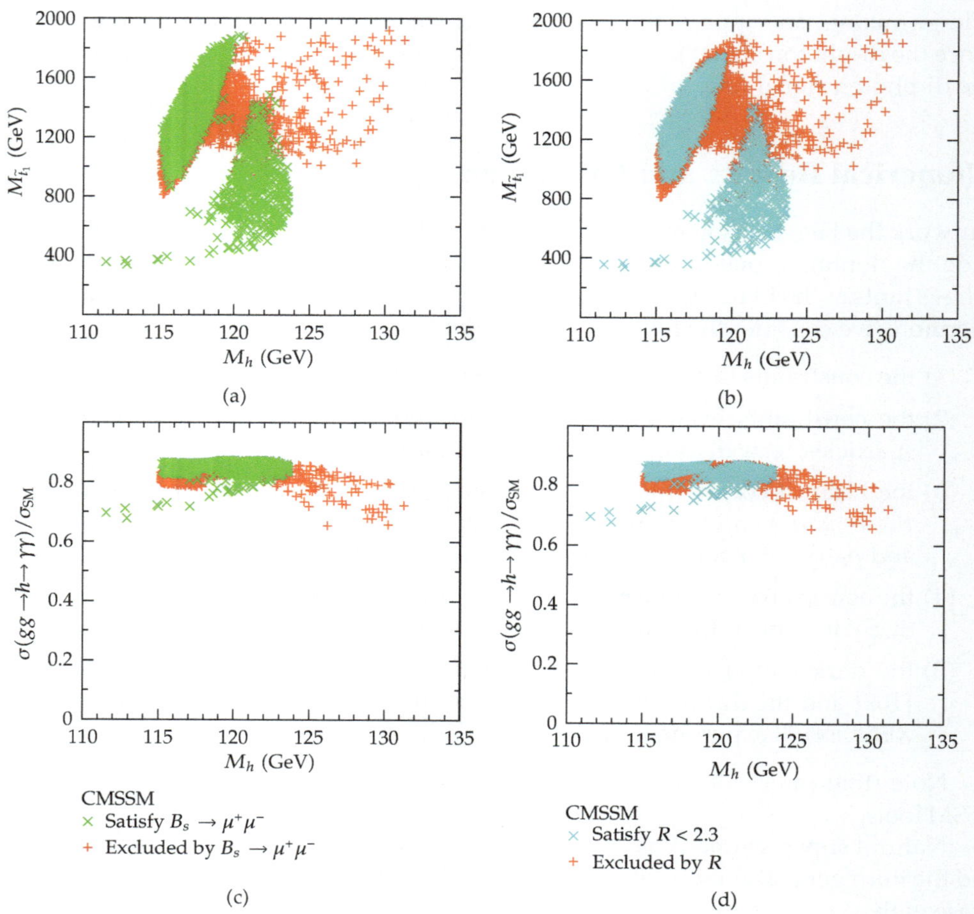

Figure 7: The scatter plots of the surviving sample in the CMSSM, displayed in the planes of the top-squark mass and the LHC di-photon rate versus the Higgs boson mass. In the left frame, the crosses (red) denote the samples satisfying all the constraints except $B_s \to \mu^+\mu^-$, and the times (green) denote those further satisfying the $\mathrm{Br}(B_s \to \mu^+\mu^-)$ constraint. In the right frame, the crosses (red) are the same as those in the left frame, while the times (sky blue) denote the samples further satisfying the R constraint (taken for [43]).

Due to the clean background, the di-photon signal is crucial for searching for the Higgs boson near 125 GeV. As discussed in the Section 2, the signal rate is relevant with the coupling $C_{h\gamma\gamma}$ and C_{hgg} and the total width of the SM-like Higgs boson. Both the coupling $C_{h\gamma\gamma}$ and C_{hgg} are affected by the contributions from the squark loops, especially the light top-squark loop, so in the Figure 3 we give the relationship between the lighter top-squark mass and the coupling $C_{h\gamma\gamma}$ and C_{hgg}, respectively. The figure shows that the light $m_{\tilde{t}_1}$ may suppress the coupling C_{hgg} significantly, especially in the NMSSM, while the light top squark has little effect on the coupling $C_{h\gamma\gamma}$ because there are additional contributions, as (2.10) and shown. As analyzed in the Section 2, light stau may enhance the coupling $C_{h\gamma\gamma}$, so in Figure 4 we give the correlation between $m_{\tilde{\tau}_1}$ and the coupling $C_{h\gamma\gamma}$ in the MSSM. The figure clearly shows that the coupling $C_{h\gamma\gamma}$ can enhance to 1.25 for $m_{\tilde{\tau}_1} \sim 100$ GeV. Figure 4 also manifests that the

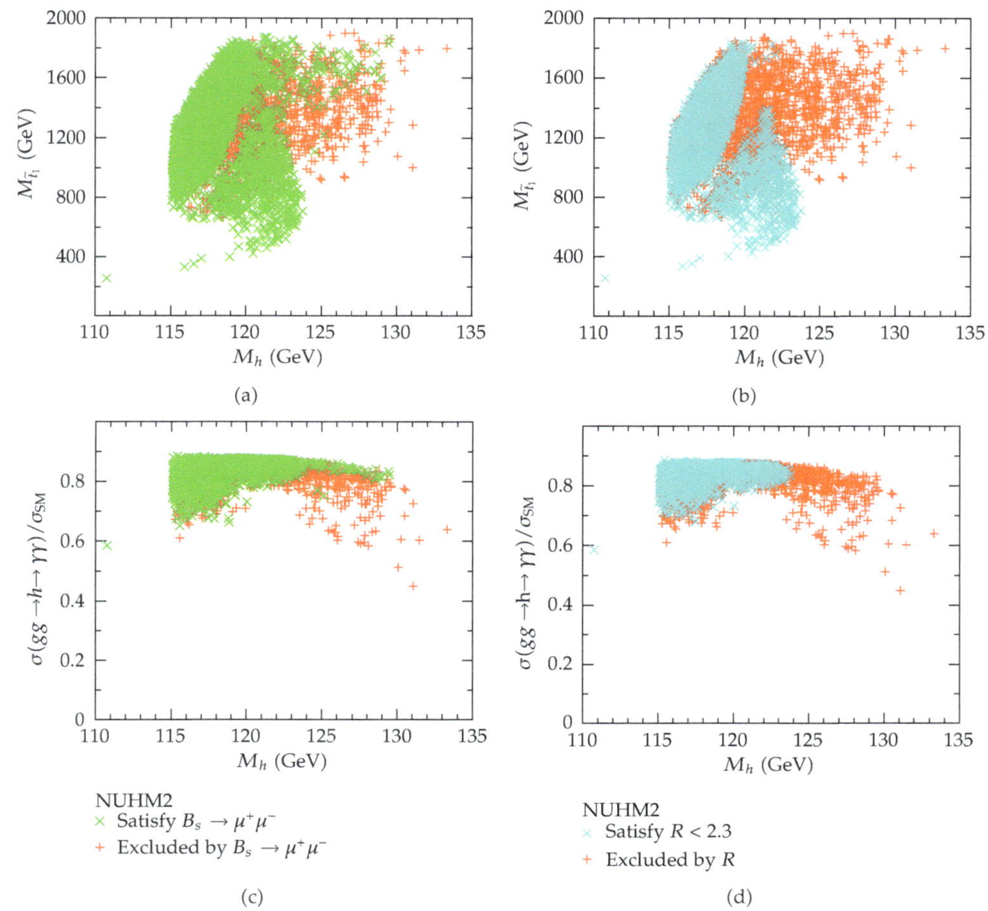

Figure 8: Same as Figure 7, but for the NUHM2 (taken for [43]).

enhancement of the coupling $C_{h\gamma\gamma}$ corresponds to large $\mu \tan\beta$, which leads to large mixing. These results exactly verify the discussions in the Section 2.

Since $h \to b\bar{b}$ is the main decay mode of the light Higgs boson, the total width of the SM-like Higgs boson may be affected by the effective $hb\bar{b}$ coupling $C_{hb\bar{b}}$, as discussed in [81]. Under the effect of the combination $C_{hgg}C_{h\gamma\gamma}/C_{hb\bar{b}}$, the di-photon Higgs signal rate may be either enhanced or suppressed, as shown in Figure 5, which also manifests that, for the signal rate larger than 1, the effective $hb\bar{b}$ coupling is enhanced a little in the MSSM, while it is suppressed significantly in the NMSSM. Therefore, we can conclude that the reason for the enhancement in the signal rate is very different between the MSSM and NMSSM. In the MSSM the enhancement of the signal is mainly due to the enhancement of the coupling $C_{h\gamma\gamma}$, while in the NMSSM it is mainly due to the suppression of the $hb\bar{b}$ coupling.

Due to the presence of the singlet field in the NMSSM, the doublet component in the SM-like Higgs boson h may be different from the case in the MSSM, which will affect the coupling $hb\bar{b}$ and accordingly affect the total width of h. At the tree-level, $C_{hb\bar{b}} = S_d/\cos\beta$. In Figure 6 we show the dependence of the signal rate $R_{\gamma\gamma}$ on S_d^2. Obviously, for the signal rate larger than 1, S_d^2 is usually very small, which leads to large suppression on the reduced coupling $hb\bar{b}$. The figure also shows that the push-up case is more effective to enhance the signal rate than the pull-down case. This is because the push-up case is easier to realize the large mixing between the singlet field and the doublet field [42].

As the case in the NMSSM, nMSSM can also accommodate a 125 GeV SM-like Higgs [81]. However, due to the peculiar property of the lightest neutralino $\tilde{\chi}_1^0$ in the nMSSM [96], the decay mode of the SM-like Higgs is very different from the case in the MSSM and NMSSM. As discussed in the Section 2, $h \to \tilde{\chi}_1^0\tilde{\chi}_1^0$ may be dominant over $h \to b\bar{b}$ in a major part of parameter space in the nMSSM [81, 105], which induces a severe suppression on the di-photon Higgs signal. Although the Higgs mass can easily reach to 125 GeV, the di-photon signal is not consistent with the LHC experiment. Therefore, the nMSSM may be excluded by LHC experiment.

We also considered the SM-like Higgs boson mass and its di-photon signal in the constrained MSSM (including CMSSM and NUHM2) under various experimental constraints, especially the limits from $B_s \to \mu^+\mu^-$. Because $\mathrm{Br}(B_s \to \mu^+\mu^-) \propto \tan^6\beta/M_A^4$, so it may provide a rather strong constraint on SUSY with large $\tan\beta$. Considering the large theoretical uncertainties for the calculation of $\mathrm{Br}(B_s \to \mu^+\mu^-)$, we use not only the LHCb data, but also the double ratio of the purely leptonic decays defined as $R \equiv \eta/\eta_{\mathrm{SM}}$ with $\eta \equiv (\mathrm{Br}(B_s \to \mu^+\mu^-)/\mathrm{Br}(B_u \to \tau\nu_\tau))/(\mathrm{Br}(D_s \to \tau\nu_\tau)/\mathrm{Br}(D \to \mu\nu_\mu))$. The surviving parameter space is plotted in Figure 7 for the CMSSM and Figure 8 for the NUHM2. It shows that both the CMSSM and NUHM2 are hard to realize a 125 GeV SM-like Higgs boson, and also the di-photon Higgs signal is suppressed relative to the SM prediction due to the enhanced $hb\bar{b}$ coupling. Therefore, the constrained MSSM may also be excluded by the LHC experiment.

4. Conclusion

In this work we briefly review our recent studies on a 125 GeV Higgs and its di-photon signal rate in the MSSM, NMSSM, nMSSM, and the constrained MSSM. Under the current experimental constraints, we find: (i) the SM-like Higgs can easily reach to 125 GeV in the MSSM, NMSSM, and nMSSM, while it is hard to satisfy in the constrained MSSM; (ii) the NMSSM may predict a lighter top squark than the MSSM, even as light as 100 GeV, which can ameliorate the fine-tuning problem; (iii) the di-photon Higgs signal is suppressed in the nMSSM and the constrained MSSM, but, in a tiny corner of the parameter space in the MSSM and NMSSM, it can be enhanced.

Acknowledgment

The work is supported by the Startup Foundation for Doctors of Henan Normal University under contract no.11108.

References

[1] G. Aad, B. Abbott, J. Abdallah et al., "Combined search for the Standard Model Higgs boson using up to 4.9 fb^{-1} of *pp* collision data at $\sqrt{s} = 7TeV$ with the ATLAS detector at the LHC," *Physics Letters, Section B*, vol. 710, no. 1, pp. 49–66, 2012.

[2] G. Aad, B. Abbott, J. Abdallah et al., "Search for the standard model Higgs boson in the diphoton decay channel with 4.9fb^{-}1 of pp collision data at $\sqrt{s} = 7TeV$ with atlas," *Physical Review Letters*, vol. 108, no. 11, Article ID 111803, 19 pages, 2012.

[3] G. Aad, B. Abbott, J. Abdallah et al., "Search for the Standard Model Higgs boson in the decay channel $H \rightarrow ZZ^{(*)} \rightarrow 4\ell$ with 4.8 fb^{-1} of pp collision data at $\sqrt{s} = 7TeV$ with ATLAS," *Physics Letters, Section B*, vol. 710, no. 3, pp. 383–402, 2012.

[4] S. Chatrchyan, V. Khachatryan, A. M. Sirunyan et al., "Search for a Higgs boson in the decay channel $H \rightarrow ZZ^{(*)} \rightarrow qq\ell^{-}\ell^{-}+$ in pp collisions at $\sqrt{s} = 7TeV$," *Journal of High Energy Physics*, vol. 2012, no. 4, article 36, 2012.

[5] S. Chatrchyan, V. Khachatryan, A. M. Sirunyan et al., "Search for the standard model Higgs boson decaying into two photons in pp collisions at $\sqrt{s} = 7TeV$," *Physics Letters, Section B*, vol. 710, no. 3, pp. 403–425, 2012.

[6] S. Chatrchyan, V. Khachatryan, A. M. Sirunyan et al., "Combined results of searches for the standard model Higgs boson in pp collisions at $\sqrt{s} = 7TeV$," *Physics Letters, Section B*, vol. 710, no. 1, pp. 26–48, 2012.

[7] S. Chatrchyan, V. Khachatryan, A. M. Sirunyan et al., "Search for the standard model Higgs boson decaying to $W^{+}W^{-}$ in the fully leptonic final state in pp collisions at $\sqrt{s} = 7TeV$," *Physics Letters, Section B*, vol. 710, no. 1, pp. 91–113, 2012.

[8] S. Chatrchyan, V. Khachatryan, A. M. Sirunyan et al., "Search for the standard model higgs boson in the decay channel $H \rightarrow ZZ \rightarrow 4I$ in pp collisions at $\sqrt{s} = 7TeV$," *Physical Review Letters*, vol. 108, no. 11, Article ID 111804, 17 pages, 2012.

[9] S. Chatrchyan, V. Khachatryan, A. M. Sirunyan et al., "Search for the standard model Higgs boson in the $H \rightarrow ZZ \rightarrow 2\ell 2\nu$ channel in pp collisions at $\sqrt{s} = 7TeV$," *Journal of High Energy Physics*, vol. 2012, no. 3, article 040, 2012.

[10] S. Chatrchyan, V. Khachatryan, A. M. Sirunyan et al., "Search for the standard model Higgs boson in the $H \rightarrow ZZ \rightarrow 1\,\ell^{+}\ell^{-}\tau^{+}\tau^{-}$ decay channel in pp collisions at $\sqrt{s} = 7TeV$," *Journal of High Energy Physics*, vol. 2012, no. 3, article 081, 2012.

[11] S. Chatrchyan, V. Khachatryan, A. M. Sirunyan et al., "Search for neutral Higgs bosons decaying to tau pairs in pp collisions at sqrt(s) = $7TeV$," *Physics Letters, Section B*, vol. 713, no. 2, pp. 68–90, 2012.

[12] S. Chatrchyan, V. Khachatryan, A. M. Sirunyan et al., "Search for the standard model Higgs boson decaying to bottom quarks in pp collisions at $\sqrt{s} = 7TeV$," *Physics Letters, Section B*, vol. 710, no. 2, pp. 284–306, 2012.

[13] D. Carmi, A. Falkowski, E. Kuflik, and T. Volansky, "Interpreting LHC higgs results from natural new physics perspective," Energy Physicsm http://arxiv.org/abs/1202.3144.

[14] S. Heinemeyer, O. Stål, and G. Weiglein, "Interpreting the LHC Higgs search results in the MSSM," *Physics Letters, Section B*, vol. 710, no. 1, pp. 201–206, 2012.

[15] A. Arbey, M. Battaglia, A. Djouadi, F. Mahmoudi, and J. Quevillon, "Implications of a 125 GeV Higgs for supersymmetric models," *Physics Letters, Section B*, vol. 708, no. 1-2, pp. 162–169, 2012.

[16] L. J. Hall, D. Pinner, and J. T. Ruderman, "A natural SUSY Higgs near 125 GeV," *Journal of High Energy Physics*, vol. 2012, no. 4, article 131, 2012.

[17] P. Draper, P. Meade, M. Reece, and D. Shih, "Implications of a 125 GeV Higgs for the MSSM and Low-Scale SUSY Breaking," High Energy Physics, http://arxiv.org/abs/1112.3068.

[18] A. Arbey, M. Battaglia, and F. Mahmoudi, "Constraints on the MSSM from the Higgs sector: a pMSSM study of Higgs searches, $B_s^0 \rightarrow \mu^{+}\mu^{-}$ and dark matter direct detection," *European Physical Journal C*, vol. 72, no. 3, article 1906, 2012.

[19] O. Buchmueller, R. Cavanaugh, A. de Roeck et al., "Higgs and supersymmetry," *European Physical Journal C*, vol. 72, no. 6, pp. 1–14, 2012.

[20] M. Kadastik, K. Kannike, A. Racioppi, and M. Raida, "Implications of the 125 GeV Higgs boson for scalar dark matter and for the CMSSM phenomenology," *Journal of High Energy Physics*, vol. 2012, no. 5, article 061, 2012.

[21] A. Arvanitaki and G. Villadoro, "A non standard model higgs at the LHC as a sign of naturalness," *Journal of High Energy Physics*, vol. 2012, no. 2, article 144, 2012.

[22] H. Baer, V. Barger, and A. Mustafayev, "Implications of a 125 GeV Higgs scalar for the LHC supersymmetry and neutralino dark matter searches," *Physical Review D*, vol. 85, no. 7, Article ID 075010, 13 pages, 2012.

[23] J. F. Gunion, Y. Jiang, and S. Kraml, "The constrained NMSSM and Higgs near 125 GeV," *Physics Letters, Section B*, vol. 710, no. 3, pp. 454–459, 2012.

[24] P. Fileviez Pérez, "SUSY spectrum and the Higgs mass in the BLMSSM," *Physics Letters, Section B*, vol. 711, no. 5, pp. 353–359, 2012.
[25] Z. Kang, J. Li, and T. Li, "On Naturalness of the (N)MSSM," *High Energy Physics*. In press http://arxiv.org/abs/1201.5305.
[26] L. Aparicio, D. G. Cerdeño, and L. E. Ibáñez, "A 119–125 GeV Higgs from a string derived slice of the CMSSM," *Journal of High Energy Physics*, vol. 2012, no. 4, article 126, 2012.
[27] C.-F. Chang, K. Cheung, Y.-C. Lin, and T.-C. Yuan, "Mimicking the standard model Higgs boson in UMSSM," *Journal of High Energy Physics*, vol. 2012, no. 6, article 51, 2012.
[28] J. Ellis and K. A. Olive, "Revisiting the Higgs mass and dark matter in the CMSSM," *European Physical Journal C*, vol. 72, no. 5, pp. 1–13, 2012.
[29] H. Baer, V. Barger, and A. Mustafayev, "Neutralino dark matter in mSUGRA/CMSSM with a 125 GeV light higgs scalar," *Journal of High Energy Physics*, vol. 2012, no. 5, article 091, 2012.
[30] N. Desai, B. Mukhopadhyaya, and S. Niyogi, "Constraints on invisible Higgs decay in MSSM in the light of diphoton rates from the LHC," High Energy Physics, http://arxiv.org/abs/1202.5190.
[31] F. Boudjema and G. D. La Rochelle, "BMSSM Higgses at 125 GeV," High Energy Physics, http://arxiv.org/abs/1203.3141.
[32] D. Albornoz Vásquez, G. Bélanger, C. Bahm, J. Da Silva, P. Richardson, and C. Wymant, "125 GeV Higgs boson in the NMSSM in light of the LHC results and astrophysics constraints," *Physical Review D*, vol. 86, no. 3, Article ID 035023, 10 pages, 2012.
[33] U. Ellwanger and C. Hugonie, "Higgs bosons near 125 GeV in the NMSSM with constraints at the GUT Scale," *Advances in High Energy Physics*, vol. 2012, Article ID 625389, 18 pages, 2012.
[34] T. Basak and S. Mohanty, "Triplet-singlet extension of the MSSM with a 125 GeV higgs and dark matter," High Energy Physics, http://arxiv.org/abs/1204.6592.
[35] N. Chen and H.-J. He, "LHC signatures of two-Higgs-doublets with fourth family," *Journal of High Energy Physics*, vol. 1204, no. 4, article 62, 2012.
[36] G. Guo, B. Ren, and X. G. He, "LHC evidence of a 126 GeV higgs boson from H to $\gamma\gamma$ with three And four generations," *High Energy Physics*, http://arxiv.org/abs/1112.3188.
[37] X.-G. He, B. Ren, and J. Tandean, "Hints of standard model Higgs boson at the LHC and light dark matter searches," *Physical Review D*, vol. 85, no. 9, Article ID 093019, 10 pages, 2012.
[38] A. Djouadi, O. Lebedev, Y. Mambrini, and J. Quevillon, "Implications of LHC searches for Higgs-portal dark matter," *Physics Letters, Section B*, vol. 709, no. 1-2, pp. 65–69, 2012.
[39] K. Cheung and T.-C. Yuan, "Could the excess seen at 124–126 GeV Be due to the Randall-Sundrum radion?" *Physical Review Letters*, vol. 108, no. 14, Article ID 141602, 5 pages, 2012.
[40] B. Batell, S. Goria, and L.-T. Wanga, "Exploring the Higgs portal with 10 fb^{-1} at the LHC," *Journal of High Energy Physics*, vol. 2012, no. 6, article 07, 2012.
[41] T. Li, J. A. Maxin, D. V. Nanopoulos, and J. W. Walker, "A Higgs mass shift to 125 GeV and a multi-jet supersymmetry signal: miracle of the flippons at the $\sqrt{s} = 7TeV$ LHC," *Physics Letters, Section B*, vol. 710, no. 1, pp. 207–214, 2012.
[42] J. Cao, Z. Heng, J. M. Yang, Y. Zhang, and J. Zhu, "A SM-like Higgs near 125 GeV in low energy SUSY: a comparative study for MSSM and NMSSM," *Journal of High Energy Physics*, vol. 2012, no. 3, article 086, 2012.
[43] J. Cao, Z. Heng, D. Li, and J. M. Yang, "Current experimental constraints on the lightest Higgs boson mass in the constrained MSSM," *Physics Letters, Section B*, vol. 710, no. 4-5, pp. 665–670, 2012.
[44] H. E. Haber and G. L. Kane, *Physics Reports*, vol. 117, p. 75, 1985.
[45] J. F. Gunion and H. E. Haber, "Higgs bosons in supersymmetric models (I)," *Nuclear Physics, Section B*, vol. 272, no. 1, pp. 1–76, 1986.
[46] "Higgs bosons in supersymmetric models (I). [Nucl. Phys. B272 (1986) 1] J.F. Gunion and H.E. Haber," *Nuclear Physics, Section B*, vol. 402, no. 1-2, pp. 567–568, 1993.
[47] A. Djouadi, "The anatomy of electroweak symmetry breaking Tome II: the Higgs bosons in the Minimal Supersymmetric Model," *Physics Reports*, vol. 459, no. 1–6, pp. 1–241, 2008.
[48] U. Ellwanger, C. Hugonie, and A. M. Teixeira, "The next-to-minimal supersymmetric standard model," *Physics Reports*, vol. 496, no. 1-2, pp. 1–77, 2010.
[49] M. Maniatis, "The next-to-minimal supersymmetric extension of the standard model reviewed," *International Journal of Modern Physics A*, vol. 25, no. 18-19, pp. 3505–3602, 2010.
[50] U. Ellwanger, "Higgs bosons in the Next-to-Minimal Supersymmetric Standard Model at the LHC," *The European Physical Journal C*, vol. 71, no. 10, p. 1782, 2011.
[51] J. Ellis, J. F. Gunion, H. E. Haber, L. Roszkowski, and F. Zwirner, "Higgs bosons in a nonminimal supersymmetric model," *Physical Review D*, vol. 39, no. 3, pp. 844–869, 1989.

[52] J. Ellis, J. F. Gunion, H. E. Haber, L. Roszkowski, and F. Zwirner, "Higgs bosons in a nonminimal supersymmetric model," *Physical Review D*, vol. 39, no. 3, pp. 844–869, 1989.
[53] M. Drees, "Supersymmetric models with extended higgs sector," *International Journal of Modern Physics A*, vol. 04, no. 14, article 3635, 1989.
[54] S. F. King and P. L. White, "Resolving the constrained minimal and next-to-minimal supersymmetric standard models," *Physical Review D*, vol. 52, no. 7, pp. 4183–4216, 1995.
[55] B. Ananthanarayan and P. N. Pandita, "The non-minimal supersymmetric standard model with tan $\beta \simeq m_t/m_b$," *Physics Letters, Section B*, vol. 353, no. 1, pp. 70–78, 1995.
[56] B. A. Dobrescu and K. T. Matchev, "Light axion within the next-to-minimal supersymmetric standard odel," *Journal of High Energy Physics*, vol. 4, no. 9, article 031, 2000.
[57] G. Hiller, "b-physics signals of the lightest CP-odd Higgs boson in the next-to-minimal supersymmetric standard model at large tan β," *Physical Review D*, vol. 70, no. 3, Article ID 034018, 10 pages, 2004.
[58] W. Wang, Z. Xiong, and J. M. Yang, "Residual effects of heavy sparticles in the bottom quark Yukawa coupling: a comparative study for the MSSM and NMSSM," *Physics Letters, Section B*, vol. 680, no. 2, pp. 167–171, 2009.
[59] J. Cao and J. M. Yang, "Anomaly of $Zb\bar{b}$coupling revisited in MSSM and NMSSM," *Journal of High Energy Physics*, vol. 2008, no. 12, article 006, 2008.
[60] J. Cao and J. M. Yang, "Current experimental constraints on the next-to-minimal supersymmetric standard model with large λ," *Physical Review D*, vol. 78, no. 11, Article ID 115001, 11 pages, 2008.
[61] J. M. Yang, "SUSY dark matter in light of CDMS/XENON limits," *International Journal of Modern Physics D*, vol. 20, no. 8, pp. 1383–1388, 2011.
[62] A. H. Chamseddine, R. Arnowitt, and P. Nath, "Locally supersymmetric grand unification," *Physical Review Letters*, vol. 49, no. 14, pp. 970–974, 1982.
[63] R. Barbieri, S. Ferrara, and C. A. Savoy, "Gauge models with spontaneously broken local supersymmetry," *Physics Letters B*, vol. 119, no. 4–6, pp. 343–347, 1982.
[64] L. Hall, J. Lykken, and S. Weinberg, "Supergravity as the messenger of supersymmetry breaking," *Physical Review D*, vol. 27, no. 10, pp. 2359–2378, 1983.
[65] N. Ohta, "Grand unified theories based on local supersymmetry," *Progress of Theoretical Physics*, vol. 70, no. 2, pp. 542–549, 1983.
[66] J. Ellis, K. A. Olive, and Y. Santoso, "The MSSM parameter space with non-universal Higgs masses," *Physics Letters, Section B*, vol. 539, no. 1-2, pp. 107–118, 2002.
[67] J. Ellis, T. Falk, K. A. Olive, and Y. Santoso, "Exploration of the MSSM with non-universal Higgs masses," *Nuclear Physics B*, vol. 652, pp. 259–347, 2003.
[68] H. Baer, A. Mustafayev, S. Profumo, A. Belyaev, and X. Tata, "Neutralino cold dark matter in a one-parameter extension of the minimal supergravity model," *Physical Review D*, vol. 71, no. 9, Article ID 095008, 5 pages, 2005.
[69] M. Carena, J. R. Espinosa, M. Quirós, and C. E. M. Wagner, "Analytical expressions for radiatively corrected Higgs masses and couplings in the MSSM," *Physics Letters, Section B*, vol. 355, no. 1-2, pp. 209–221, 1995.
[70] U. Ellwanger and C. Hugonie, "The upper bound on the lightest higgs mass in the NMSSM revisited," *Modern Physics Letters A*, vol. 22, no. 21, pp. 1581–1590, 2007.
[71] U. Ellwanger and C. Hugonie, "Masses and couplings of the lightest Higgs bosons in (M + 1)SSM," *European Physical Journal C*, vol. 25, no. 2, pp. 297–305, 2002.
[72] U. Ellwanger, "Higgs bosons in the Next-to-Minimal Supersymmetric Standard Model at the LHC," *The European Physical Journal C*, vol. 71, no. 10, article 1782, 2011.
[73] P. H. Chankowski, J. Ellis, and S. Pokorski, "The fine-tuning price of LEP," *Physics Letters, Section B*, vol. 423, no. 3-4, pp. 327–336, 1998.
[74] R. Barbieri and A. Strumia, "About the fine-tuning price of LEP," *Physics Letters, Section B*, vol. 433, no. 1-2, pp. 63–66, 1998.
[75] G. L. Kane and S. F. King, "Naturalness implications of LEP results," *Physics Letters, Section B*, vol. 451, no. 1-2, pp. 113–122, 1999.
[76] S. Moretti and S. Munir, "Di-photon Higgs signals at the LHC in the next-to-minimal supersymmetric standard model," *European Physical Journal C*, vol. 47, no. 3, pp. 791–803, 2006.
[77] K. Hsieh and C. P. Yuan, "Lone Higgs boson at the CERN LHC," *Physical Review D*, vol. 78, no. 5, Article ID 053006, 19 pages, 2008.
[78] I. Low and S. Shalgar, "Implications of the Higgs discovery in the MSSM golden region," *Journal of High Energy Physics*, vol. 2009, no. 4, article 091, 2009.

[79] U. Ellwanger, "Enhanced di-photon Higgs signal in the next-to-minimal supersymmetric standard model," *Physics Letters, Section B*, vol. 698, no. 4, pp. 293–296, 2011.

[80] U. Ellwanger, "A Higgs boson near 125 GeV with enhanced di-photon signal in the NMSSM," *Journal of High Energy Physics*, vol. 2012, no. 3, article 044, 2012.

[81] J. Cao, Z. Heng, T. Liu, and J. M. Yang, "Di-photon Higgs signal at the LHC: a comparative study in different supersymmetric models," *Physics Letters, Section B*, vol. 703, no. 4, pp. 462–468, 2011.

[82] C. Panagiotakopoulos and K. Tamvakis, "Stabilized NM SSM without domain walls," *Physics Letters, Section B*, vol. 446, no. 3-4, pp. 224–227, 1999.

[83] C. Panagiotakopoulos and K. Tamvakis, "New minimal extension of MSSM," *Physics Letters, Section B*, vol. 469, no. 1–4, pp. 145–148, 1999.

[84] C. Panagiotakopoulos and A. Pilaftsis, "Higgs scalars in the minimal nonminimal supersymmetric standard model," *Physical Review D*, vol. 63, no. 5, Article ID 055003, 33 pages, 2001.

[85] A. Dedes, C. Hugonie, S. Moretti, and K. Tamvakis, "Phenomenology of a new minimal supersymmetric extension of the standard model," *Physical Review D*, vol. 63, no. 5, Article ID 055009, 9 pages, 2001.

[86] A. Menon, D. E. Morrissey, and C. E. M. Wagner, "Electroweak baryogenesis and dark matter in a minimal extension of the MSSM," *Physical Review D*, vol. 70, no. 3, Article ID 035005, 20 pages, 2004.

[87] V. Barger, P. Langacker, and H.-S. Lee, "Lightest neutralino in extensions of the MSSM," *Physics Letters, Section B*, vol. 630, no. 3-4, pp. 85–99, 2005.

[88] C. Balázs, M. Carena, A. Freitas, and C. E. M. Wagner, "Phenomenology of the nMSSM from colliders to cosmology," *Journal of High Energy Physics*, vol. 2007, no. 6, article 066, 2007.

[89] J. Cao, H. E. Logan, and J. M. Yang, "Experimental constraints on the nearly minimal supersymmetric standard model and implications for its phenomenology," *Physical Review D*, vol. 79, no. 9, Article ID 091701, 5 pages, 2009.

[90] J. Cao, Z. Heng, and J. M. Yang, "Rare Z-decay into light *CP*-odd Higgs bosons: a comparative study in different new physics models," *Journal of High Energy Physics*, vol. 2010, no. 11, article 110, 2010.

[91] V. Barger, P. Langacker, H.-S. Lee, and G. Shaughnessy, "Higgs sector in extensions of the minimal supersymmetric standard model," *Physical Review D*, vol. 73, no. 11, Article ID 115010, 31 pages, 2006.

[92] M. Carena, S. Gori, N. R. Shah, and C. E. M. Wagner, "A 125 GeV SM-like Higgs in the MSSM and the $\gamma\gamma$ rate," *Journal of High Energy Physics*, vol. 2012, no. 3, article 014, 2012.

[93] S. F. King, M. Mühlleitner, and R. Nevzorov, "NMSSM Higgs benchmarks near 125 GeV," *Nuclear Physics B*, vol. 860, no. 2, pp. 207–244, 2012.

[94] R. Dermíšek and J. F. Gunion, "Escaping the large fine-tuning and little hierarchy problems in the next to minimal supersymmetric model and $h \to aa$ decays," *Physical Review Letters*, vol. 95, no. 4, Article ID 041801, 4 pages, 2005.

[95] U. Ellwanger, G. Espitalier-Noel, and C. Hugonie, "Naturalness and fine tuning in the NMSSM: implications of early LHC results," *Journal of High Energy Physics*, vol. 2011, no. 9, article 105, 2011.

[96] J. Cao, Z. Heng, and J. M. Yang, "Rare Z-decay into light *CP*-odd Higgs bosons: a comparative study in different new physics models," *Journal of High Energy Physics*, vol. 2010, no. 11, article 110, 2010.

[97] U. Ellwanger, J. F. Gunion, and C. Hugonie, "NMHDECAY: a fortran code for the Higgs masses, couplings and decay widths in the NMSSM," *Journal of High Energy Physics*, vol. 0502, article 066, 2005.

[98] U. Ellwanger and C. Hugonie, "NMHDECAY 2.1: an updated program for sparticle masses, Higgs masses, couplings and decay widths in the NMSSM," *Computer Physics Communications*, vol. 175, no. 4, pp. 290–303, 2006.

[99] G. Altarelli and R. Barbieri, "Vacuum polarization effects of new physics on electroweak processes," *Physics Letters, Section B*, vol. 253, no. 1-2, pp. 161–167, 1991.

[100] M. E. Peskin and T. Takeuchi, "Estimation of oblique electroweak corrections," *Physical Review D*, vol. 46, no. 1, pp. 381–409, 1992.

[101] M. Davier, A. Hoecker, B. Malaescu, C. Z. Yuan, and Z. Zhang, "Reevaluation of the hadronic contribution to the muon magnetic anomaly using new $e^+e^- \to \pi^+\pi^-$ cross section data from BABAR," *European Physical Journal C*, vol. 66, no. 1, pp. 1–9, 2010.

[102] J. Dunkley, E. Komatsu, M. R. Nolta et al., "Five-year *wilkinson microwave anisotropy probe* observations: likelihoods and parameters from the *WMAP* data," *Astrophysical Journal, Supplement Series*, vol. 180, no. 2, pp. 306–329, 2009.

[103] E. Aprile, K. Arisaka, F. Arneodo et al., "Dark matter results from 100 live days of XENON100 data," *Physical Review Letters*, vol. 107, no. 13, Article ID 131302, 6 pages, 2011.

[104] H. Baer, V. Barger, P. Huang, and X. Tata, "Natural supersymmetry: LHC, dark matter and ILC searches," *Journal of High Energy Physics*, vol. 2012, no. 5, article109, 2012.
[105] J. Cao, Z. Heng, J. M. Yang, and J. Zhu, "Higgs decay to dark matter in low energy SUSY: is it detectable at the LHC?" *Journal of High Energy Physics*, vol. 2012, no. 6, article 34, 2012.

10

Modified Chaplygin Gas Cosmology

H. B. Benaoum

Mathematics and Natural Sciences Department, Prince Mohammad Bin Fahd University, Al-Khobar 31952, Saudi Arabia

Correspondence should be addressed to H. B. Benaoum, hbenaoum@pmu.edu.sa

Academic Editor: Vlad Popa

Modified Chaplygin gas as an exotic fluid has been introduced by H. B. Benaoum (2002). Essential features of the modified Chaplygin gas as a cosmological model are discussed. Observational constraints on the parameters of the model have been included. The relationship between the modified Chaplygin gas and a homogeneous minimally coupled scalar field is reevaluated by constructing its self-interacting potential. In addition, we study the role of the tachyonic field in the modified Chaplygin gas cosmological model and the mapping between scalar field and tachyonic field is also considered.

1. Introduction

Astronomical and cosmological observations, such as type Ia supernovae (SNe Ia) [1–4], large-scale redshift surveys structure (LSS) [5, 6], the cosmic microwave background (CMB) [7, 8], and Wilkinson Microwave Anisotropy Probe (WMAP) [9, 10], indicate that the observable universe experiences an accelerated expansion. These observations suggest also that the universe is nearly flat and dominated by a nonbaryonic substratum. The source of this acceleration is usually attributed to an exotic type of fluid with negative pressure called commonly dark energy.

Various kinds of dark energy models have been proposed such as cosmological constant [11], quintessence [12–14], *k*-essence [15–17], tachyon [18–20], phantom [21–23], Chaplygin gas [24], quintom [25], holographic dark energy [26], and extra dimensions [27]. The nature of the dark sector of the universe (i.e., dark energy and dark matter) remains a mystery. An economical and attractive idea to unify the dark sector of the universe is to consider it as a single component that acts as both dark energy and dark matter. One way to achieve the unification of dark energy and dark matter is by using the so-called Chaplygin gas. The pure Chaplygin gas or generalized Chaplygin gas is a perfect fluid which behaves like a pressureless fluid at an early stage and a cosmological constant at a later stage.

Pure Chaplygin gas with an exotic equation of state is characterized by a negative pressure [24],

$$p = -\frac{B}{\rho}, \tag{1.1}$$

where p is the pressure, ρ is the energy density, and B is a positive parameter.

The pure Chaplygin gas has been extended to the so-called generalized Chaplygin gas with the following equation of state [28]:

$$p = -\frac{B}{\rho^{\alpha}}, \quad \text{where } 0 \leq \alpha \leq 1. \tag{1.2}$$

It is clear that the pure Chaplygin gas is recovered for the case $\alpha = 1$.

The interesting feature of the Chaplygin gas is its connection to the string theory. It can be obtained from the Nambu-Goto action for a D-brane moving in a $(D+2)$-dimensional spacetime in the light cone parametrization [29–33].

The outline of this paper is as follows. In the next section, we study the cosmological model of the modified Chaplygin gas which was introduced by the author in [34]. We show that the modified Chaplygin gas model interpolates between an epoch with a soft equation of state and a de Sitter phase. In Section 3, the relationships between the modified Chaplygin gas and a cosmological scalar field are reevaluated. In Section 4, a tachyonic field is considered as a candidate for the modified Chaplygin gas model. The correspondence between minimally coupled scalar field and the tachyon field is also investigated.

2. FRW Cosmology from the Modified Chaplygin Gas Model

Within the framework of Friedmann-Robertson-Walker (FRW) cosmology, a model called modified Chaplygin gas has been proposed by the author [34]. This model includes an initial phase of radiation and it is based on the following equation of state:

$$p = A\rho - \frac{B}{\rho^{\alpha}}, \tag{2.1}$$

where A, B, and α are constant parameters.

When $B = 0$, we recover the equation of state of perfect fluid, that is, $p = A\rho$. For $A = 0$, it reduces to the generalized Chaplygin gas.

In (2.1), the two terms start to be of the same order when the pressure vanishes (i.e., $p = 0$). In this case, the fluid has pressureless density ρ_0, corresponding to some scale factor a_0,

$$\rho_0 = \rho^{\alpha+1}(a_0) = \frac{B}{A}. \tag{2.2}$$

The metric of D-dimensional FRW spacetime is

$$ds^2 = -dt^2 + a^2(t)d\Omega_k^2, \tag{2.3}$$

where $a(t)$ is the scale factor and $d\Omega_k^2$ is the metric of the maximally symmetric $(D-1)$-space with curvature $k = 0, \pm 1$.

The Friedmann equation which governs the evolution of the scale factor is given by

$$H^2 = \left(\frac{\dot{a}}{a}\right)^2 = \frac{2\rho}{(D-1)(D-2)} - \frac{k}{a^2}, \tag{2.4}$$

where H is the Hubble parameter.

In the FRW framework, a fluid with an energy density ρ and a pressure p must satisfy the conservation law:

$$\dot{\rho} + (D-1)H(\rho + p) = 0. \tag{2.5}$$

The latter two equations imply that

$$\frac{\ddot{a}}{a} = \dot{H} + H^2 = -\frac{(D-1)p + (D-3)\rho}{(D-1)(D-2)}. \tag{2.6}$$

By defining $W = a^{(D-1)(A+1)}$ and a rescaled density $\overline{\rho} = \rho W$, (2.5) becomes

$$\dot{\overline{\rho}} - \frac{B}{A+1}\frac{W^\alpha \dot{W}}{\overline{\rho}^\alpha} = 0. \tag{2.7}$$

This equation can be easily integrated leading to

$$\frac{\overline{\rho}^{\alpha+1}}{\alpha+1} = \frac{B}{A+1}\frac{W^{\alpha+1}}{\alpha+1} + \frac{C}{\alpha+1}, \tag{2.8}$$

where C is an integration constant.

The density will be

$$\rho = \left(\frac{B}{A+1} + \frac{C}{W^{\alpha+1}}\right)^{1/(\alpha+1)}. \tag{2.9}$$

The constant of integration C can be expressed in terms of the cosmological scale a_0, (i.e., $W_0 = a_0^{(D-1)(A+1)}$) where the fluid has a vanishing pressure:

$$C = \frac{B}{A+1}\frac{W_0^{\alpha+1}}{A}. \tag{2.10}$$

The energy density ρ will be

$$\rho = \left(\frac{B}{A+1}\right)^{1/(\alpha+1)} \left(1 + \frac{1}{AW_r^{\alpha+1}}\right)^{1/(\alpha+1)}, \tag{2.11}$$

where $W_r = W/W_0$.

For large-scale factor a, that is, $W_r \gg 1$, we have

$$\begin{aligned} \rho &\simeq \left(\frac{B}{A+1}\right)^{1/(\alpha+1)}, \\ p &\simeq -\left(\frac{B}{A+1}\right)^{1/(\alpha+1)} = -\rho, \end{aligned} \tag{2.12}$$

which correspond to an empty universe with a cosmological constant $(B/(A+1))^{1/(\alpha+1)}$ (i.e., a de Sitter space).

Also for small-scale factor a, that is, $W_r \ll 1$, we have

$$\begin{aligned} \rho &\simeq \left(\frac{B}{A(A+1)}\right)^{1/(\alpha+1)} \frac{1}{W_r}, \\ p &\simeq A\left(\frac{A}{B(B+1)}\right)^{1/(\alpha+1)} \frac{1}{W_r} = A\rho, \end{aligned} \tag{2.13}$$

which correspond to universe dominated by an equation of state $p = A\rho$. This shows that this model interpolates between a universe dominated by matter phase with equation of state $p = A\rho$ and a de Sitter phase $p \simeq -\rho$.

Moreover, expanding (2.11) and (2.1) to the subleading terms at large cosmological constant, we obtain the following expressions for the energy and the pressure:

$$\begin{aligned} \rho &\simeq \left(\frac{B}{A+1}\right)^{1/(\alpha+1)} \left(1 + \frac{1}{(\alpha+1)AW_r^{\alpha+1}}\right), \\ p &\simeq \left(\frac{B}{A+1}\right)^{1/(\alpha+1)} \left(-1 + \frac{A+\alpha(A+1)}{(\alpha+1)AW_r^{\alpha+1}}\right). \end{aligned} \tag{2.14}$$

These correspond to the mixture of a cosmological constant $(B/(A+1))^{1/(\alpha+1)}$ and a type of matter described by an equation of state:

$$p = (\alpha + A(\alpha+1))\rho. \tag{2.15}$$

The equation of state parameter takes the form:

$$\omega = \frac{p}{\rho} = -\frac{1 - 1/W_r^{\alpha+1}}{1 + 1/AW_r^{\alpha+1}}, \tag{2.16}$$

which ranges over $-1 < \omega < A$, depending on the cosmological scale a,

$$\omega = \begin{cases} -1 & \text{for } W_r \gg 1, \\ 0 & \text{for } W_r = 1, \\ A & \text{for } W_r \ll 1. \end{cases} \tag{2.17}$$

The speed of sound c_s is defined as

$$c_s^2 = \frac{\partial p}{\partial \rho} = \frac{\dot{p}}{\dot{\rho}}. \tag{2.18}$$

Now by computing $\dot{\omega}$, we obtain

$$\dot{\omega} = \left(c_s^2 - \omega\right)\frac{\dot{\rho}}{\rho}, \tag{2.19}$$

which for the modified Chaplygin gas gives the following expression for the speed of sound:

$$c_s^2 = \omega + \rho\frac{d\omega}{d\rho} = \frac{\alpha + (\alpha+1)A + 1/W_r^{\alpha+1}}{1 + 1/AW_r^{\alpha+1}}, \tag{2.20}$$

implying that c_s^2 is always positive and hence there is no concern about imaginary speed of sound.

Moreover, it has the following asymptotic limit,

$$c_s^2 = \begin{cases} \alpha + (\alpha+1)A & \text{for } W_r \gg 1, \\ (\alpha+1)A & \text{for } W_r = 1, \\ A & \text{for } W_r \ll 1. \end{cases} \tag{2.21}$$

The speed of sound never exceeds that of light for smaller-scale a or scale of the order of a_0 where the pressure vanishes, provided that $A < 1/(\alpha+1)$ and will exceed it for large scale compared to a_0.

The constraints from the astrophysical and cosmological observables on the modified Chaplygin gas have been studied by many authors [35–40]. The permissible values of the parameters A, B, and α have been explored from the observed data.

Using data from different observations, namely, observational Hubble Data (OHD), baryon acoustic oscillation (BAO), and CMB shift parameters data, the allowed values for some of the parameters of the modified Chaplygin gas have been extracted. In the light of the 182 Gold SNe Ia, the 3-year WMAP and the SDSS data, the best fits correspond to $B = -0.085$ and $\alpha = 1.724$ [35]. This result was obtained by decomposing the modified Chaplygin gas into two components, that is, dark matter and dark energy component.

However, by using Markov chain Monte Carlo method with the observational data from the SN Ia Union 2, OHD, cluster X-ray gas mass fraction (CBF), BAO, and CMB data, the best fits of the modified Chaplygin gas parameters give $B = 0.00189$ and $\alpha = 0.1079$ [36].

Moreover, a perturbative analysis of the modified Chaplygin gas shows that the power spectrum observational data restricts the value of $|B| < 10^{-6}$ [40] such that the modified Chaplygin gas is disfavored.

3. Modified Chaplygin Gas as a Scalar Field

Following [41–45], we describe the modified Chaplygin gas cosmological model by introducing a scalar fields having a self-interacting potential $U(\varphi)$ with the Lagrangian:

$$\mathcal{L}_{\varphi} = \frac{\dot{\varphi}^2}{2} - U(\varphi). \tag{3.1}$$

Both the energy density and the pressure of the modified Chaplygin gas can be related to the scalar φ through the following transformation equations:

$$\begin{aligned} \rho_{\varphi} &= \frac{\dot{\varphi}^2}{2} + U(\varphi) = \rho, \\ p_{\varphi} &= \frac{\dot{\varphi}^2}{2} - U(\varphi) = A\rho - \frac{B}{\rho^{\alpha}}. \end{aligned} \tag{3.2}$$

The kinetic energy of the scalar field φ and its corresponding potential $U(\varphi)$ are

$$\begin{aligned} \dot{\varphi}^2 &= (1+\omega_{\varphi})\rho_{\varphi}, \\ U(\varphi) &= \frac{1}{2}(1-\omega_{\varphi})\rho_{\varphi}, \end{aligned} \tag{3.3}$$

where $\omega_{\varphi} = p_{\varphi}/\rho_{\varphi}$.

Now since $\dot{\varphi} = \varphi' \dot{W}_r$ where the prime denotes derivation with respect to W_r and $\dot{W}_r = (D-1)(A+1)HW_r$, we get

$$\varphi'^2 = \frac{D-2}{2(D-1)(A+1)^2}\frac{1+\omega_{\varphi}}{W_r^2}. \tag{3.4}$$

Here, we have used (2.4) for the Hubble constant and guided by the cosmic microwave background CMB data which is strongly consistent with a flat universe and restricted ourselves to the flat case $k = 0$.

By using (2.16), we get

$$\begin{aligned} \varphi' &= \sqrt{\frac{D-2}{2(D-1)A(A+1)}}\frac{1}{W_r^{(\alpha+3)/2}\sqrt{1+1/AW_r^{\alpha+1}}}, \\ U(\varphi) &= \frac{1}{2}\left(\frac{B}{A+1}\right)^{1/2}\frac{2+(1-A)/AW_r^{\alpha+1}}{\sqrt{1+1/AW_r^{\alpha+1}}}. \end{aligned} \tag{3.5}$$

The first equation can be integrated easily which gives

$$AW_r^{\alpha+1} = \frac{1}{\sinh^2(\varpi(\alpha+1)\Delta\varphi)}, \tag{3.6}$$

where $\varpi = \sqrt{(D-1)(A+1)/2(D-2)}$ and $\Delta\varphi = \varphi - \varphi_0$.

We note that for larger scales, the scalar field asymptotically approaches to the constant field φ_0 and becomes infinite (i.e., $\phi \rightarrow +\infty$) for small scales.

Next, by substituting the latter expression in (3.2), we can write all our physical quantities ρ_φ, p_φ, and ω_φ in terms of the scalar field φ as

$$\begin{gathered}
\rho_\varphi = \left(\frac{B}{A+1}\right)^{1/(\alpha+1)} \cosh^{2/(\alpha+1)}(\varpi(\alpha+1)\Delta\varphi), \\
p_\varphi = \left(\frac{B}{A+1}\right)^{1/(\alpha+1)} \left[A\cosh^{2/(\alpha+1)}(\varpi(\alpha+1)\Delta\varphi) - \frac{A+1}{\cosh^{2\alpha/(\alpha+1)}(\varpi(\alpha+1)\Delta\varphi)}\right], \\
\omega_\varphi = -\frac{1 - A\sinh^2(\varpi(\alpha+1)\Delta\varphi)}{\cosh^2(\varpi(\alpha+1)\Delta\varphi)}.
\end{gathered} \tag{3.7}$$

Notice that these physical quantities do not depend on the intermediate constant W_0 (i.e., constant of integration C).

Finally, we get the following potential which has a simple form:

$$U(\varphi) = \frac{1}{2}\left(\frac{B}{A+1}\right)^{1/(\alpha+1)} \left[\frac{1+A}{\cosh^{2\alpha/(\alpha+1)}(\varpi(\alpha+1)\Delta\varphi)} + (1-A)\cosh^{2/(\alpha+1)}(\varpi(\alpha+1)\Delta\varphi)\right]. \tag{3.8}$$

4. Modified Chaplygin Gas as Tachyonic Field

The importance of tachyon in cosmology is inspired by string theory [18, 19]. The action of the homogeneous tachyon condensate of string theory in a gravitational background is given by

$$S = \int d^D x \sqrt{-g}\left(R + \mathcal{L}_{\text{tachyon}}\right), \tag{4.1}$$

where R is the scalar curvature. For a tachyonic field T with tachyonic potential $V(T)$, the relativistic Lagrangian can be expressed as

$$\mathcal{L}_{\text{tachyon}} = -V(T)\sqrt{1 + g^{\mu\nu} \cdot \nabla_\mu T \nabla_\nu T}. \tag{4.2}$$

The corresponding energy-momentum tensor for the tachyonic field is

$$T_{\mu\nu} = -\frac{2\delta S}{\sqrt{-g}\delta g^{\mu\nu}} = p_T g_{\mu\nu} + (p_T + \rho_T) u_\mu u_\nu, \tag{4.3}$$

where the velocity u_μ is

$$u_\mu = \frac{-\nabla_\mu T}{\sqrt{-g^{\mu\nu}\nabla_\mu T \nabla_\nu T}} \tag{4.4}$$

with $u^\mu u_\mu = -1$.

It follows that the energy density ρ_T and the pressure p_T of the tachyonic field are given by

$$\begin{aligned} \rho_T &= \frac{V(T)}{\sqrt{1-\dot{T}^2}}, \\ p_T &= -V(T)\sqrt{1-\dot{T}^2}. \end{aligned} \tag{4.5}$$

The equation of state parameter is

$$\omega_T = \frac{p_T}{\rho_T} = -\left(1-\dot{T}^2\right). \tag{4.6}$$

The condition of accelerating universe (i.e., $\ddot{a}/a > 0$) requires that

$$\dot{T}^2 < \frac{2}{D-1}, \tag{4.7}$$

and the rolling tachyon has an interesting equation of state whose parameter ω_T interpolates between -1 and $(-D+3)/(D-1)$.

The evolution of the tachyonic field is driven by

$$\ddot{T} + \left(1-\dot{T}^2\right)\left((D-1)H\dot{T} + \frac{1}{V(T)}\frac{dV(T)}{dT}\right) = 0 \tag{4.8}$$

with the constraint equation for the Hubble parameter given by

$$\begin{aligned} H^2 &= \frac{2}{(D-1)(D-2)}\frac{V(T)}{\sqrt{1-\dot{T}^2}}, \\ \dot{H} &= -\frac{1}{D-1}\frac{\dot{T}^2 V(T)}{\sqrt{1-\dot{T}^2}}. \end{aligned} \tag{4.9}$$

By combining the last two equations, the tachyonic field T and the potential $V(T)$ can be expressed as

$$\begin{aligned} T(t) &= \int dt \left(-\frac{2}{D-1}\frac{\dot{H}}{H^2}\right)^{1/2}, \\ V(T) &= \frac{(D-1)(D-2)}{2} H^2 \left(1 + \frac{2}{D-1}\frac{\dot{H}}{H^2}\right)^{1/2}. \end{aligned} \tag{4.10}$$

Note that the knowledge of H and $\dot{H}$ (i.e., cosmological scale factor $a(t)$) completely determines the tachyonic field T and its corresponding potential $V(T)$.

By mapping the pressure p_φ and the energy density ρ_φ for the scalar field with the corresponding tachyon field p_T and ρ_T, we obtain that

$$\begin{aligned} dT &= \frac{d\varphi}{\sqrt{\rho_\varphi}}, \\ V(T) &= \sqrt{-\omega_\varphi \rho_\varphi}. \end{aligned} \tag{4.11}$$

Cosmological correspondence between the tachyonic field and a minimally coupled scalar field has been noted by [34, 46]. Such correspondence was also investigated by [47] where it was demonstrated explicitly that distinct scalar field and tachyonic models may give rise to the same cosmological evolution for a particular choice of initial conditions.

Now by using (3.7), an exact integration to the first equation in (4.11) can be performed without any approximation as follows:

$$T - T_0 = \frac{\cosh^{\alpha/(\alpha+1)}(\varpi(\alpha+1)\Delta\varphi)}{\alpha\varpi(B/(A+1))^{1/2(\alpha+1)}} \, {}_2F_1\left[\begin{matrix} \frac{1}{2}, \frac{1}{2} - \frac{1}{2(\alpha+1)} \\ \frac{3}{2} - \frac{1}{2(\alpha+1)} \end{matrix} ; -\cosh^2(\varpi(\alpha+1)\Delta\varphi)\right], \tag{4.12}$$

where ${}_2F_1$ is the hypergeometric function given by

$${}_2F_1\left[\begin{matrix} a, b \\ c \end{matrix} ; x\right] = \sum_{k=0}^{[\infty} \frac{(a)_k (b)_k}{(c)_k} \frac{x^k}{k!} \tag{4.13}$$

with $(a)_k = a(a+s)\ldots(a+(n-1)s)$ as the Pochhammer symbol.

Furthermore, by replacing the first equation in (3.7), (4.12) for tachyonic field T can be written in terms of the energy density ρ_φ as

$$T - T_0 = \frac{\rho_\varphi^{\alpha/2}}{\alpha\varpi(B/(A+1))^{1/2}} \, {}_2F_1\left[\begin{matrix} \frac{1}{2}, \frac{1}{2} - \frac{1}{2(\alpha+1)} \\ \frac{3}{2} - \frac{1}{2(\alpha+1)} \end{matrix} ; -\left(\frac{A+1}{B}\right)\rho_\varphi^{\alpha+1}\right]. \tag{4.14}$$

To find the scalar field φ and its potential $U(\varphi)$ in terms of the tachyon field and its potential $V(T)$, one has to perform the inverse transformation:

$$\begin{aligned} d\varphi &= \sqrt{\rho_T} dT, \\ U(\varphi) &= \frac{1}{2}(1 - \omega_T)\rho_T. \end{aligned} \tag{4.15}$$

In the slow-rolling approximation, the tachyon field potential $V(T)$ and the scalar field potential $U(\varphi)$ are approximately the same. To see this, we expand (4.10) up to the first order in $\dot{H}/H^2$ and use (2.4) and (2.6) to get:

$$V(T) \simeq \frac{(D-1)(D-2)}{2}\left(H^2 + \frac{1}{D-1}\dot{H}\right) = \frac{1}{2}(1 - \omega_\varphi)\rho_\varphi = U(\varphi). \tag{4.16}$$

Now to rewrite the potential $V(T)$ in terms of T, we first expand (4.14) to the first order in ρ_φ:

$$T - T_0 \simeq \frac{\rho_\varphi^{\alpha/2}}{\alpha\varpi(B/(A+1))^{1/2}} \tag{4.17}$$

and then substitute it in (4.16) to finally get:

$$V(T) \simeq \frac{c_1}{(T - T_0)^2} + c_2(T - T_0)^2, \tag{4.18}$$

where c_1 and c_2 are constant parameters depending on D, α, A, and B.

5. Conclusions

One approach in modern cosmology consists in supposing that dark energy and dark matter are different manifestations of a single entity. Following such an idea, this work (see also [34]) presents a cosmological model based on the modified Chaplygin gas that acts as a single component. It is shown that the equation of state of the modified Chaplygin gas interpolates from matter-dominated era to a cosmological constant-dominated era. Astronomical and cosmological constraints on the parameters A, B, and α of the modified Chaplygin have been investigated and still no basic agreement is reached on their values.

In addition, from the theoretical point of view, the modified Chaplygin gas model is equivalent to that of a scalar field φ having self-interacting potential $U(\varphi)$. Such a description has been explored and its self-interacting potential has been determined.

Furthermore, it has been shown that the modified Chaplygin gas can also be described by a tachyonic field T having a potential $V(T)$. Correspondence between scalar field and tachyonic field descriptions has been investigated. The transformations between the scalar field and tachyonic field and between their corresponding potentials have been determined. Such a correspondence has been applied to derive the exact expression of the tachyonic field T in terms of the scalar field φ for the modified Chaplygin gas. Finally, in the slow-rolling approximation, we have expressed the the potential $V(T)$ in terms of the tachyonic field T.

Acknowledgment

The author would like to express his gratitude to the anonymous referees for very helpful and detailed comments.

References

[1] S. Perlmutter, G. Aldering, M. Della Valle et al., "Discovery of a supernova explosion at half the age of the Universe," *Nature*, vol. 391, no. 6662, pp. 51–54, 1998.
[2] S. Perlmutter, G. Aldering, G. Goldhaber et al., "Measurements of Ω and Λ from 42 high-redshift Supernovae," *Astrophysical Journal Letters*, vol. 517, no. 2, part 1, pp. 565–586, 1999.
[3] A. G. Riess, A. V. Filippenko, P. Challis et al., "Observational evidence from supernovae for an accelerating universe and a cosmological constant," *Astronomical Journal*, vol. 116, no. 3, pp. 1009–1038, 1998.
[4] A. G. Riess, L.-G. Sirolger, J. Tonry et al., "Type Ia supernova discoveries at $z > 1$ from the *hubble* space telescope: evidence for past deceleration and constraints on dark energy evolution," *Astrophysical Journal Letters*, vol. 607, no. 2, pp. 665–687, 2004.
[5] N. A. Bahcall, J. P. Ostriker, S. Perlmutter, and P. J. Steinhardt, "The cosmic triangle: revealing the state of the universe," *Science*, vol. 284, no. 5419, pp. 1481–1488, 1999.
[6] M. Tegmark, M. A. Strauss, M. R. Blanton et al., "Cosmological parameters from SDSS and WMAP," *Physical Review D*, vol. 69, no. 10, Article ID 103501, 26 pages, 2004.
[7] A. D. Miller, R. Caldwell, M. J. Devlin et al., "A measurement of the angular power spectrum of the cosmic microwave background from $l = 100$ to 400," *Astrophysical Journal Letters*, vol. 524, no. 1, part 2, pp. L1–L4, 1999.
[8] C. L. Bennet, M. Halpern, G. Hinshaw et al., "First year Wilkinson Microwave Anisotropy Probe (WMAP) observations: preliminary maps and basic results," *The Astrophysical Journal Supplement Series*, vol. 148, no. 1, 2003.
[9] S. L. Bridle, O. Lahav, J. P. Ostriker, and P. J. Steinhardt, "Precision cosmology? Not just yet...," *Science*, vol. 299, no. 5612, pp. 1532–1533, 2003.
[10] D. N. Spergel, L. Verde, H. V. Peiris et al., "First-year Wilkinson Microwave Anisotropy Probe (WMAP) observations: determination of cosmological parameters," *Astrophysical Journal, Supplement Series*, vol. 148, no. 1, pp. 175–194, 2003.
[11] P. J. E. Peebles and B. Ratra, "The cosmological constant and dark energy," *Reviews of Modern Physics*, vol. 75, no. 2, pp. 559–606, 2003.
[12] B. Ratra and P. J. E. Peebles, "Cosmological consequences of a rolling homogeneous scalar field," *Physical Review D*, vol. 37, p. 3406, 1988.
[13] R. R. Caldwell, R. Dave, and P. J. Steinhardt, "Cosmological imprint of an energy component with general equation of state," *Physical Review Letters*, vol. 80, no. 8, pp. 1582–1585, 1998.
[14] M. Sami and T. Padmanabhan, "Viable cosmology with a scalar field coupled to the trace of the stress tensor," *Physical Review D*, vol. 67, no. 8, Article ID 083509, 10 pages, 2003.
[15] C. Armendariz-Picon and V. Mukhanov, "Essentials of *k*-essence," *Physical Review D*, vol. 63, no. 10, Article ID 103510, 10 pages, 2001.
[16] T. Chiba, "Tracking *k*-essence," *Physical Review D*, vol. 66, no. 6, Article ID 063514, 4 pages, 2002.
[17] R. J. Scherrer, "Purely kinetic *k* essence as unified dark matter," *Physical Review Letters*, vol. 93, no. 1, Article ID 011301, 1 pages, 2004.
[18] A. Sen, "Rolling tachyon," *Journal of High Energy Physics*, vol. 04, article 048, 2002.
[19] A. Sen, "Time evolution in open string theory," *Journal of High Energy Physics*, vol. 07, article 065, 2002.
[20] G. W. Gibbons, "Cosmological evolution of the rolling tachyon," *Physics Letters B*, vol. 537, no. 1-2, pp. 1–4, 2002.
[21] R. R. Caldwell, "A phantom menace? Cosmological consequences of a dark energy component with super-negative equation of state," *Physics Letters, Section B*, vol. 545, no. 1-2, pp. 23–29, 2002.
[22] E. Elizalde, S. Nojiri, and S. D. Odintsov, "Late-time cosmology in a (phantom) scalar-tensor theory: dark energy and the cosmic speed-up," *Physical Review D*, vol. 70, no. 4, Article ID 43539, 20 pages, 2004.
[23] J. M. Cline, S. Jeon, and G. D. Moore, "The phantom menaced: constraints on low-energy effective ghosts," *Physical Review D*, vol. 70, no. 4, Article ID 43543, 4 pages, 2004.

[24] A. Kamenshchik, U. Moschella, and V. Pasquier, "An alternative to quintessence," *Physics Letters, Section B*, vol. 511, no. 2–4, pp. 265–268, 2001.
[25] B. Feng, M. Li, Y.-S. Piao, and X. Zhang, "Oscillating quintom and the recurrent universe," *Physics Letters, Section B*, vol. 634, no. 2-3, pp. 101–105, 2006.
[26] P. Hořava and D. Minic, "Probable values of the cosmological constant in a holographic theory," *Physical Review Letters*, vol. 85, no. 8, pp. 1610–1613, 2000.
[27] M. Rogatko, "Uniqueness theorem for generalized Maxwell electric and magnetic black holes in higher dimensions," *Physical Review D*, vol. 70, no. 4, Article ID 044023, 5 pages, 2004.
[28] M. C. Bento, O. Bertolami, and A. A. Sen, "The Revival of the unified dark energy—dark matter model?" *Physical Review D*, vol. 70, Article ID 083519, 2004.
[29] J. C. Fabris, S. V. B. Gonçalves, and P. E. De Souza, "Density perturbations in a universe dominated by the Chaplygin gas," *General Relativity and Gravitation*, vol. 34, no. 1, pp. 53–63, 2002.
[30] N. Bili, G. B. Tupper, and R. D. Viollier, "Unification of dark matter and dark energy: the inhomogeneous Chaplygin gas," *Physics Letters, Section B*, vol. 535, no. 1–4, pp. 17–21, 2002.
[31] A. Hayashigaki, "Chaplygin-like gas and branes in black hole bulks," *Physics Letters, Section B*, vol. 487, no. 1-2, pp. 7–13, 2000.
[32] R. Jackiw, "A particle field theorist's lectures on supersymmetric, non-Abelian fluid mechanics and d-branes," http://arxiv.org/abs/physics/0010042.
[33] N. Ogawa, "Remark on the classical solution of the Chaplygin gas as d-branes," *Physical Review D*, vol. 62, no. 8, Article ID 085023, 8 pages, 2000.
[34] H. B. Benaoum, "Accelerated universe from modified chaplygin gas and tachyonic fluid," http://arxiv.org/abs/hep-th/0205140.
[35] J. Lu, L. Xu, J. Li, B. Chang, Y. Gui, and H. Liu, "Constraints on modified Chaplygin gas from recent observations and a comparison of its status with other models," *Physics Letters, Section B*, vol. 662, no. 2, pp. 87–91, 2008.
[36] J. Lu, L. Xu, Y. Wu, and M. Liu, "Combined constraints on modified Chaplygin gas model from cosmological observed data: markov Chain Monte Carlo approach," *General Relativity and Gravitation*, vol. 43, no. 3, pp. 819–832, 2011.
[37] L. Xu, Y. Wang, and H. Noh, "Modified Chaplygin gas as a unified dark matter and dark energy model and cosmic constraints," *European Physical Journal C*, vol. 72, no. 3, article 1931, 2012.
[38] P. Thakur, S. Ghose, and B. C. Paul, "Modified Chaplygin gas and constraints on its B parameter from cold dark matter and unified dark matter energy cosmological models," *Monthly Notices of the Royal Astronomical Society*, vol. 397, no. 4, pp. 1935–1939, 2009.
[39] B. C. Paul, P. Thakur, and A. Saha, "Modified Chaplygin gas in Horava-Lifshitz gravity and constraints on its *B* parameter," *Physical Review D*, vol. 85, no. 2, Article ID 024039, 9 pages, 2012.
[40] J. C. Fabris, H. E. S. Velten, C. Ogouyandjou, and J. Tossa, "Ruling out the Modified Chaplygin Gas cosmologies," *Physics Letters, Section B*, vol. 694, no. 4-5, pp. 289–293, 2011.
[41] J. D. Barrow, "Graduated inflationary universes," *Physics Letters B*, vol. 235, no. 1-2, pp. 40–43, 1990.
[42] J. D. Barrow and P. Saich, "Scalar-field cosmologies," *Classical and Quantum Gravity*, vol. 10, no. 2, pp. 279–283, 1993.
[43] A. A. Starobinsky, "How to determine an effective potential for a variable cosmological term," *JETP Letters*, vol. 68, no. 10, pp. 757–763, 1998.
[44] D. Huterer and M. S. Turner, "Prospects for probing the dark energy via supernova distance measurements," *Physical Review D*, vol. 60, no. 8, Article ID 081301, 5 pages, 1999.
[45] T. Nakamura and T. Chiba, "Determining the equation of state of the expanding Universe: inverse problem in cosmology," *Monthly Notices of the Royal Astronomical Society*, vol. 306, no. 3, pp. 696–700, 1999.
[46] T. Padmanabhan, "Accelerated expansion of the universe driven by tachyonic matter," *Physical Review D*, vol. 66, no. 2, Article ID 021301, 4 pages, 2002.
[47] V. Gorini, A. Kamenshchik, U. Moschella, and V. Pasquier, "Tachyons, scalar fields, and cosmology," *Physical Review D*, vol. 69, no. 12, Article ID 123512, 16 pages, 2004.

Permissions

The contributors of this book come from diverse backgrounds, making this book a truly international effort. This book will bring forth new frontiers with its revolutionizing research information and detailed analysis of the nascent developments around the world.

We would like to thank all the contributing authors for lending their expertise to make the book truly unique.They have played a crucial role in the development of this book. Without their invaluable contributions this book wouldn't have been possible. They have made vital efforts to compile up to date information on the varied aspects of this subject to make this book a valuable addition to the collection of many professionals and students.

This book was conceptualized with the vision of imparting up-to-date information and advanced data in this field. To ensure the same, a matchless editorial board was set up. Every individual on the board went through rigorous rounds of assessment to prove their worth. After which they invested a large part of their time researching and compiling the most relevant data for our readers. Conferences and sessions were held from time to time between the editorial board and the contributing authors to present the data in the most comprehensible form. The editorial team has worked tirelessly to provide valuable and valid information to help people across the globe.

Every chapter published in this book has been scrutinized by our experts. Their significance has been extensively debated. The topics covered herein carry significant findings which will fuel the growth of the discipline. They may even be implemented as practical applications or may be referred to as a beginning point for another development. Chapters in this book were first published by Hindawi Publishing Corporation; hereby published with permission under the Creative Commons Attribution License or equivalent.

The editorial board has been involved in producing this book since its inception. They have spent rigorous hours researching and exploring the diverse topics which have resulted in the successful publishing of this book. They have passed on their knowledge of decades through this book. To expedite this challenging task, the publisher supported the team at every step. A small team of assistant editors was also appointed to further simplify the editing procedure and attain best results for the readers.

Our editorial team has been hand-picked from every corner of the world. Their multi-ethnicity adds dynamic inputs to the discussions which result in innovative outcomes. These outcomes are then further discussed with the researchers and contributors who give their valuable feedback and opinion regarding the same. The feedback is then collaborated with the researches and they are edited in a comprehensive manner to aid the understanding of the subject.

Apart from the editorial board, the designing team has also invested a significant amount of their time in understanding the subject and creating the most relevant covers. They scrutinized every image to scout for the most suitable representation of the subject and create an appropriate cover for the book.

The publishing team has been involved in this book since its early stages. They were actively engaged in every process, be it collecting the data, connecting with the contributors or procuring relevant information. The team has been an ardent support to the editorial, designing and production team. Their endless efforts to recruit the best for this project, has resulted in the accomplishment of this book. They are a veteran in the field of academics and their pool of knowledge is as vast as their experience in printing. Their expertise and guidance has proved useful at every step. Their uncompromising quality standards have made this book an exceptional effort. Their encouragement from time to time has been an inspiration for everyone.

The publisher and the editorial board hope that this book will prove to be a valuable piece of knowledge for researchers, students, practitioners and scholars across the globe.

List of Contributors

C. Broggini
INFN, Sezione di Padova, Via F. Marzolo 8, 35131 Padova, Italy

C. Giunti
INFN, Sezione di Torino, Via P. Giuria 1, 10125 Torino, Italy

A. Studenikin
Department of Theoretical Physics, Moscow State University, Moscow 119991, Russia
Joint Institute for Nuclear Research, Dubna 141980, Moscow, Russia

K. Nozari, F. Moafi and F. Rezaee Balef
Department of Physics, Islamic Azad University, Sari Branch, Sari, Iran

Chee Sheng Fong and Enrico Nardi
INFN Laboratori Nazionali di Frascati, Via Enrico Fermi 40, 00044 Frascati, Italy

Antonio Riotto
Department of Theoretical Physics, University of Geneva and Center for Astroparticle Physics (CAP), 24 quai E. Ansermet, 1211 Geneva 4, Switzerland

M. Dima
Department of Computational Physics, Horia Hulubei National Institute of Physics and Nuclear Engineering, Atomistilor Street 407, P.O. Box MG-6, 76900 Bucharest, Romania

M. K. Pavićević and G. Amthauer
Division of Material Sciences and Physics, Department of Mineralogy, University of Salzburg, Hellbrunner Street 34, 5020 Salzburg, Austria

F. Bosch
GSI Helmholtzzentrum für Schwerionenforschung, Planck Straße 1, 64291 Darmstadt, Germany

I. Aničin and V. Pejović
Institute of Physics, University of Belgrade, Pregrevica 118, 11000 Belgrade, Serbia

B. Boev
Faculty of Mining and Geology, Goce Delˇcev University of ˇ Stip, Goce Delˇcev 89, 92000 ˇ Stip, Macedonia

W. Brü chle
GSI Helmholtzzentrum für Schwerionenforschung, Planck Straße. 1, 64291 Darmstadt, Germany

V. Cvetković and R. Jelenković
Faculty of Mining and Geology, University of Belgrade, D− uˇsina 7/II, 11000 Belgrade, Serbia

Z. Djurčić
Physics Division, Argonne National Laboratory (ANL), 9700 South Cass Avenue, Argonne, IL 60439, USA

W. F. Henning
Physics Department E12, Technical University of Munich, James-Franck Street, 85748 Garching, Germany

A. Weiss
Max-Planck Institute for Astrophysics, Karl-Schwarzschild Street 1, 85741 Garching, Germany

Fu-Hu Liu, Cai-Xing Tian, Mai-Ying Duan and Bao-Chun Li
Institute of Theoretical Physics, Shanxi University, Shanxi, Taiyuan 030006, China

Murat Korunur and Irfan Açıkgöz
Department of Physics, Faculty of Science, Dicle University, 21280 Diyarbakır, Turkey

Vladimir V. Skobelev
Moscow State Industrial University, Faculty of Applied Mathematics and the Technical Physics, 115280, Russia

Zhaoxia Heng
Department of Physics, Henan Normal University, Xinxiang 453007, China

H. B. Benaoum
Mathematics and Natural Sciences Department, Prince Mohammad Bin Fahd University, Al-Khobar 31952, Saudi Arabia